AF443709

Wireless Physical Layer Network Coding

Discover a fresh approach for designing more efficient and cooperative wireless communications networks with this systematic guide. Covering everything from fundamental theory to current research topics, leading researchers describe a new, network-aware coding strategy that exploits the signal interactions that occur in dense wireless networks directly at the waveform level. Using an easy-to-follow, layered structure, this unique text begins with a gentle introduction for those new to the subject, before moving on to explain key information-theoretic principles and establish a consistent framework for wireless physical layer network coding (WPNC) strategies. It provides a detailed treatment of Network Coded Modulation, covers a range of WPNC techniques such as Noisy Network Coding, Compute and Forward, and Hierarchical Decode and Forward, and explains how WPNC can be applied to parametric fading channels, frequency selective channels, and complex stochastic networks. This is essential reading whether you are a researcher, graduate student, or professional engineer.

Jan Sykora is a professor in the Faculty of Electrical Engineering at the Czech Technical University in Prague, and a consultant for the communications industry in the fields of advanced coding and signal processing.

Alister Burr is Professor of Communications in the Department of Electronic Engineering at the University of York.

Wireless Physical Layer Network Coding

JAN SYKORA

Czech Technical University in Prague

ALISTER BURR

University of York

CAMBRIDGE
UNIVERSITY PRESS

University Printing House, Cambridge CB2 8BS, United Kingdom

One Liberty Plaza, 20th Floor, New York, NY 10006, USA

477 Williamstown Road, Port Melbourne, VIC 3207, Australia

314–321, 3rd Floor, Plot 3, Splendor Forum, Jasola District Centre,
New Delhi – 110025, India

79 Anson Road, #06–04/06, Singapore 079906

Cambridge University Press is part of the University of Cambridge.

It furthers the University's mission by disseminating knowledge in the pursuit of
education, learning, and research at the highest international levels of excellence.

www.cambridge.org
Information on this title: www.cambridge.org/9781107096110
DOI: 10.1017/9781316156162

First published 2018

Printed in the United Kingdom by Clays, St Ives plc

A catalogue record for this publication is available from the British Library.

ISBN 978-1-107-09611-0 Hardback

Contents

Preface

About the Book

The book addresses strategies and principles of physical layer coding and signal processing that fully respect and utilize knowledge of the structure of a wireless network. This technique substantially increases the overall network throughput, efficiency, and reliability. Wireless Physical Layer Network Coding (WPNC) (a.k.a. Physical Layer Network Coding (PLNC)) is a general framework for physical (PHY) layer coding and processing strategies in which PHY behavior at a given node depends on its position in the network topology, and the signal-level processing/decoding exploits multiple paths between source and destination. We introduce the concept of Network Coded Modulation (NCM) as a network-structure-aware signal space code, which processes a (hierarchical) joint function of source data. At intermediate nodes NCM utilizes hierarchical decoding, and it is also designed to allow unambiguous decoding at the final destination using multiple hierarchical observations, arriving via different routes. The book addresses the fundamental principles of WPNC in the context of network information theory, and provides a comprehensive classification of the strategies. It also covers advanced design and techniques, including particular coding and processing designs and their respective properties. We also address selected hot research topics and open problems.

Motivation for the Book

It is becoming widely accepted that the most significant future developments in the physical layer of wireless communication systems will not take place in the PHY layer of individual communication links, but rather in the context of complete wireless networks, especially as the density of wireless networks continues to increase. Over the past decade or so there have been significant developments in network information theory; these have shown that very significant overall performance gains are available compared with the conventional paradigm in which PHY techniques are applied to individual links only, leaving network aspects to be dealt with only at higher layers of the protocol stack. One such new research field is network coding, in which coding techniques are applied to multiple data streams at intermediate nodes in a network, rather than only to individual streams on single links. This can exploit network topology to significantly improve throughput in multi-user networks. However, in its original form it operates at the level of data streams, rather than signal waveforms, and hence is not well suited to the inherently broadcast nature of wireless networks. Wireless physical layer network coding (WPNC) allows it to be applied directly to wireless networks, with a further significant improvement in efficiency. The key advance on conventional PHY techniques is that the nodes are aware of the network topology and their place within it, and both signaling waveforms and node signal processing exploit this knowledge to improve overall network throughput.

Book Scope and Organization

The book is carefully balanced, being divided into several "layers" giving different depths of information for audiences with various levels of background knowledge. Part I gives a gentle introduction to the key concept with the explanation kept in accessible form. Part II presents fundamental principles in more detail, but still using a "big picture" global perspective. Part III addresses a mosaic of various particular design techniques and principles that can practically fulfill the general principles of Part II. The Appendix provides some background material for readers with a weaker background in communication, signal processing, and information theory.

Throughout the book, we maintain a strong emphasis on the proper classification and structuring of the problems, techniques, and particular coding, processing, and decoding schemes under discussion. This will help readers to properly orient themselves in the complex landscape of the different individual approaches. In the currently available literature these frequently overlap, and suffer from rather "fuzzy" terminology. This may lead to incorrect comparisons due to the high complexity of the field and the ambiguity and inconsistency of the terminology. (Terminology also changes rapidly due to the rapid progress of the research community.)

The book is not primarily intended as a university course textbook but rather as a reference source for researchers, PhD students, and engineers who would like to understand the principles of WPNC in the context of other techniques or would like to start their own research work in this field. Therefore the book is a highly structured set of Parts–Chapters–Sections, which are intended, as far as possible, to be read in a self-contained manner.

Jan Sykora and Alister Burr

Mathematical Symbols

Basic Symbols, Sets

$\mathbb{N}$	positive integers
$\mathbb{N}_0$	non-negative integers
$\mathbb{R}$	real numbers
$\mathbb{C}$	complex numbers
$\mathbb{Z}$	integer numbers
$\mathbb{Z}_\mathrm{j} = \{a + \mathrm{j}\,b : a, b \in \mathbb{Z}\}$	complex (Gaussian) integers
$\mathcal{A}_1 \times \mathcal{A}_2$	Cartesian product of sets
$\mathcal{A}^N = \underbrace{\mathcal{A} \times \cdots \times \mathcal{A}}_{N \text{ times}}$	Cartesian product of sets
$|\mathcal{A}|$	size (cardinality) of the set $\mathcal{A}$
$\emptyset$	empty set
$\cup, \cap$	union and intersection operators for the sets
$\setminus$	set difference (set minus)
$\{c_i\}_i$	set of variables c_i for all feasible indices i
$\tilde{a} = \{a_1, \ldots, a_n\}$	set of all components
$[a, b) = \{x : a \leq x < b\}$	semiopen interval
$[k_1 : k_2]$	integer interval $\{k_1, k_1 + 1, \ldots, k_2\}$, $k_1, k_2 \in \mathbb{Z}$
$f : \mathcal{A} \mapsto \mathcal{B}$	f mapping from domain $\mathcal{A}$ to codomain (range) $\mathcal{B}$
$F[\cdot]$	operator F
$\exists$	there exists
$\forall$	for all
$\triangleq$	equal by definition
$\equiv$	equivalent, defines equivalence class
$\approx$	approximately or asymptotically equal
$\lesssim, \gtrsim$	approximately less than and greater than
$\lesssim, \gtrsim$	asymptotically less than and greater than
$\Rightarrow, \Leftrightarrow$	implication and equivalence
$\{a\}$	the set of all values the variable a can take
$\delta(t)$	Dirac delta function (continuous time)
$\delta[k]$	Kronecker delta (discrete time)
$\sup$	supremum
$\mathrm{sinc}(x) = \sin(\pi x)/(\pi x)$	sampling function

$\lg x = \log_2 x$	binary logarithm
a^*	complex conjugation
$U(x)$	unit step function
$(x)^+ = \max(0, x)$	positive part function
$\wedge, \vee$	Boolean "and", "or"
$j = \sqrt{-1}$	imaginary unit
e	base of the natural logarithm
$\sphericalangle z$	angle of complex number
$\partial f / \partial x$	standard partial derivative of the function f over variable x
$\tilde{\partial} f / \tilde{\partial} z$	generalized partial derivative of complex valued function over complex valued variable
$\int (.)\, dx$	abbreviated form for the integration over the whole domain of variable x, e.g. $\int_{-\infty}^{\infty} (.)\, dx$
$\sum_{x:g} f(x)$	sum over the set of all x consistent with condition g
$\sum_{x:g(x)=y} f(x)$	sum over the set of all x consistent with explicit condition $g(x) = y$

Number Theory, Vectors, Matrices, Inner-Product Spaces, Lattices

$\langle . ; . \rangle$	inner product
$\mathbf{a}$	vector (all vectors are column vectors)
$\mathbf{1} = [1, \ldots, 1]^{\mathrm{T}}$	unity vector
$\mathbf{I}, \mathbf{I}_N$	identity matrix with size defined by context, $N \times N$ identity matrix
$\mathrm{diag}(\mathbf{a})$	diagonal matrix with the components of $\mathbf{a}$ on the main diagonal
$\mathbf{A} \in \mathbb{C}^{m \times n}$	(m,n) matrix of complex numbers
$[\mathbf{A}]_{i,j}$	element of the matrix on the ith row and jth column
$\mathbf{A} \geq \mathbf{B}$	$\mathbf{A}$–$\mathbf{B}$ matrix is positive semi-definite
$(.)^{\mathrm{T}}$	transposed matrix or vector
$(.)^{\mathrm{H}}$	Hermitian transpose
$\mathbf{A}^{-1}$	matrix inverse
$\mathbf{A}^{\dagger} = (\mathbf{A}^{\mathrm{H}} \mathbf{A})^{-1} \mathbf{A}^{\mathrm{H}}$	matrix pseudoinverse
$\det \mathbf{A}$	determinant of matrix $\mathbf{A}$
$\boxtimes$	Kronecker matrix product
$\boxdot$	element-wise Hadamard product of two matrices/vectors
$\mathbb{S}_M$	finite ring
$\mathbb{F}_{p^m}$	Galois (finite) extended field with characteristic p
$\mathbb{F}_M^N, \mathbb{F}_M^{N_1 \times N_2}$	N-dimensional vector and $N_1 \times N_2$ matrix on $\mathbb{F}_M$ GF
$\oplus, \otimes$	addition and multiplication on GF (this explicit notation is used only when we need explicitly to distinguish it, otherwise ordinary "plus" and "times" operators are also used)
$\mathcal{E}$	energy

$c(t) = a(t) * b(t)$	convolution in continuous time $\int a(t - \tau)b(\tau)\, \mathrm{d}\tau$
$\mathbf{c} = \mathbf{a} * \mathbf{b}$	convolution in discrete time $c_n = \sum_k a_{n-k}b_k$, $\mathbf{a} = [\dots, a_0, a_1, \dots]^\mathrm{T}$ and similarly for $\mathbf{b}, \mathbf{c}$
$\mathbf{c} = \mathbf{a} \circledast \mathbf{b}$	cyclic convolution $c_n = \sum_{k=0}^{N-1} a_{(n-k) \bmod N}b_k$, $\mathbf{a} = [a_0, a_1, \dots, a_{N-1}]^\mathrm{T}$ and similarly for $\mathbf{b}, \mathbf{c}$
$x \perp y$	orthogonal x and y, i.e. $\langle x; y \rangle = 0$ for some inner product definition
Λ	lattice
$\mathcal{V}_0(\Lambda_s)$	fundamental Voronoi cell of lattice Λ_s
Λ_c/Λ_s	quotient group for lattices Λ_c, Λ_s

Random Variables, Processes, and Information Theory

X, x	strict notation for random variable and its particular realization
$y, y^{(i)}$	alternative (relaxed) form of notation (identified by its context) for random variable and its particular realization
$x^K = \{x_1, x_2, \dots, x_K\}$	a sequence (a tuple) of variables
$X^K = \{X_1, X_2, \dots, X_K\}$	a sequence (a tuple) of random variables
$x(\mathcal{S}) = \{x_k : k \in \mathcal{S}\}$	set (a tuple) of variables with indices given by $\mathcal{S}$
$\Pr\{\cdot\}$	probability
$p(x), p_X(x), p_x(x)$	PDF (PMF) with implicit and explicit denotation of random variable
$p(x\|z), p_{X\|Z}(x\|z), p_{x\|z}(x\|z)$	conditional PDF (PMF) with implicit and explicit denotation of random variables
$x \sim p(x)$	drawn according to the given PDF/PMF
$A \perp\!\!\!\perp B$	independent random variables
$A \perp\!\!\!\perp B\|C$ (or $(A \perp\!\!\!\perp B)\|C$	A and B conditionally independent given C
$\mathcal{U}(\mathcal{S})$	uniform distribution over the set $\mathcal{S}$
$\mathcal{N}(\mathbf{m}, \mathbf{C})$	Gaussian distribution with mean vector $\mathbf{m}$ and variance matrix $\mathbf{C}$
$\mathrm{E}[\cdot]$	ensemble domain expectation operator
$\mathrm{E}_x[\cdot], \mathrm{E}_{p(x)}[\cdot]$	expectation over explicit random variable or distribution
$\mathcal{H}[X]$	entropy of random variable X
$\mathcal{H}[X\|Y]$	conditional entropy of X conditioned by Y
$I(X; Y)$	mutual information between X and Y
$I(X; Y\|Z)$	conditional mutual information between X and Y given Z
$\mathrm{H}(p)$	binary entropy function
$A \mapsto B \mapsto C$	Markov chain variables
$(x, y) \in \mathcal{T}$	x and y are jointly typical
$\mathcal{R}(S_1, S_2)$	rate region for independent codebooks with S_1, S_2 random channel symbols

Abbreviations

2WRC	2-Way Relay Channel
AF	Amplify and Forward
AWGN	Additive White Gaussian Noise
BC	Broadcast Channel
BPSK	Binary Phase Shift Keying
CF	Compute and Forward
CpsF	Compress and Forward
CRLB	Cramer–Rao Lower Bound
CSE	Channel State Estimation
DF	Decode and Forward
DFT	Discrete Fourier Transform
GF	Galois Field
H-	Hierarchical
H-BC	Hierarchical BC
H-constellation	Hierarchical Constellation
H-decoding	Hierarchical Decoding
HDF	Hierarchical Decode and Forward
HI	Hierarchical Information
H-Ifc	Hierarchical Interference
H-MAC	Hierarchical MAC
HNC map	Hierarchical Network Code map
H-NTF	Hierarchical Network Transfer Function
H-NTM	Hierarchical Network Transfer Matrix
H-PEP	Hierarchical Pairwise Error Probability
H-SCFD	Hierarchical Successive CF Decoding
HSI	Hierarchical Side-Information
H-SODEM	Hierarchical Soft-Output Demodulator
Ifc	Interference
iff	if and only if
IH-codebook	Isomorphic H-codebook
IID	Independent and Identically Distributed
JDF	Joint Decode and Forward
LHS	left-hand side
MAC	Multiple Access Channel
MAP	Maximum A posteriori Probability

MIMO	Multiple-Input Multiple-Output
ML	Maximum Likelihood
MMSE	Minimum Mean Square Error
MPSK	M-ary Phase Shift Keying
MSE	Mean Square Error
NCM	Network Coded Modulation
NC	Network Coding
NC-JDF	Network Coding over JDF
NNC	Noisy Network Coding
OFDM	Orthogonal Frequency Division Multiplexing
PDF	Probability Density Function
PMF	Probability Mass Function
PSK	Phase Shift Keying
QAM	Quadrature Amplitude Modulation
QF	Quantize and Forward
QPSK	Quadriphase Phase Shift Keying
RHS	right-hand side
Rx	Receiver
SF	Soft Forward
SNR	Signal-to-Noise Ratio
SODEM	Soft-Output Demodulator
UMP	Uniformly Most Powerful
WPNC	Wireless Physical Layer Network Coding
w.r.t.	with respect to
Tx	Transmitter
WCC	Wireless Cloud Coding
XOR	eXclusive OR operation

Part I

Motivation and Gentle Introduction

1 Introduction

1.1 Introduction

Wireless networks are becoming more and more ubiquitous in the modern world, and more and more essential to today's society. In 30 years they have progressed from the province of a tiny minority of the world's population in only the most developed nations, to the point where there are very nearly as many wireless subscriptions as people in the world [24]. The services offered have extended from very limited speech services at the introduction of first-generation mobile systems in 1985, to broadband Internet access and full motion video today. Moreover, we are at the point where wireless networks will extend beyond connecting people (of whom there are a limited number), to connecting their devices – an effectively unlimited number. Some believe that there are already more devices than people connected to the Internet, and predictions that 50 billion or more devices will be connected by 2020 are circulating widely [60]. Of course, that is only the start.

All this implies that the *density* of wireless networks will inevitably increase. To provide telecommunication services to the human populations of our cities, at continually increasing data rates, will require increasing numbers of access points, for which backhaul will become an increasing problem, and require more widespread use of wireless backhaul. The devices will also form a network many times as dense as any current wireless networks, also likely to require connection to the core network. In both cases it is likely that the current point-to-multipoint architecture of wireless networks, exemplified by both cellular and WiFi systems, will be replaced by a multi-hop mesh network architecture.

The concept of the *mobile ad-hoc network* (MANET), one of the best-established concepts in wireless mesh networking, has been in existence for many years [9], yet has not really fulfilled its predicted potential. There are very few wireless networks in use today that implement a truly multi-hop networking approach. There seems to be a barrier to the practical implementation of multi-hop wireless networking that will surely have to be overcome in order to implement the ultra-dense wireless networks that are likely to be required in the near future.

Perhaps the most fundamental basis for such a barrier is that described by Gupta and Kumar in their well-known paper [20]. They show that for a conventional approach to wireless networking, in which transmissions from other nodes in the network are treated as interference, the total capacity of the network scales as the square root of the number of nodes – that is, the capacity per node decreases as the size of the network increases. Hence as networks become denser, and more hops are required, the capacity available to each terminal will decrease.

This interference problem has become widely recognized as the most significant problem limiting the performance of future wireless networks, including point-to-multipoint networks as well as multi-hop. Traditionally it has been mitigated by means of the cellular paradigm, which limits interference by ensuring that a certain *re-use distance* is respected. Increased density is accommodated by using smaller and smaller cells with greatly reduced transmit power, but this approach is now reaching its limit, both because of the large numbers of radio access points it requires and the resulting backhaul problem, and because cell sizes are becoming comparable in size with buildings and other city features.

All this suggests that it is time for a completely new paradigm in wireless networking, and a major objective of this book is to lay the foundations for such a paradigm, which we call the "Network-Aware Physical Layer."

1.2   The "Network-Aware Physical Layer"

Since the 1970s the design of communications networks has been based upon a *layered* paradigm, in which network functions are divided between protocol layers, each assumed to be transparent to the ones above it. The original layered model, dating from the late 1970s, was of course the OSI seven-layer model [2], but recently the layers implicitly defined in the TCP-IP protocol suite [1] have been more influential. In either case, the lower layers – the *network layer*, the *link layer*, and the *physical layer* – are of most interest to us here, since they provide the most basic functions of a communication network, namely routing, multiple access and error control, and modulation and coding, respectively.

Of these layers, the physical layer is the one that handles the signals which are actually transmitted over the communication medium: in our case these are related to the electromagnetic fields that form the radio waves. In the traditional layered paradigm the physical layer receives a signal from the medium and converts it to a bit stream, which is then passed to the link layer. However, this has the fundamental disadvantage that information is lost in the process that might improve the performance of functions which are located in higher layers. For example, it is well known that error correction is less efficient when operating on a bit stream (corresponding to *hard decision decoding*) than when it has access to a *soft decision metric*, which is usually obtained from the signal.

Moreover, it also means that signals from nodes other than the transmitter of interest must be treated as interference, which conveys no useful information but degrades the performance of the receiver in decoding the wanted signal. This arises because the traditional physical layer is assumed to operate on only one point-to-point link, which means signals on other links are interference (and vice versa). This is illustrated in Figure 1.1. The figure illustrates a multi-hop network in which data can travel from source to destination via two routes. We focus on the link of interest marked: in the traditional

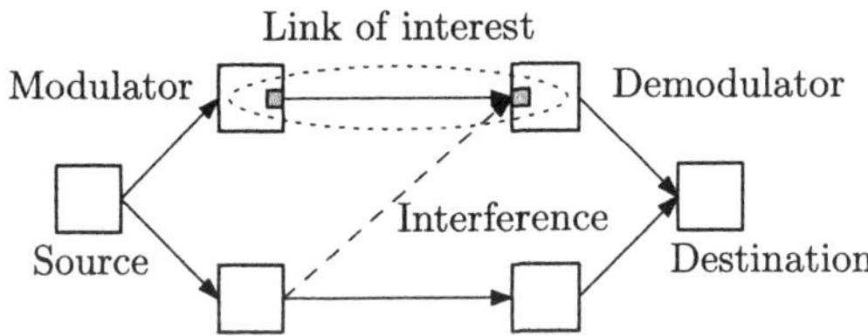

Figure 1.1 Traditional physical layer in a network.

paradigm the physical layer consists of the modulator at the transmitting node, the radio link between them, and the demodulator in the receiving node: that is, it relates to that link only, in isolation from the rest of the network. Thus a signal from another transmitter must be treated as interference (as shown), even though it carries information from the same original source, and could in principle be exploited to improve the reception of the data of interest.

Because interference is deleterious, it must usually be avoided wherever possible in traditional networks. This means that each node must transmit as far as possible on a channel orthogonal to the channel assigned to every other node – typically in a different time-slot or at a different frequency. This divides the resource available to the network and greatly reduces its efficiency. Again, information theory teaches us that greater capacity can often be achieved when multiple sources are allowed to transmit at the same time in non-orthogonal channels: for example, the capacity region of the multiple access channel (MAC) is achieved when the sources transmit simultaneously in the same channel, and is greater than the rate achieved by time-sharing of the channel.

The "network-aware" physical layer, on the other hand, does not need to nominate one node as transmitter of interest and hence treat all other signals but this one as interference. A network-aware receiver is aware – at the physical layer – of its location in the network, and what signals it may expect to receive in a given channel or time-slot. It is therefore able to determine what processing to apply to the composite signal formed by the mixture of all these signals. Similarly a network-aware transmitter is aware what effect its transmitted signals will have on other receivers, and can tailor the transmission in such a way that the received combination can also be processed as required.

Simply, if multiple interacting signals are unavoidable (e.g. due to the physical density of the network), it is better to make them useful one to each other as much as possible, instead of avoiding them. We do that directly on the signal level by properly constructing the transmitted coded signals and properly processing and decoding the received signals. This allows multiple nodes to transmit on the same channel, and avoids the division of resources. A receiver may even benefit from receiving combined signals rather than separate signals. It means that fewer signals have to be treated as deleterious interference, and any that do are typically weaker signals that have little effect.

This paradigm is not entirely novel: some functions which might be regarded as belonging to the link layer have already been implemented in the physical layer. One example is multiple access, which in computer networks is commonly implemented at the link layer by using protocols such as ALOHA or CSMA (Carrier Sense Multiple

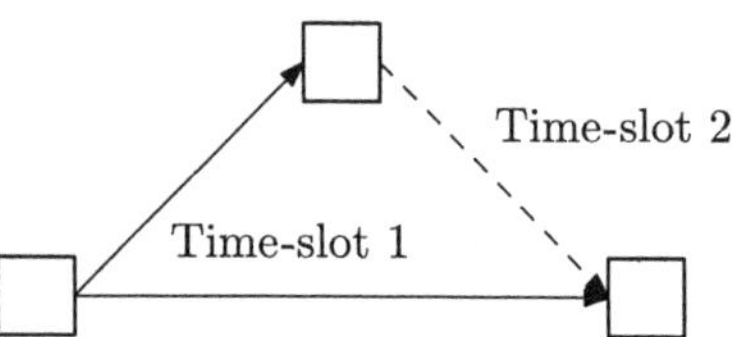

Figure 1.2 A simple cooperative communication system.

Access), or else is scheduled by using time-division or frequency-division multiple access (TDMA or FDMA). However code-division multiple access (CDMA), widely used in third-generation (3G) mobile systems, uses channels (corresponding to spreading codes) that are typically not fully orthogonal, and hence requires processing of the received mixed signal, which must be carried out at the physical layer, to separate the data. Similarly error control: while forward error correction (FEC) coding is conventionally regarded as part of the physical layer, retransmission protocols such as ARQ (Automatic Repeat reQuest) have traditionally been implemented at the link layer. However, recently hybrid FEC/ARQ schemes have somewhat blurred this distinction, since they require combining of signals transmitted in the course of multiple retransmissions.

Until recently, however, the functions of the network layer, specifically routing, have been excluded from the physical layer. This began to change about a decade ago with the introduction of *cooperative communications* [32]. Cooperative systems involve at least one relay node as well as the source and destination nodes (Figure 1.2), to assist the transmission of the source's data. Typically it receives the source signal in one time-slot, and retransmits it in some form in a subsequent slot. In most cases the processing within the relay is entirely at the physical layer, and frequently it is the original signal or some function of it that is retransmitted, without being converted to bits first. This is perhaps the simplest example of the physical layer being extended over a network involving multiple hops, beyond the simple link between one transmitter and one receiver.

This is, however, a very rudimentary version of routing. In this book we consider a much more general scenario involving multiple sources and multiple destinations, and multi-hop relaying between them. Thus routing is an essential element. The approach we will use, however, differs from routing in the conventional layered paradigm in two respects. The first is that it resembles cooperative communications in that processing within the relay takes place at the physical layer, involving signals directly. Unlike a bridge or a router in a conventional network, the relay does not decode the source data and transfer it to the link or network layer, but rather processes the received signals and forwards some function of them. The second is that what it forwards may not be a representation of data from a single source, but rather some function of data from several sources – a "mixture" of data from multiple sources to be separated at a later stage and delivered to the required destination. Thus it may no longer be possible to identify distinct routes for individual data streams, as is conventionally assumed.

This latter aspect can also be applied at the network layer of a multi-hop network, and corresponds to a technique introduced at the beginning of this century, known as *network coding*, which we will now discuss.

1.3   Network Coding at the Network Layer

Network layer *network coding* (NC) [5] addresses a network modeled as a directed graph connecting source nodes to destination nodes via a set of relaying nodes. In general there may be multiple sources and multiple destinations. The edges of the graph represent discrete links between pairs of nodes. This is clearly a good model of a data communications network with wired connections, such as the Internet, though we will see later that it does not represent a wireless network so well. For a unicast network, in which there is only one source and one destination, it can be proven that the maximum data flow rate is given by the *max-flow, min-cut theorem* [14]. However, Ahlswede *et al.* [5] showed that in the multicast case, where multiple destinations wish to receive the same data, the maximum flow rate cannot be achieved if relaying nodes operate simply as switches, connecting data flows on incoming links to outgoing links. Instead nodes should apply network coding, in which the symbols on an outgoing link are generated by some function of the symbols on two or more incoming links. This may be illustrated by the network shown in Figure 1.3, known as the *butterfly network*. The figure shows two versions of a network, in which two data sources each wish to send their data to both of two destinations, over a network in which all links have unit capacity. Figure 1.3a represents a conventional network in which the nodes can only switch a data stream from an incoming link onto an outgoing edge, or duplicate it and send on more than one outgoing edge. Thus the upper of the two relay nodes (which are marked as circles) can only select either stream A or stream B to send on its outgoing link (here it selects A). This is duplicated by the lower relay node, and hence the right-hand destination node can receive both streams, but the left-hand one receives only A. Figure 1.3b shows a network employing network coding. Here the upper relay node computes the exclusive OR (XOR) function (or modulo-2 sum) of the symbols in the data streams, and forwards the result. The lower relay node duplicates this to both destinations, and they can each recover both streams, because one is directly available, and the other can be reconstructed by reversing the network coding function applied at the relay node with the aid of the directly available stream. Thus the left-hand destination can now reconstruct stream B by applying $A \oplus (A \oplus B) = B$.

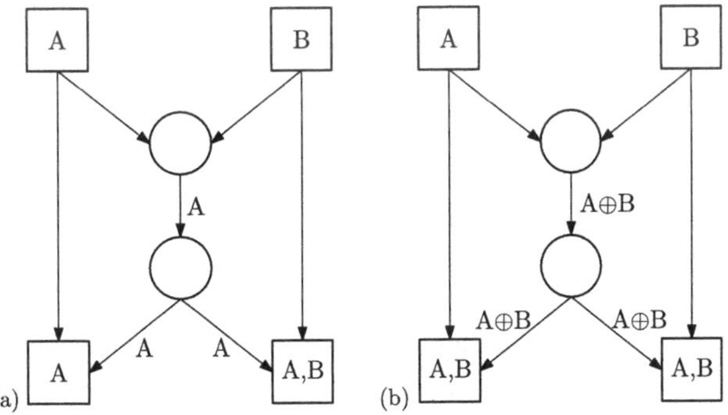

Figure 1.3 Butterfly network.

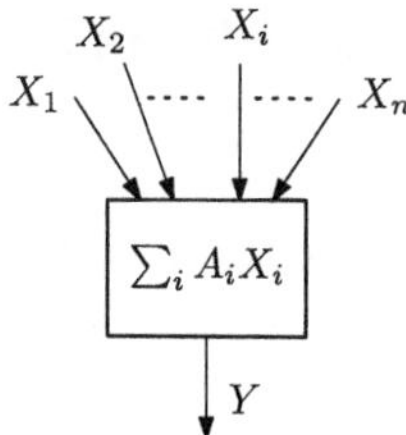

Figure 1.4 Linear network coding.

We will revisit this network topology later, in a slightly different context, but of course this principle also applies to much more complex networks, including networks containing cycles. Also in this case very simple coding is applied at the relay – simply the bit-by-bit XOR – but in general more complex encoding is required. There exists a wide variety of forms of coding, but [27] showed that linear coding over the finite field $\mathbb{F}_{2^m}$ is effective: in fact [34] had already shown that linear coding can achieve the maximum flow in a multicast network. Figure 1.4 illustrates this coding applied to a node: the output symbol Y is given by the formula in the diagram, in two different notations. In the first form $\otimes$ and $\oplus$ represent multiplication and addition within $\mathbb{F}_{2^m}$; in the second this is simply represented as a summation. The symbols on the incoming links are symbols in $\mathbb{F}_{2^m}$: they are drawn from an alphabet whose size is a power of 2, and can in fact be represented as length m binary strings. The coefficients A_i, $i = 1 \ldots n$ are also elements of $\mathbb{F}_{2^m}$, and again can be represented as length m binary strings. The addition operation is in fact simple bit-by-bit modulo-2 addition, but multiplication is more complicated: it is usually defined using primitive element operations on finite field (see Section A.2.1 or [8]).

It is clear that if all nodes apply a linear function of this sort, with symbols and coefficients from the same field, then the vector of output symbols across all relay nodes may be related to the vector of source symbols by a matrix. Equally clearly, for the destination nodes to reconstruct the source data this matrix must be full rank. We will revisit this model more rigorously later in the book.

1.4   Wireless Physical Layer Network Coding

The network model implicit in the conception of network coding, as illustrated in Figures 1.3 and 1.4, has one important deficiency as a representation of a wireless network. It assumes that the incoming links are *discrete*, and the symbols they carry are *separately* available to the network coding function in the node. This is a valid model of a wired network, but a wireless network does not have defined, discrete connections between nodes in the same way. Rather the electromagnetic fields due to signals transmitted simultaneously from two nodes will add together at the antenna of a receiving node, resulting in a superimposition of the two signals. Moreover they may be attenuated and/or phase shifted due to the wireless channel in largely unpredictable ways. In

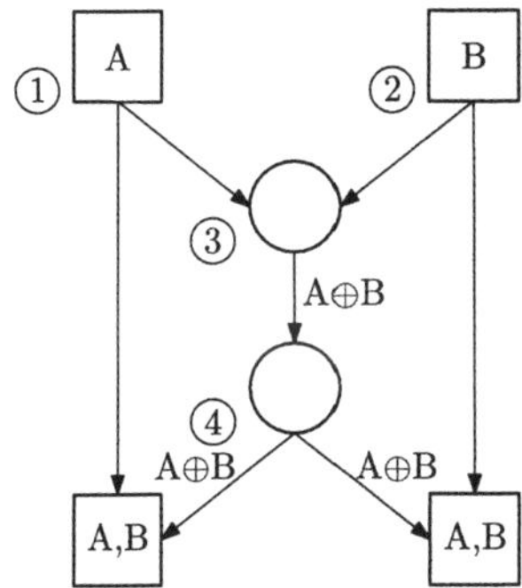

Figure 1.5 Network coded butterfly network with schedule.

the classical paradigm they are subject to fading and cause mutual interference to one another.

However, there are two approaches by which such discrete links can be emulated in a wireless network. The first is straightforward: separate orthogonal channels are provided for each link. In principle any means of orthogonalization could be used: different time-slots, different frequency channels, or different orthogonal bearer waveforms. For simplicity we will here assume that different time-slots are used: that the links are orthogonal in the time domain. Considering the network coded butterfly network in Figure 1.3b, this would require four time-slots per pair of source symbols to deliver the data to both destinations, as shown in Figure 1.5. This clearly reduces the efficiency of the network.

This also illustrates a general point about wireless networks that will be important in this book. Wireless devices are typically subject to the *half-duplex constraint*: that is, they cannot transmit and receive simultaneously on the same channel or in the same time-slot. There has been recent work on the implementation of full duplex wireless nodes, but that is beyond the scope of this book, in which for the most part we will assume the half-duplex constraint must be respected. This constraint immediately implies that a relay node can transmit in at most half of the time-slots.

As mentioned previously, information theory shows that transmission on orthogonal channels is not the optimum way of signaling from multiple source nodes to a single destination or relay node. In information theoretic terms this is known as the *multiple access channel* (MAC). The capacity of a MAC is defined by its *rate region*, as illustrated in Figure 1.6, for a two-user MAC. The left of the diagram illustrates the scenario: two sources, S_1 and S_2, transmit at rates R_1 and R_2 respectively to a common destination. The region within the solid line in the graph on the right denotes the rate region: the set of rate pairs that can be achieved with low error rate. Note that it implies that three limits operate: a limit on the rates R_1 and R_2 that each source can transmit independently plus a limit on the *sum rate* $R_1 + R_2$.

Note, however, that a conventional system using TDMA (i.e. using orthogonal time-slots) would be restricted to the triangular region shown by the dashed line — since any increase in the rate from one source would always have to be exactly balanced by

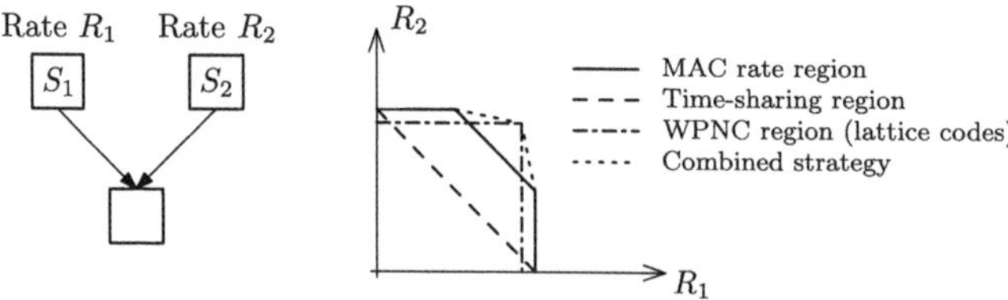

Figure 1.6 Rate region for two-user MAC.

a reduction in the rate from the other. The system can do better than time-sharing by allowing both sources to transmit simultaneously, and at the receiver to first decode one, then cancel the interference it causes and decode the other. This allows an increase in the sum rate significantly above the time-sharing rate. Thus in the network coded butterfly network we could allow sources A and B to transmit simultaneously, merging time slots 1 and 2 in the schedule shown in Figure 1.5, and increasing the network throughput.

However, this still constitutes a bottleneck in the network, because it requires symbols from both sources to be decoded even though what is required is only the one symbol formed by combining them with the network code function. Taking this into account, it is possible (as we will see later) to establish what we will call the *WPNC region*, which is the set of source rates which allows this symbol to be decoded. This is shown by the dash-dotted lines in Figure 1.6, and allows rates outside the conventional two-user MAC region. It is achievable e.g. by the use of nested lattice codes, as will be discussed in Chapter 5.

To achieve a rate outside the MAC region requires that rather than being obtained by decoding the two sources separately, and then applying network coding at the network layer (a strategy we will call *joint decoding*), the network coded symbol must be decoded directly from the received signal at the physical layer – in other words by *physical layer network coding* (PLNC). In this book we refer to the technique as *wireless physical layer network coding* (WPNC), and it is the main topic of the book. The term "wireless" is used here because the inherent superposition of wireless signals mentioned above means that this form of network coding is essential in wireless systems to obtain all the information available. There will of course be much more detail to come, and in particular there will be a "gentle" introduction to the main principles in the next chapter, so here we will restrict ourselves to a very simple example of how this might work and how it can enhance capacity.

Figure 1.7 shows the scenario. Two terminals transmit uncoded BPSK, taking signal values ±1 over channels with the same attenuation and phase shift to a relay. We assume that the relay applies network coding using the XOR function. At the relay the signals add, resulting in the values ±2 and 0. A joint detection strategy would need to decode the two sources separately, and this is clearly not possible if the value 0 is received, since it might represent the data 01 or 10. WPNC, on the other hand, has only to detect which network coded symbol the received signal corresponds to. This avoids the problem, since 01 and 10 both correspond to the network coded symbol 1. Thus the received signal can be interpreted as a constellation in which both the signals marked with white circles

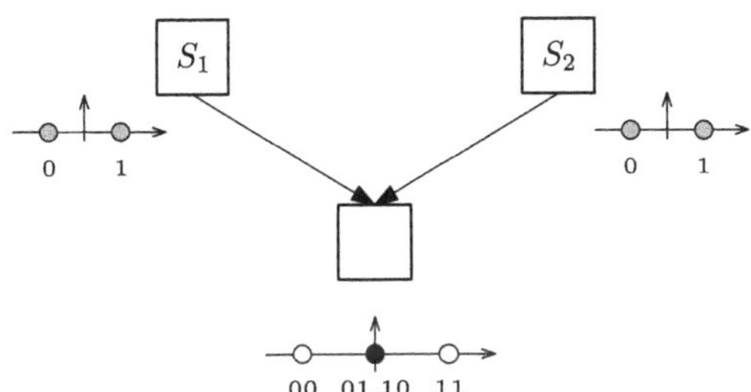

Figure 1.7 Illustration of PNC operation.

correspond to (network coded) 0, while the black circle corresponds to 1. This clearly increases capacity compared to both the joint decoding approach and the network coding approach.

1.5 Historical Perspective

At this point we will take a break from the technical details of WPNC to discuss how we reached this point, and the initial development of WPNC up to the present. We have already discussed some of the information theoretic background, and have mentioned the development of network coding. It is worth noting, however, that many of the theoretical foundations of multi-user information theory were laid in the 1970s – including analysis of the multiple access channel [4], [35], of the broadcast channel [11], and of the relay channel [12]. However, there has been little practical implementation of these concepts even up to today, although that is now changing, notably because of the pressures on wireless networks noted above, and also because multiple antenna systems have important synergies with the MAC and broadcast channels, which have led to the introduction of multi-user MIMO (MU-MIMO) systems in fourth-generation wireless systems. Multi-user information theory can now be seen as an important step towards the development of network information theory in the past decade or so, extending these concepts beyond single-hop multi-user networks. Both network coding and WPNC occupy the field of network information theory, and many concepts from it underlie the work in this book.

WPNC itself was discovered independently by three research groups, who approached it from slightly different angles, resulting in distinct approaches that, however, are clearly based on the same principles. Zhang, Liew, and Lam [64], of the Chinese University of Hong Kong, were probably motivated by concepts from network coding. They introduced the application of WPNC to the two-way relay channel, which we will review in the next chapter but which is quite similar to the butterfly network we have already seen. They also generalized it to a multi-hop chain network.

Popovski and colleagues at the University of Aalborg introduced an analog version of WPNC at the same time [49], based on earlier work applying network coding to the two-way relay channel [33]. They subsequently extended this to a scheme they refer to as *denoise and forward* [28]. Around the same time other work, e.g. [50], discussed other strategies for the two-way relay channel (though without proposing the full WPNC concept), and this was also the emphasis of the work by Popovski *et al.*

The third group was Nazer and Gastpar, then both at University of California Berkeley. Their earliest published work dates from 2005 [43], and was framed more as an approach to a new information-theoretic problem: that of decoding functions of symbols from multiple sources, rather than the sources themselves. However, this is evidently directly relevant to WPNC if the functions are those required in network coding, and leads to an approach called *compute and forward*. Their subsequent work and the work of other workers inspired by it has moved into the area of lattice coding, as a useful basis for the functions, and has retained a strong algebraic flavor.

Lattice coding is itself a field with a long history. It is based on the mathematical theory of lattice constructions, especially in more than three dimensions, but is connected with group theory as well as the physics and chemistry of crystals, going back to the middle of the nineteenth century. Its application to coding theory was extensively discussed in the 1980s in the classic reference on the topic, [10]. However, more recently it has undergone something of a renaissance, especially since it has been demonstrated that lattice codes with lattice decoding can also approach the Shannon capacity [16]. The work of Nazer and Gastpar [45] also used it to establish achievable regions for compute and forward.

Since this fundamental work the field has remained very active. Much of the early work continued to focus on the two-way relay channel, but recently this has been extended to other topologies, such as the multi-way relay channel, multiple relay networks, and multi-hop networks. Early work also focussed on information theoretic aspects, with little attention to practical implementation, but more recently more practical aspects have been investigated, such as the use of practical coding schemes, synchronization, performance on realistic wireless channels, etc. Recently also practical test-beds for the concept have been implemented [3, 38]. Of course, much of this work will feature in the remainder of this book.

1.6   Practical Usage Scenarios

We have already described the developments in wireless communications that provide the practical drivers for the move toward the network-aware physical layer in general, and the implementation of WPNC in particular. Here we will look in a little more detail at some specific scenarios in which it might be applied. The drivers we have considered include both conventional wireless broadband services via cellular and WiFi networks, and machine-type communications, including the "Internet of Things." However, these two different application areas may give rise to different network topologies, so we will discuss them separately here.

As mentioned above, access networks for cellular mobile networks are becoming denser in order to support rapidly increasing capacity density requirements arising from both increasing numbers of users and increasing data rate demand per user. To mitigate the interference problems this causes, the concept of *network MIMO* or *cooperative multipoint* (CoMP) has been introduced. In this approach several base stations cooperate to serve a user terminal, instead of each one being served by a single base station, so that signals received by another base station must be treated as interference. The network then exploits signals that would otherwise be interference, which can then enhance performance rather than degrading it. However, this requires that signals are no longer decoded only in one base station, and also implies that digitized signals rather than only user data should be transmitted between base stations and the core network. More recently the *cloud radio access network* (C-RAN) concept has been introduced, in which base station sites, containing baseband signal processing and higher-layer networking functions, are replaced by *remote radio units* (RRU) containing only the antennas, RF processing, and signal sampling and digitization. Baseband processing and all higher-layer functions for a large number of these RRUs are then concentrated in centralized *baseband units* (BBU). This clearly enables base station cooperation of the sort required by network MIMO to be more readily implemented. The connection between the RRU and the BBU is then known as *fronthaul* rather than backhaul, because it carries signal information rather than user data. The concept is illustrated in Figure 1.8. The major disadvantage of C-RAN is that the capacity required for the fronthaul is typically many times the total user data being transmitted, since it is a digitized signal rather than the actual data, and therefore typically requires longer sample words than the number of information bits per symbol to represent the signal at sufficient precision. It has therefore usually been assumed that optical fiber would need to be used to provide fronthaul connections (as opposed to wireless), which would greatly increase the cost of the network.

WPNC provides a potential alternative, which greatly reduces fronthaul load, potentially allowing it to be implemented over wireless. As in the example illustrated in Figure 1.7 above, a base station receiving signals simultaneously from two terminals might decode a network coded function of the two, rather than attempting to decode one in the presence of interference from the other. Thus it exploits all signals received from a terminal just as network MIMO does, and it achieves a performance that is similar in the sense that it provides the same diversity order, albeit typically with a small degradation in terms of required signal to noise ratio. However, because the network coded signal in principle contains the same number of symbols as each of the user data streams, it

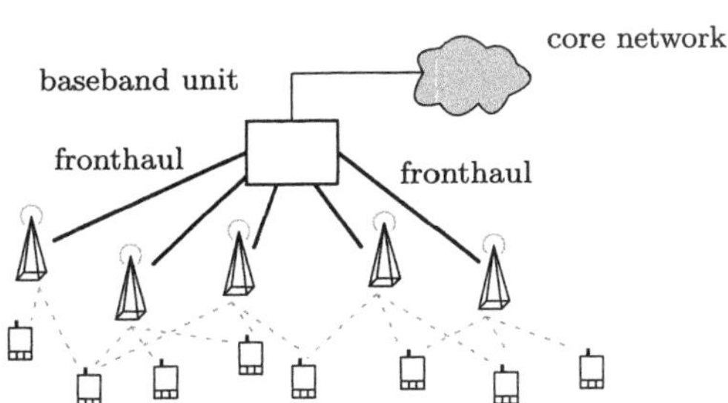

Figure 1.8 Cloud Radio Access Network.

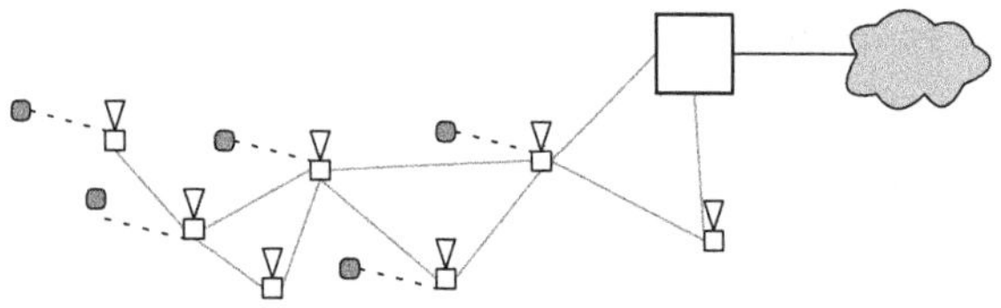

Figure 1.9 Mesh network for "Internet of Things" applications.

requires no expansion of the fronthaul load compared to the total user data rate. This might well allow wireless links to be used, with the potential to reduce network costs.

Machine-type communications, on the other hand, are likely to call for a different network structure. Potential applications include sensor networks, industrial process control, "smart grid" and "smart city" systems, to name just a few. These have in common that they are likely to involve very large numbers of devices, widely distributed across a service area, with very low power. This may mean that it is not feasible to provide a dense enough access network to serve all these devices directly, so these applications are likely to lead to a mesh network topology based on device-to-device communications and low-cost relay nodes to provide links back to the core network, as illustrated in Figure 1.9. In many cases the data rate per device is relatively small and occurs in the form of small packets, but there are large numbers of devices and large numbers of packets in total. In addition many applications are highly time-critical and require very low latency.

We have already reviewed the limitations of multi-hop mesh network topologies when the conventional network paradigm is used, especially the capacity bottleneck that results from interference between links, and this will clearly apply in many of these applications. Moreover, the conventional paradigm tends to result in packet collisions, requiring retransmission at intermediate hops that potentially increases end-to-end delay. Thus WPNC is very well suited to these applications, since its exploitation of otherwise interfering signals has the potential to overcome the capacity bottleneck in multi-hop networks. Similarly it can exploit colliding packets to extract information that can be further forwarded through a network, minimizing the need for retransmissions.

Both of these application areas are examples of the current developments in wireless communications towards *ultra-dense* networks, in which it is no longer feasible to avoid interference between different links within the same network. The paradigm of the "network-aware physical layer," which we have introduced in this chapter, and will explore in the remainder of this book, is therefore extremely timely.

2 Wireless Physical Layer Network Coding: a Gentle Introduction

2.1 The 2-Way Relay Channel

In this chapter we begin to describe the principles of WPNC, taking a more "gentle" approach than we do in the remainder of the book, minimizing the use of mathematics in favor of more descriptive and graphical approaches as a means to explain these principles. We will see that the simple example described in Section 1.4 already captures some of the important issues, but we will begin the process of generalizing it and setting it in the context of a complete network, albeit a very simple one.

Accordingly we focus on the *2-way relay channel* (2WRC)[1] as a very simple example of a complete network (in fact the simplest possible, as we will see in a moment) in which WPNC can be applied. The 2WRC is illustrated in Figure 2.1. The basic idea is that two terminals each have data to exchange with the other, but (perhaps because the distance between them is too great for a direct link) they wish to use an intermediate *relay* node for the purpose.

The reason for focussing on the 2WRC is that it provides a simple example of a multi-hop wireless network supporting multiple data flows, as well as being an example that demonstrates the benefits of WPNC particularly clearly and one that is of some practical interest. In fact, as mentioned in Section 1.5, a large proportion of the work in the field in the past decade has exclusively addressed this network.

We emphasize here, following on from Section 1.2, that WPNC applies to wireless *networks*, not to individual point-to-point links – this is the essence of the "network-aware physical layer." Such networks must necessarily involve more than one wireless "hop" between transmitter and receiver, and hence must include a relay node as well as source and destination terminal nodes. They must also necessarily involve more than one data source, leading to multiple data flows through the network that also interact at some point within it. On this basis the 2WRC, containing two terminal nodes and one relay and involving two flows each originating at one of the terminals, is in fact the simplest possible example.

We will begin by comparing the WPNC approach to the 2WRC with two previous approaches: the conventional one and one based on network coding at the network layer, showing the potential benefits of WPNC over both of these. We will then describe and compare some alternative schemes which can all in some sense be labeled as WPNC.

---

[1] Sometimes, it is also abbreviated as TWRC (Two-Way Relay Channel).

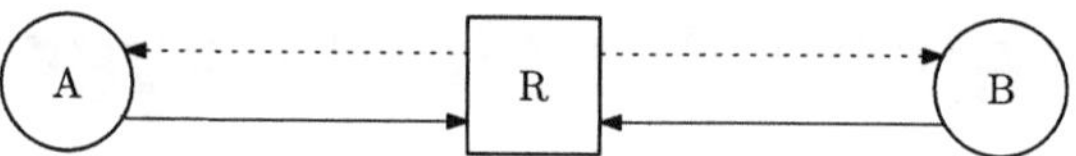

Figure 2.1 2-way relay channel.

This will lead us to one of the common requirements of these schemes: the need for *unambiguous decodability*, that is, that the original data can be recovered at the destination without ambiguity and therefore with certainty. We will also introduce the concept of *hierarchical side information*, and describe its role in unambiguous decoding.

Up to this point we will assume that BPSK modulation is used, as in our example in Section 1.4, but we will next extend our consideration to QPSK modulation. As we will then see, this introduces additional problems that do not arise with BPSK as a result of the unpredictable parameters of the channel – primarily the effect of *fading*. This causes phase shifts and amplitude variations in the signal that in general are unknown to the transmitter. It is in particular the relative values of these parameters between the channels from the two sources and the relay that influence the behavior of the network.

Finally we will extend our consideration to other example network topologies, and in particular to what we refer to as the *hierarchical wireless network*, where a set of source nodes are connected to the destination via one or more layers of relays. We will see how similar considerations apply in such networks as in the 2WRC.

Note that in this chapter, for simplicity in explaining the basic principles of WPNC, we assume uncoded transmission in most of the text (with the exception of Section 2.5). In later chapters an important theme will be how forward error correction (FEC) coding can be incorporated into the scheme.

2.2   Conventional, Network-Layer Network Coding, and WPNC Approaches

The 2WRC can be operated in a variety of modes, involving different schedules for the activation of the nodes that comprise it. These are illustrated in Figure 2.2. The conventional approach using a point-to-point physical layer would require four time-slots, or *phases*, for a complete cycle of transmissions. First terminal A transmits to the relay R, then R retransmits A's data to terminal B. Next B transmits to R, and R retransmits B's data to A. In the conventional paradigm none of these phases can take place concurrently, either because the transmissions would then interfere at the relay, or because of the half-duplex constraint on the relay.

In the network-layer network coding (NC) approach, illustrated in Figure 2.2b, the relay is no longer restricted to simply forwarding data it has received. Instead it calculates a function of the data of both A and B, which we refer to as the network code function or *mapping*. In our present example, because the data are binary, the function is the exclusive OR (XOR) function, but in the general case a wide range of other options are possible, as we will see. This then allows a three-phase schedule, as shown in the figure. Terminal A transmits its data to the relay in the first phase, then terminal B transmits its data in the second phase. The relay then forms the function A $\oplus$ B, and transmits this simultaneously to terminals A and B in the third phase. This procedure

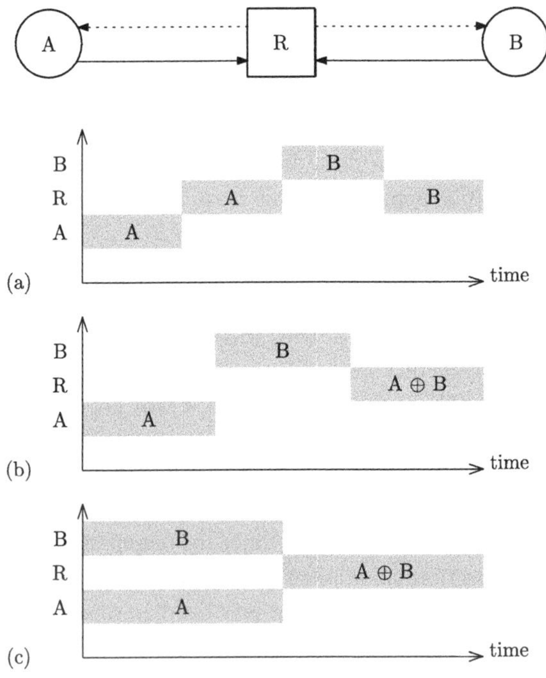

Figure 2.2 Activation schedules for the 2-way relay channel: (a) conventional, four-phase; (b) network-layer NC, three-phase; (c) WPNC, two-phase.

works because each terminal has available the data it originally transmitted, and can decode the data from the other terminal by applying a second XOR function, as we will see in Section 2.4 below. We will refer to information like data B in this case, which assists a terminal in recovering its data of interest even though it does not itself depend on that data, as *hierarchical side-information* (HSI). The rationale for this terminology will be explained in Chapter 3. Of course terminal A can perform an equivalent process.

We may note that in terms of data flows the 2WRC is equivalent to the "butterfly network" discussed in Section 1.3 above, illustrated in Figure 1.3b. Here the upper of the two nodes in the center of the diagram represents the application of the XOR function to the incoming data, while the lower represents the broadcast of the network coded (i.e. XORed) data. The links directly joining sources and destinations represent the HSI which the source in each terminal makes available to the network decoding function, carried out in the nodes at the bottom of the diagram. This diagram has the advantage of making the transfer of the HSI explicit.

Note that the 2WRC is equivalent to a butterfly network in which the HSI transfer is perfect, because the source and destination reside in the same terminal. Later in the book we will consider another example in which these links may not be perfect, because source and destination may be separated. This clearly has implications for the operation of the network, as we will see.

We noted in Section 1.3 when considering the application of NC (at the network layer) to the butterfly network that the NC model effectively assumes that the data flows from the two sources arrive over discrete links, which we noted was not naturally the case in wireless networks. However, the schedule shown in Figure 2.2b overcomes this by separating the two links in two time-slots, in other words by applying time-division multiple access (TDMA) over the wireless medium to provide orthogonal channels for the links. The data on these links can then be decoded separately before the network code function is applied. For this reason it must be treated as a form of network-layer NC, rather than WPNC.

The approach shown in Figure 2.2c, however, reduces the schedule to two phases. Now terminals A and B transmit simultaneously in the same time-slot (and in the same frequency channel). Thus their signals arrive at the relay as a superposition of the electromagnetic waves of the two wireless signals, so that the signals are no longer readily separable at the relay, and so it will not be easy (unless using *coded* signals and multi-user decoding which, however, imposes some limitations on the rates as will be described later) to decode their data separately. However, the relay does not necessarily need to do so: all it requires to do is to extract the network code function from the superposed received signal. Since the output of the function has less entropy (that is, contains less information) than the combined information of the original data sequences, in general this may be an easier process than separate decoding. This question will be addressed much more rigorously in later chapters of this book.

However, the very simple example of WPNC that we gave in Section 1.4 shows how in some circumstances it may be impossible to regenerate the original data sequences but still readily possible to obtain the network coded data. The example is illustrated in Figure 1.7, where it is assumed that both sources transmit BPSK to the relay over channels that happen to have the same phase shift and attenuation. Thus the signals combine to give a constellation with three signal points rather than four, which we have labelled -2, 0 and 2. Note that -2 and $+2$ correspond to the case where the two sources transmit $(0,0)$ and $(1, 1)$, respectively, while 0 occurs with either $(0, 1)$ or $(1, 0)$. Hence if this point is received at the relay it cannot with certainty decide which of these two pairs of data symbols was received. However, since these two pairs both result in the same network coded symbol, namely 1 (since $1 \oplus 0 = 0, 0 \oplus 1 = 1$), it is able to decode this symbol with certainty. And of course if either -2 or $+2$ is received, this will be decoded as network coded 0, since $1 \oplus 1 = 1, 0 \oplus 0 = 0$. (Note that while it is very unlikely that the two channels will be exactly the same, as required by this example, nevertheless if they are close, so that the pairs $(0, 1)$ and $(1, 0)$ produce very similar signals, in the presence of noise it will still be very difficult to distinguish them, but remain easy to obtain the network coded symbol.)

Note, however, that this direct extraction of the network code function must necessarily take place at the physical layer, since the information must be obtained from the received signal, which is only available at the physical layer. It cannot in general be separated into decoding of source data symbols followed by network coding applied at the network layer. However, it must be a physical layer that is aware of the nature of the superposed signals it will receive: both their statistical characteristics (especially the combined constellation they may form) and their significance as a representation of different combinations of source data. In this sense the physical layer must be "network aware," as discussed in Section 1.2.

The example discussed above provides only one of several ways of processing the received signal and retransmitting some function of it. In the next section we compare it with some alternative strategies. In the remainder of the chapter (and indeed the remainder of the book) we will for the most part focus on the two-phase protocol of Figure 2.2. We will often refer to the first phase (sources to relay) as the *multiple access channel* (MAC) phase, and the second (relay to destinations) as the *broadcast channel* (BC) phase, because the phases involve many-to-one and one-to-many transmission, respectively, like the corresponding channels.

2.3   WPNC Relay Strategies

Here we consider the case of WPNC as applied to the 2WRC (that is, where a two-phase schedule is applied, as illustrated in Figure 2.2c), and especially some alternative strategies available to the relay.

The fundamental requirement that the relay must fulfill is to transmit some function of the two data symbols which is such that the required data can be unambiguously decoded at the destination, given the appropriate HSI. In the next section we will consider in more detail the requirements placed on the relay function by this unambiguous decodability criterion, but here we will consider some simple functions and strategies to obtain them.

The simplest such strategy is for the relay to directly store the received signal, amplify it and retransmit it. This is known as *amplify and forward* (AF). The destination in each of the two terminals can recover the required data, assuming that its own data and information about the channels between both terminals and the relay are available to it, by subtracting the interference at the relay due to those data. The disadvantage of AF is that the noise at the relay receiver is also amplified, and adds to the noise on the relay–destination link. However, provided both channels and data are perfectly known, the effect of the second signal at the relay can be completely eliminated. In terms of the rate region illustrated in Figure 2.3, this means that the rate region is rectangular, since the data flow from one source to destination is completely unaffected by the flow from the other. Once the interference has been removed, the end-to-end link can be represented by a single equivalent channel whose noise is given by the sum of the noise at the final destination and the noise at the relay amplified and transmitted over the relay–destination link. Therefore the capacity of each user and hence the size of the region is reduced because noise is greater than on either of the channels on their own. The rate region is shown by the solid line in Figure 2.3. Note that the regions shown in this diagram are intended to be illustrative only, not exact results for any specific channel.

The second strategy is to apply multiple access techniques at the relay to first decode each source separately, then apply the network code function to the decoded symbols, and broadcast the resulting network coded symbol to both destinations. As previously mentioned, the classical way to do this is to first decode the lower-rate source, which is able to use a more powerful error correction code, estimate the interference this causes and subtract it, so that the higher-rate source is able to decode as if it were operating on

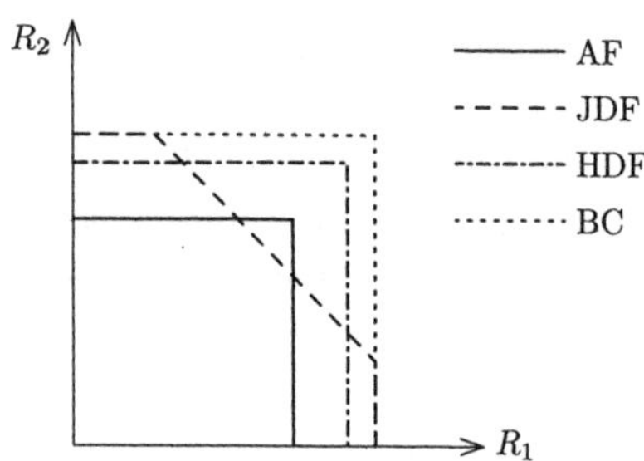

Figure 2.3 Rate regions for 2WRC: AF, JDF, HDF, and BC.

an interference-free channel. The network code function which then operates on the two decoded symbols must be chosen in such a way that it can be decoded at the destination given its own source data. This will be discussed in the next section, but we will note here that an advantage of this *joint decode and forward* (JDF) strategy, if it turns out to be possible, is that we are free to choose any network code function that fulfills this requirement. It also has the advantage compared with AF that each node decodes in the presence only of its own noise: we do not encounter the cascade combination of noise from successive links that occurs in AF. The rate region now has to be evaluated for the two phases of the network's schedule separately, whereupon the overall rate is the smaller of the two, since the phase with lower capacity will act as a bottleneck. Here the rate region of the MAC phase is just the expected rate region for a MAC (as shown in Figure 2.3, where the dashed line is the JDF rate region, and also previously discussed in Figure 1.6), because as for the MAC both sources have to be decoded at the relay. The individual rate limits for R_1 and R_2 that bound the rate region arise from the cancellation of interference due to the other flow, and thus are simply the capacity bound for the corresponding point-to-point link. In the broadcast phase also the two links each function like a point-to-point link, in which the rate of one does not affect the other, and so the rate region is rectangular. Figure 2.3 addresses the case where the channels between the two terminals and the relay are balanced in terms of propagation, and hence the rates for the two users for the broadcast channel are the same as the individual rate limits for the MAC channel. Hence in this case the MAC rate region lies within the BC rate region, and hence in this and many other cases it is the MAC phase that gives rise to a bottleneck and defines the overall rate region for the network. Moreover, because the MAC rate region is pentagonal rather than rectangular, it may also mean that the corner point of the AF region extends outside it, as shown in Figure 2.3, so that it is possible to achieve higher rates for the two users simultaneously by using AF than JDF, although its individual rate limits are lower than in JDF.

The third strategy is the one we have already described above, and illustrated in the previous chapter. The relay decodes the network coded function directly from the received signal. Thus it does not necessarily need to decode the two source symbols separately, but only determine which value the network coded function should take. This in general is an easier decoding task than the joint decoding described above, because the function is a many-to-one function, and it requires fewer values to be distinguished. For example, in the case discussed in Section 2.2 above, and illustrated in Figure 1.7,

Table 2.1 Summary of multi-source/node PHY techniques – classical single-user point-to-point (P2P), classical multi-user (MU), network-level NC, and native WPNC.

	P2P PHY	MU PHY	NC	WPNC
Topology: *direct neighbors* signal interaction	−	+	−	+
Topology: *full network* structure	−	−	+	+
Signal structure: *constellation (signal) space* level	+	+	−	+
Relay Tx signal codeword map: a *function* of data	−	−	+	+
Relay Rx signal codeword map: a *function* of data	−	−	−	+

the received signal could in principle take four values, but the network code function (the XOR function) takes only two values. Thus the decoder at the relay needs only distinguish between two pairs of signals. In the example given in Figure 1.7 one of these pairs contains two fully coincident points, and so, as already mentioned, it would be impossible to decode both sources separately,[2] but nevertheless the network code can be decoded. We refer to the sets of points from the full received constellation that correspond to the same network code value as *clusters*: in the example illustrated, the distance between the nearest points in the two clusters is in fact the same as it would be in the received constellation from a single source without interference but, in general, if points do not coincide the inter-cluster distance will be smaller than in the interference-free constellation. For this reason the limitations on the rates of the individual sources are a little lower, and hence the rate region is smaller than the BC region, although again, because it is rectangular, its corner may project beyond the MAC rate region, as shown in Figure 2.3. It is, however, larger than for AF, because the noise is smaller. We refer to this strategy as *hierarchical decode and forward* (HDF), because what is decoded is a hierarchical function of the source symbols, although in this very simple example the hierarchy contains only a single level (see Chapter 3, which explains the hierarchical principle in detail). Similarly the constellation of the received signal is a hierarchical constellation, consisting of a hierarchy of clusters and points.

In terms of relative performance, Figure 2.3 shows the comparison between the three approaches we have discussed. As mentioned, the figure is of course only illustrative, and the exact numerical comparison depends on the details of the channels and signal-to-noise ratios involved. However, it is clear that at least potentially HDF can outperform the other schemes in terms of overall sum rate, even if JDF can achieve a higher rate for the individual sources. In terms of complexity, AF is undoubtedly the simplest to implement, especially at the relay, since very little baseband processing except storage is required. In principle JDF may require a maximum likelihood (ML) or maximum a posteriori probability (MAP) detector, with complexity proportional to the received constellation size, and therefore exponential with the rate. The implementation of HDF, and conditions under which it may be simplified, will be an important theme of this book. Table 2.1 shows a summary of processing aspects for various classes of PHY techniques used in multi-node and multi-source networks.

---

[2] For simplicity we refer here to uncoded transmission. In coded systems the codebook structure might help to distinguish these points.

2.4   Unambiguous Decoding and Hierarchical Side-Information

If the relay transmits a function of the source symbols back to the terminals (rather than the symbols themselves), it is clearly essential that the terminals are able to recover the original data symbols that are of interest to them: in other words to decode the network code function applied at the relay. More formally, we say that the symbol received at the destination must allow *unambiguous decoding* of the source of interest. Unambiguous decoding is possible provided the combination of network coded symbols received at a given destination corresponds only to one possible symbol from the source of interest. Otherwise an ambiguity remains about the source symbol after the network coded symbol has been received, and information is lost. As we will see, however, the destination terminals require additional information to allow them to decode; we have already referred to this as hierarchical side-information (HSI). We must ensure that unambiguous decoding is possible when the HSI and the network coded symbol, which we call *hierarchical information* (HI), are both available at the destination.

In our example using the 2WRC unambiguous decoding is very easy to achieve. As we have seen, the relay obtains the XOR function $A \oplus B$ of the two source data symbols, and forwards it to both destinations, where it provides HI about the source symbol of interest. In this case the destinations also have as HSI the data symbol transmitted in the previous time-slot by the source collocated in the same terminal. This does not itself contain any information about the source symbol of interest (that from the other terminal), but it does help to decode that symbol. For example terminal B combines the data $A \oplus B$ received from the relay with its own data, forming $(A \oplus B) \oplus B = A \oplus (B \oplus B) = A \oplus 0 = A$, and thus recovers the data A that it requires.

To generalize this somewhat, let us suppose that the data symbols from the two sources, which we will denote as b_A and b_B, are drawn from an alphabet $\mathcal{A}$ of size M (we say that they have *cardinality M*). The network code or mapping function applied at the relay is denoted as $\chi(b_A, b_B)$. In order unambiguously to decode data symbol b_A at terminal B we require that the combination of the network coded symbol $\chi(b_A, b_B)$ and the source symbol b_B should uniquely define the symbol b_A from source A, for all possible b_A and b_B. This requires that the combination is different if b_A is different, that is, that

$$\{\chi(b_A, b_B), b_B\} \neq \{\chi(b'_A, b_B), b_B\}, \quad \forall b_B, b_A, b'_A \neq b_A \tag{2.1}$$

or, more simply,

$$\chi(b_A, b_B) \neq \chi(b'_A, b_B), \quad \forall b_B, b_A, b'_A \neq b_A. \tag{2.2}$$

This is commonly called the *exclusive law*. Conversely, for unambiguous decoding of b_B at terminal A we require

$$\chi(b_A, b_B) \neq \chi(b_A, b'_B), \quad \forall b_A, b_B, b'_B \neq b_B. \tag{2.3}$$

Note that this form of the requirement for unambiguous decoding applies specifically to the 2WRC: for other topologies it should be modified, as we will see in Section 2.7 of this chapter.

Table 2.2 Table to define mapping function.

b_B / b_A	0	1	...	$M-1$
0	0	1	...	$M-1$
1	$M-1$	0	...	$M-2$
$\vdots$	$\vdots$	$\vdots$	$\ddots$	$\vdots$
$M-1$	1	2	...	0

This requirement in its turn imposes requirements on the mapping function. These requirements can be expressed in various ways, just as the mapping function can be defined in different ways. A general way to define the mapping, at least for small numbers of arguments, is by means of a table, as illustrated in Table 2.2. Once again, this table is intended to illustrate principles: except as discussed below the particular content of the table is not intended to be prescriptive.

This table exhaustively lists the output value of the function $b_{AB} = \chi(b_A, b_B)$ for all combinations of input, and thus allows us to define an arbitrary (discrete) function of the two arguments. The approach can also be extended, in principle, to functions of more than two arguments by increasing the number of dimensions of the table, but this clearly is not necessary for the 2WRC. Note that the cardinality of the output alphabet of the function, $M_{AB} = |\mathcal{A}_{AB}|$, $b_{AB} \in \mathcal{A}_{AB}$, need not be the same as that of its arguments, and indeed the cardinalities of the two inputs, $M_A = |\mathcal{A}_A|$ and $M_B = |\mathcal{A}_B|$, $b_A \in \mathcal{A}_A$, $b_B \in \mathcal{A}_B$ do not need to be the same.

We observe that if the output cardinality of the function is equal to the total size of the table, i.e. $M_{AB} = M_A M_B$, then the function may be unambiguously decodable even without any HSI, since each entry can be mapped unambiguously to the corresponding pair of source symbols, provided no symbol is repeated within the table. This is referred to as *full* cardinality. However, in many ways it would nullify the benefits of the 2WRC, so for HDF we prefer a function with lower cardinality than this. We may observe from the table illustrated in Table 2.2 that symbol b_A can be unambiguously decoded provided any symbol occurs only once on any given column of the table, so that if b_B is known (which defines the column), the coded symbol unambiguously defines the row, and hence b_A. This requires that $M_{AB} \geq M_A$. Similarly b_B can be decoded if any symbol occurs only once in a row, which requires that $M_{AB} \geq M_B$. Hence correct operation of the 2WRC requires that $M_{AB} \geq \max(M_A, M_B)$. The equality in this expression defines what is known as *minimal* cardinality. Any value between this minimum and full cardinality will be referred to as *extended* cardinality.

There are other, less general ways of defining the function. In particular we have already noted that network coding functions which are *linear* on some algebraic field are used. We note that linearity may also be defined on a ring as well as a field, but for brevity we refer here primarily to the field. The function may then be defined in the form

$$\chi(b_A, b_B, \ldots) = a_A \otimes b_A \oplus a_B \otimes b_B \oplus \cdots \tag{2.4}$$

where the symbols $b_A, b_B, \ldots$ and the coefficients $a_A, a_B, \ldots$ belong to the same field, and $\oplus$ and $\otimes$ denote addition and multiplication in the field, respectively.

If such a function is applied in the 2WRC, it is easy to see that b_A can be unambiguously decoded provided the corresponding coefficient a_A has a unique inverse in the field, since at destination B the term $a_B b_B$ can be subtracted and the residue multiplied by the inverse of a_A (and conversely for b_B). Because in a field all elements except 0 have unique inverses, this is always possible provided both coefficients are non-zero (that is, the function depends on both its arguments).

In the binary case we have been considering so far the table definition of the function as described above is 2×2, and its entries are 1s and 0s. Since there must be one "1" and one "0" on each row and each column, the table must take the form of the XOR function (or its inverse). It is therefore also a linear function, whose symbols and coefficients are in $\mathbb{F}_2$. The argument above also shows that both coefficients must be "1"; thus our binary 2WRC example leaves us no options in the choice of network code function.

In Section 2.7 of this chapter we will extend these concepts to a more general network topology, but at this point it is worth noting that the considerations we have dealt with here create conditions on the design of the network code functions for a WPNC network that apply to the whole network. In the next section, on the other hand, we will encounter conditions on the function that apply at an individual relay node.

2.5   Achievable Rates of HDF and JDF

Among all strategies for multi-user and multi-node wireless networks, the HDF (as one particular example of a *PHY-native* WPNC technique) and JDF (as a more traditional approach) are the ones sharing some important commonalities, namely in processing a hierarchical many-to-one function of the data streams at the relay. The JDF does that by concatenating the traditional multi-user decoding of all individual data streams, whereupon the discrete network-level NC is subsequently applied. In contrast, HDF decodes the mapping function directly using the signal space observation. The example cases treated so far, have assumed uncoded transmission or kept the statements at a quite generic qualitative level for the sake of simplicity.

However, the performance comparison of HDF and JDF is of such importance that we now expose the coded case in a slightly more exact form. More elaborate mathematical treatment will serve as a gentle introduction to the information-theoretic style of analyzing WPNC systems used in the rest of the book. Particular numerical results will also serve as a justification of the HDF-based approach and as a motivation for the rest of the book.

We will consider a very simple scenario for the hierarchical MAC channel where two sources communicate at the same time and frequency (with mutually interfering signals) with one relay that aims to decode a hierarchical many-to-one data mapping function. We will assume *coded* transmission and compare the achievable rates of HDF and JDF. There are many additional conditions and constraints under which the following statements hold and these are treated in detail in the rest of the book. For the sake of clarity, we will not state them explicitly now and we urge the reader to check them carefully in order to avoid misinterpretations. We also consider the simplistic case of two BPSK sources in a *real-valued* AWGN hierarchical MAC channel. Even though this example has very little practical relevance, and it still does not allow closed-form mathematical results (they must be determined numerically), the treatment is relatively simple and prepares the ground for the more complex expositions used later in the book.

2.5.1 Two-Source BPSK Hierarchical MAC

We assume two coded sources with messages $b_A \in [1 : 2^{NR_A}]$, $b_B \in [1 : 2^{NR_B}]$, where N is the codeword length, and source codebooks $\mathcal{C}_A, \mathcal{C}_B$ with identical code rates $R_A = R_B$. In information theory, the message is frequently described by a scalar index drawn from some discrete value range. It stresses the fact that the form of the information is irrelevant and the only important aspect is the total number of message values. It also has a nice interpretation as line index numbers of the codebook. The total number of codebook lines is $M_i = 2^{NR_i}$, $i \in \{A, B\}$, where R_i is the so-called rate. The codeword length N is the length of the line in the codebook. The rate of the code is the binary-base logarithm of the codebook size per codesymbol, $R_i = \lg M_i / N$, i.e. how many binary symbols are represented by one codesymbol.

The codesymbols $c_{A,n}, c_{B,n} \in \{0, 1\}$ use the BPSK channel alphabet $s_{A,n}, s_{B,n} \in \{\pm 1\}$, with size $M = 2$, mapped symbol-wise to the codesymbols. The observation model is a *real-valued* AWGN channel

$$x_n = s_{A,n}(c_{A,n}) + s_{B,n}(c_{B,n}) + w_n \tag{2.5}$$

where the noise has σ_w^2 variance per dimension and its probability density function (PDF) is

$$p_w(w) = \frac{1}{\sqrt{2\pi\sigma_w^2}} \exp\left(-\frac{w^2}{2\sigma_w^2}\right). \tag{2.6}$$

The SNR is defined as

$$\gamma = \frac{\mathrm{E}[|s_i|^2]}{\sigma_w^2}. \tag{2.7}$$

The hierarchical mapping function is XOR

$$c_n = \chi_c(c_{A,n}, c_{B,n}) = c_{A,n} \oplus c_{B,n} \tag{2.8}$$

and thus it has the minimal cardinality $c_n \in \{0, 1\}$. Under a number of specific assumptions treated later (e.g. isomorphic layered code, regular and symbol-wise independent and identically distributed (IID) perfect random codebooks, etc.; see Sections 5.7, 5.7.3, 5.7.4, and Chapter 4), we can assess the coded system performance using *single channel symbol* information-theoretic properties. The isomorphic assumption implies that we can uniquely decode the hierarchical map of the information data messages. The hierarchical data map is $b = \chi(b_A, b_B)$ and $b \in [1 : 2^{NR}]$ where R is the hierarchical data rate.

2.5.2   JDF Strategy

The JDF strategy is *limited* by a *classical multi-user rate region*. Both data streams must be first reliably *individually* decoded before they can be used in the network-level NC. The achievable rates are given in terms of mutual information expressions

$$R_A < I(C_A; X|C_B), \tag{2.9}$$

$$R_B < I(C_B; X|C_A), \tag{2.10}$$

$$R_A + R_B < I(C_A, C_B; X). \tag{2.11}$$

We dropped the sequence index n from the notation. All following statements refer to a single symbol.

The mutual information between a pair of random variables describes how much the outcome uncertainty of one of them is reduced after observing the other one. The conditional mutual information assumes that the stochastic behavior of the variables involved is conditioned by the knowledge of the conditioning variable (e.g. the codesymbol is known). In the case where it is not clear from the context, or when we need to distinguish it explicitly, we use capital letters to denote the random variables and lower-case letters to denote their particular values.

The achievable rate is the rate of some given codebook construction that can be decoded with some given decoding strategy with error probability approaching zero for $N \to \infty$. The achievable rate is typically determined by some function containing mutual information expressions. Under common memoryless channel and so-called IID random codebook assumptions, the involved mutual information expressions are related to *individual* symbols. The random IID codebook is an abstraction in constructing the *hypothetical idealized* codebook that makes the information theoretic proofs of coding theorems possible; see Section A.4 for details.

Owing to the symmetry of the channel and the symmetry of the codebooks, the achievable rates have the first-order limit

$$R_A = R_B < I_1 = I(C_A; X|C_B) = I(C_B; X|C_A) \tag{2.12}$$

and the second-order limit

$$R_A = R_B < I_2/2 \tag{2.13}$$

where

$$I_2 = I(C_A, C_B; X). \tag{2.14}$$

Thy symmetry of the system and the minimal cardinality map then implies $R = R_A = R_B$.

The first-order limits are essentially the single-user rates

$$I_1 = \mathcal{H}[X'] - \mathcal{H}[X'|C]$$
$$= \mathcal{H}[X'] - \mathcal{H}[W] \tag{2.15}$$

where

$$\mathcal{H}[W] = \frac{1}{2} \lg(2\pi \, e \, \sigma_w^2) \tag{2.16}$$

is the AWGN entropy and $X' = S_A(C_A) + W$ is the effective single-user channel model with the second source removed to equivalently model the conditioning in the mutual information. The effective observation has PDF

$$p(x') = \sum_{c_A} p(x'|s_A(c_A))p(c_A)$$

$$= \sum_{c_A} p_w(x' - s_A(c_A))p(c_A) \tag{2.17}$$

and entropy

$$\mathcal{H}[X'] = -\mathrm{E}_{p(x')}\left[\lg p(x')\right]. \tag{2.18}$$

The second-order limit is

$$I_2 = \mathcal{H}[X] - \mathcal{H}[X|C_A, C_B]$$

$$= \mathcal{H}[X] - \mathcal{H}[W] \tag{2.19}$$

where the observation entropy is

$$\mathcal{H}[X] = -\mathrm{E}_{p(x)}\left[\lg p(x)\right] \tag{2.20}$$

and

$$p(x) = \sum_{c_A,c_B} p(x|s_A(c_A), s_B(c_B))p(c_A)p(c_B)$$

$$= \sum_{c_A,c_B} p_w(x - s_A(c_A) - s_B(c_B))p(c_A)p(c_B). \tag{2.21}$$

All codesymbols have uniform a priori probability mass function (PMF) $p(c_A) = 1/M$ and $p(c_B) = 1/M$.

2.5.3 HDF Strategy

The HDF strategy, in contrast to JDF, directly decodes the hierarchical data map. The achievable hierarchical rate, under some conditions (e.g. using regular isomorphic layered NCM, etc.; see Sections 5.7 and 5.7.4 for details), is given by the hierarchical mutual information

$$R < I(C; X) = I_H. \tag{2.22}$$

Notice that we do *not* need to explicitly decode the individual source data streams, and that the hierarchical data rate is directly given by the single-symbol information-theoretic limit. The symmetry of the scenario and the minimal cardinality map again implies

$$R = R_A = R_B. \tag{2.23}$$

The hierarchical mutual information $I(C; X)$ evaluation requires the knowledge of the hierarchical channel symbol conditional PDF. It describes the observed received signal from the perspective of the hierarchical channel symbol which is, in turn, mapped to the hierarchical codebook encoded message b. The conditional PDF for our minimal cardinality map with uniformly distributed symbols is (see details in Section 4.4)

$$p(x|c) = \frac{1}{M} \sum_{c_A,c_B:c} p\left(x|s_A(c_A), s_B(c_B)\right), \tag{2.24}$$

where the summation set $c_A, c_B : c$ is the summation over all c_A, c_B consistent with hierarchical symbol c, i.e. such that $c = \chi_c(c_A, c_B)$. The hierarchical mutual information is then

$$I_H = \mathcal{H}[X] - \mathcal{H}[X|C]. \tag{2.25}$$

Unlike the JDF case, the conditioning in $\mathcal{H}[X|C]$ still leaves some ambiguity because of the many-to-one hierarchical mapping function property. The conditional entropy thus needs to be explicitly evaluated using

$$\mathcal{H}[X|C] = -\,\mathrm{E}_{p(x,c)}\left[\lg p(x|c)\right] \tag{2.26}$$

where $p(x|c)$ is given above and $p(x, c) = p(x|c)p(c)$ where hierarchical symbols have uniform PMF $p(c) = 1/M$.

2.5.4   Achievable Rates

The achievable rates for JDF and HDF strategies are now evaluated numerically. The integrals of the expectations in the entropies do not have closed-form solutions; however, it is a relatively easy numerical task. We first visualize them by plotting (in Figure 2.4) the achievable hierarchical rate for JDF, which is given by the bottleneck of the first- and the second-order rate limits

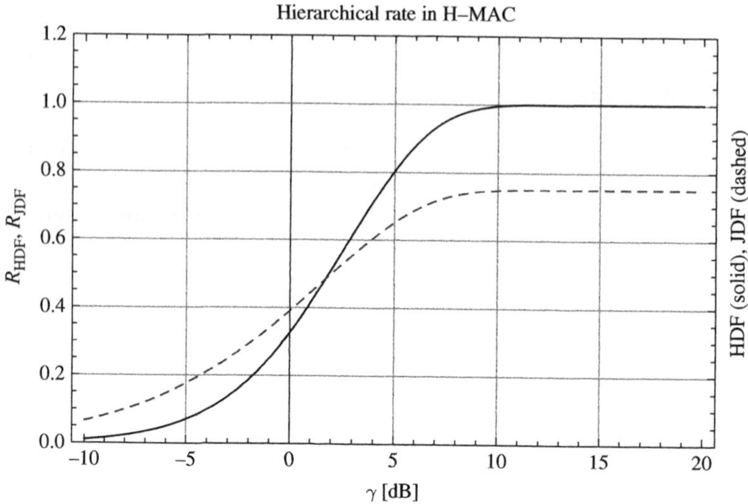

Figure 2.4 Hierarchical rate achievable for two BPSK sources and relay with JDF (dashed line) and HDF (solid line) strategies.

$$R_{\text{JDF}} < \min\left(I_1, \frac{I_2}{2}\right),\tag{2.27}$$

and for HDF, which is given by the hierarchical mutual information

$$R_{\text{HDF}} < I_H.\tag{2.28}$$

As we see, the performance is clearly alphabet-limited at high SNR values, where it saturates at the fixed ceiling. This ceiling is, however, higher for the HDF strategy, where it is given by $\lg M$, and it provides the single-user level performance as if there was no interference at all. It is, however, in contrast to JDF, where the performance is given by the interference limited regime of the second-order rate region condition. Even with zero noise, the JDF cannot support the $\lg M$ hierarchical rate. This comparison exactly shows the performance advantage where WPNC (HDF in this example) technique demonstrates its supremacy and it also justifies our aim of turning the interference into a "friendly" form.

The low SNR region is dominated by the influence of the noise and the actual interaction of the coded signals remains less significant. We call this region the noise-limited region. The advantage of HDF, which can effectively cope with the interference by turning it into a "friendly" interaction that reduces the cardinality of codewords that need to be distinguished, does *not* help now. The specific hierarchical constellation shape, namely the fact that *two* points (± 2) belong to the same codesymbol map, now makes the situation slightly worse for low SNR. This will be explicitly treated in Section 4.5. In the noise limited region, JDF outperforms HDF.

Figure 2.5 shows the rate region from the perspective of both sources. The HDF strategy has a rectangular region since both the rates R_A and R_B are equal, provided that they are less than I_H. In contrast with that, the JDF strategy has the classical multi-user MAC shaped region. The region has close-to-rectangular shape for low SNR – the noise-limited regime. The interference limited regime for high SNR makes the second-order limit the dominant one. The symmetric rate $R_A = R_B$ is thus limited by the second-order limit $I_2/2$. The pair of lines for $\gamma = 5\,[\text{dB}]$ nicely demonstrates that the "corner" point of HDF can be *outside* the JDF region, while the JDF itself provides slightly *greater first-order rate* limits.

The trade-off between noise limitation and interference limitation can be nicely seen when evaluating the ratio $(I_2/2)/I_1$ (Figure 2.6). It describes how much the second-order limit influences the symmetric rate. The second-order limit captures how the performance is affected by the presence of the other user. The first-order limit captures the stand-alone single-user behavior and thus captures the noise-related performance.

2.6 2WRC with QPSK: the Problem of Channel Parametrization

So far our examples have been restricted to the binary case: BPSK modulation. For most practical applications it will be necessary to extend to higher-order modulation. In

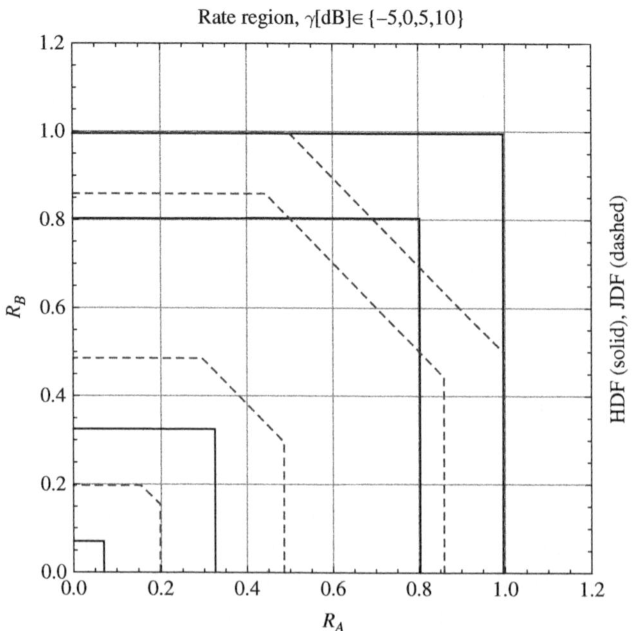

Figure 2.5 Achievable rate regions for two BPSK sources and relay with JDF and HDF strategies. Each pair of solid (HDF) and dashed (JDF) lines corresponds to one SNR value γ. High SNR values correspond to outer pairs.

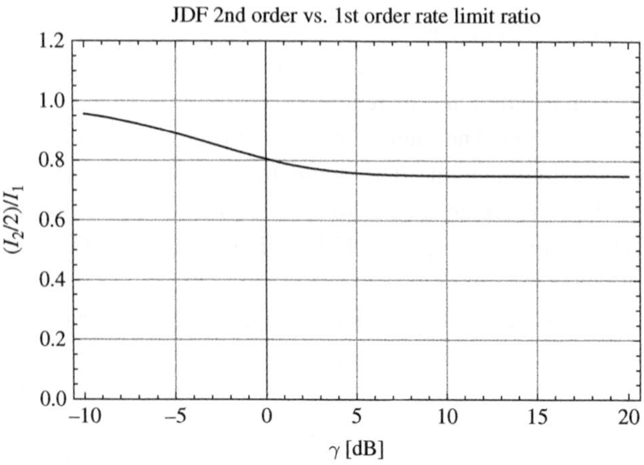

Figure 2.6 Second-order vs. first-order limit ratio $(I_2/2)/I_1$ for BPSK sources and relay with JDF strategy.

this section we will consider QPSK modulation, and we will see that this raises further issues about the choice of mapping function. First, it allows more options: for linear functions because there are more coefficients to choose from, and for the more general table representation because the table is larger and allows more permutations. Secondly, we find that problems arise for some values of channel parameters (that is, amplitude and phase of channel fading) that do not arise with BPSK. Note here that for most of this book we will assume that channels are subject to quasi-static flat fading, and thus that a wireless channel can be defined by its amplitude and phase, usually defined by the complex coefficient h.

In the simple example of WPNC with BPSK described in Figure 1.7 in Section 1.4 we have assumed that the two channels have the same parameters: they are subject to exactly the same fading. This of course is unlikely in practice, but it results in the received constellation shown, which contains only three points since two of the combinations of source data symbols ("01" and "10") result in the same signal at the receiver – namely zero. This is a state we describe as *singular fading*, defined as follows for the case of two source nodes received at one relay. The full details will be given in Section 3.5.3. Here we present only a simplified case for two source nodes and uncoded signals.

Singular fading occurs if the channel fading parameters are such that two different combinations of source symbols transmitted from two nodes result in the same received signal at a relay, neglecting the effect of noise. Mathematically it means

$$\exists (s_A, s_B) \neq (s'_A, s'_B) : \quad u_{AB} = h_A s_A + h_B s_B = u'_{AB} = h_A s'_A + h_B s'_B. \tag{2.29}$$

That is

$$h_A \left(s_A - s'_A \right) = h_B \left(s'_B - s_B \right) \tag{2.30}$$

and

$$(s'_B - s_B)h = (s_A - s'_A), \quad h = \frac{h_B}{h_A} \tag{2.31}$$

for some $(s_A, s_B) \neq \left(s'_A, s'_B \right)$, where s_A, s_B, s'_A, s'_B are transmitted signals corresponding to symbols b_A, b_B, b'_A, b'_B, and h denotes the relative fading of the two channels. It will already have been obvious that the shape of the constellation depends only on the ratio of the two channel coefficients, since any common factor of the two will result only in a phase/amplitude shift of the whole received constellation. We refer to symbol combinations that result in the same signal as *clashes*.

In the case of BPSK, since the ss take only two possible values, there are only two values of h that give rise to singular fading: $+1$ and -1. We might say there also exist two further such values, 0 and ∞, in which one channel or the other is completely faded, but these are not of interest for the 2WRC since they would in any case prevent the network from operating, in the same way they would in a conventional relay network. All other relative fade coefficients will yield a constellation with four distinct points, as shown in Figure 2.7. Singular fade states such as these are important for two reasons: firstly because they represent channel conditions under which joint decoding will not operate, and secondly because if WPNC is to operate correctly clashing symbol combinations should encode to the same network coded symbol.[3] If this is the case we say that the clash is *resolved*, and if all clashes corresponding to a singular fade state are resolved,

---

[3] Both these points hold for a simple uncoded case. For the coded case with properly constructed codebooks (e.g. the idealized abstraction of random IID codebook), the unresolved singular fading only reduces the achievable rates (see Part III for more details).

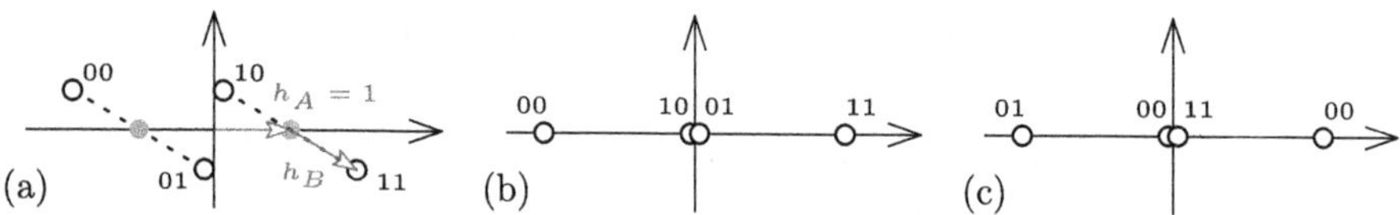

Figure 2.7 Receive constellations at relay for BPSK in 2WRC with different fade states: (a) non-singular fading; (b) singular fade with $h = 1$; (c) singular fade with $h = -1$.

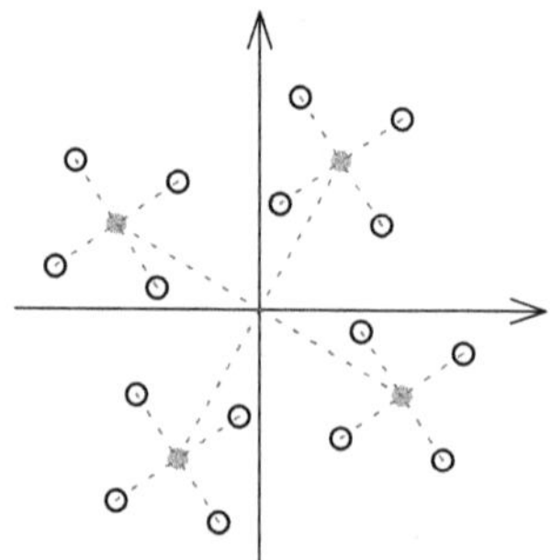

Figure 2.8 Receive constellation for QPSK.

we say that the singular fade state is itself resolved. An unresolved clash will mean that the relay is unable to decode the corresponding network coded symbol, since the received signal will correspond with equal probability to (at least) two network coded symbols. We note that for the binary 2WRC both singular fade states are resolved by the XOR function, which is fortunate, because we have also shown above that this too is the only function that will allow unambiguous decoding.

As we have already noted, it is vanishingly improbable that an exactly singular fade state will occur. However, in the presence of noise, fading that is close to singular will also prevent reliable decoding, since the noise will result in a high error rate in distinguishing between the two network coded symbols.

For QPSK, however, the situation becomes more complex. The four transmitted signals now take the values $\pm 1 \pm j$, resulting (for general non-singular fading) in 16 points in the received constellation, as illustrated in Figure 2.8. Excluding the values 0 and ∞, singular fading now occurs for

$$h \in \{\pm 1, \pm j, \pm 1 \pm j, (\pm 1 \pm j)/2\}. \tag{2.32}$$

Figure 2.9 shows the received constellations for three representative cases from these (or rather for fading close to these states, so the separate labels can more easily be seen). Note that the binary labels shown for the constellation points are formed by concatenating the two-bit binary labels (using conventional Gray code labeling) for the symbols from the two sources.

We now consider mapping functions that can resolve the clashes that occur in these fade states. We will look for linear functions for this purpose. Perhaps the obvious approach is to apply the XOR function to each bit of the binary label separately (in

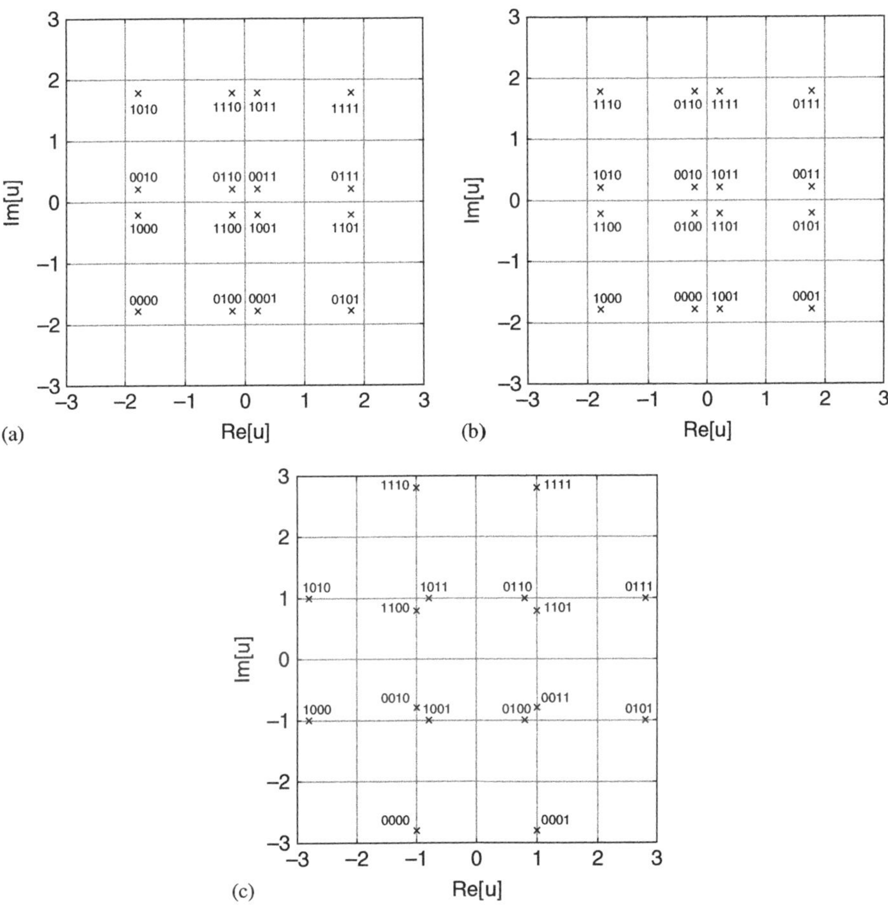

Figure 2.9 (Nearly) singular fading for (a) $h \approx 1$, (b) $h \approx j$, (c) $h \approx 1 + j$.

terms of the labels shown in Figure 2.9 this means the network coded label is formed by two XOR functions, first of bits 1 and 3 and second of bits 2 and 4). For $h = 1$, as in Figure 2.9a, we can see that this results in the same two-bit binary label for all the (nearly) coincident points in the constellation. However, for $h = j$, as in Figure 2.9b, we observe that the clashes are not resolved: for example of the four coincident points around the origin two will be labelled "10," and the other two "01." But if instead the two functions XOR first bits 1 and 4, and secondly bits 2 and 3 of the composite label, we will label all four of these points "01," and similarly the other four clashes in the constellation will also be resolved.

This highlights an important general issue that arises with any modulation scheme with more than two points in its constellation. Unlike the BPSK case, where we observed that the XOR function resolved both singular fade states and therefore could be used for any fading, for QPSK (and any other non-binary modulation), different mapping functions are required in order to resolve all fade states. Hence adaptive mapping is required at the relay.

Incidentally these mapping functions can also be represented by using a binary matrix notation. We represent the M-ary source symbols b_A, b_B as length m vectors $\mathbf{b}_A, \mathbf{b}_B$, where $M = 2^m$ and concatenate them. The linear function may then be represented as multiplication by a binary matrix

$$\mathbf{b}_{AB} = \chi_{\mathbf{b}}\,(\mathbf{b}_A, \mathbf{b}_B) = \mathbf{G} \begin{bmatrix} \mathbf{b}_A \\ \mathbf{b}_B \end{bmatrix}. \tag{2.33}$$

The mapping function we invoked for Figure 2.9a can then be represented by the matrix

$$\mathbf{G} = \begin{bmatrix} 1 & 0 & 1 & 0 \\ 0 & 1 & 0 & 1 \end{bmatrix}, \tag{2.34}$$

and for Figure 2.9b by

$$\mathbf{G} = \begin{bmatrix} 1 & 0 & 0 & 1 \\ 0 & 1 & 1 & 0 \end{bmatrix}. \tag{2.35}$$

Considering the third singular fade state, $h = 1 + j$, for which the received constellation is illustrated in Figure 2.9c, we observe that neither of the two functions so far discussed will resolve any of the four clashes. There is, however, a similar function that will resolve this state (and others like it). We may use a pair of XOR functions which combine both bits of one symbol label with one each of the bits of the other symbol – that is, the first function XORs bits 1, 2, and 3 of the composite label, while the second XORs bits 1, 2, and 4. We observe that this resolves the four clashes in the constellation shown, and similar functions can resolve clashes in the other equivalent fade states (namely $h = \pm 1 \pm j$ and $h = (\pm 1 \pm j)/2$). In this case the mapping matrix is

$$\mathbf{G} = \begin{bmatrix} 1 & 1 & 1 & 0 \\ 1 & 1 & 0 & 1 \end{bmatrix}. \tag{2.36}$$

However unfortunately this function fails the exclusive law mentioned in the previous section, and hence does not ensure unambiguous decodability at both destinations. Since both XOR functions combine both bits of the label of b_A, there are at least two different b_As (e.g. "01" and "10") which yield the same network code for given b_B, and hence the destination is unable to distinguish them. In fact it can be shown that there is no quaternary network code function (i.e. giving a four-level result, equivalent to two bits) that resolves the fade states $h = \pm 1 \pm j$ and $h = (\pm 1 \pm j)/2$ and allows unambiguous decoding at both destinations in the 2WRC. This underlines the point that the end-to-end performance of a network using WPNC needs to be considered: it is not sufficient to choose mapping functions at relays only on the basis that they resolve the singular fading encountered there.

2.7   Hierarchical Wireless Network Example

Finally we consider a second, a little more complicated, example network. We describe it as *hierarchical wireless network* (HWN) because it models a hierarchy of nodes from

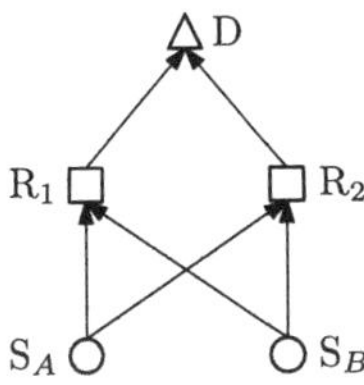

Figure 2.10 Hierarchical wireless network example.

the source terminals at the lowest level, via a series of layers of relay nodes, to a *hub* node, which is the final destination for all data. This could be a model for the uplink of a radio access network in which terminals communicate with a series of small, local access points, which then forward data via backhaul links to a concentrator node that is connected to the core network. Here we consider the simplest possible example of such a network (illustrated in Figure 2.10), consisting of two sources S_A and S_B transmitting symbols b_A and b_B to two relays R_1 and R_2, which then forward network coded symbols $b_1 = \chi_1(b_A, b_B)$ and $b_2 = \chi_2(b_A, b_B)$ to one destination D. We assume that both relays receive signals from both sources, via channels subject to independent fading. Note that in this network there is no HSI: the data from both relays constitute HI, since they both depend on source data which is of interest to the destination.

The same issues of singular fading and unambiguous decodability arise in this network. At each relay the mapping function should adapt to the fading of the channels to resolve as far as possible any singular fade states. But the resulting mapping functions should combine at the destination to enable the destination to deduce unambiguously which combination of symbols was sent. We can define a new version of the exclusive law to cover this case (see Section 3.4 for a general treatment)

$$(b_1, b_2) = (\chi_1(b_A, b_B), \chi_2(b_A, b_B)) \neq (b_1', b_2') = \left(\chi_1\left(b_A', b_B'\right), \chi_2\left(b_A', b_B'\right)\right),$$
$$\forall (b_A, b_B), (b_A', b_B') : (b_A, b_B) \neq (b_A', b_B'). \quad (2.37)$$

That is, any two different pairs of source symbols must result in a different pair of network coded symbols at the relays. We can treat the pair of mapping functions at the two relays as a single joint mapping function, which can be tabulated in the same way as in Table 2.2. Table 2.3 illustrates such a table – again the content of the table is illustrative only. In this case the entries of the table are pairs of symbols from the two relays, and each pair must be distinct, corresponding unambiguously to a pair of source symbols.

The number of distinct pairs of symbols (b_1, b_2) must be at least as great as the number of pairs (b_A, b_B), that is $M_1 M_2 \geq M_A M_B$ (where $M_1 = |\mathcal{A}_1|$, $b_1 \in \mathcal{A}_1$, $M_2 = |\mathcal{A}_2|$, $b_2 \in \mathcal{A}_2$). Once again, the cardinality of the outputs of the mapping functions do not need to have the same cardinality as the inputs. In this network there is in fact no lower limit on the cardinality of one relay, provided that of the other is sufficient to compensate. If one relay has full cardinality (i.e. $M_A M_B$), then the other is not needed at all (although we may treat it as having cardinality 1).

Table 2.3 Table for joint mapping function from two relays in HWN.

b_A \ b_B	0	1	...	$M-1$
0	$(0,0)$	$(1,M-1)$	...	$(M-1,1)$
1	$(1,1)$	$(2,0)$	...	$(0,2)$
$\vdots$	$\vdots$		$\ddots$	$\vdots$
$M-1$	$(M-1,M-1)$	$(0,M-2)$	...	$(M-2,0)$

It is also clear that the functions at the two relays must be different. In fact a stronger condition is required: whenever two different pairs of source symbols give the same output for one function, they must produce a different result for the other function:

$$\forall (b_A, b_B), (b'_A, b'_B) : (b_A, b_B) \neq (b'_A, b'_B), \; \chi_i(b_A, b_B) = \chi_i\left(b'_A, b'_B\right),$$
$$\text{it must hold that } \chi_{\bar{i}}(b_A, b_B) \neq \chi_{\bar{i}}\left(b'_A, b'_B\right) \quad (2.38)$$

where $i \in \{1, 2\}$, $\bar{i} = 3 - i$. The table formulation, as before, can be used for any arbitrary pair of discrete functions. If we restrict ourselves to linear functions, then the pair of output symbols from the relays can be written

$$(b_1, b_2) = (a_{1A} \otimes b_A \oplus a_{1B} \otimes b_B, \; a_{2A} \otimes b_A \oplus a_{2B} \otimes b_B). \quad (2.39)$$

This may also be written in matrix form, as

$$\mathbf{b}^r = \mathbf{A}\mathbf{b}^s \quad (2.40)$$

where $\mathbf{b}^s = [b_A, b_B]^{\mathrm{T}}$, $\mathbf{b}^r = [b_1, b_2]^{\mathrm{T}}$, and

$$\mathbf{A} = \begin{bmatrix} a_{1A} & a_{1B} \\ a_{2A} & a_{2B} \end{bmatrix}. \quad (2.41)$$

Then the condition for unambiguous decodability becomes simply that $\mathbf{A}$ is invertible, that is, that its rows and columns be linearly independent. This, of course, also implies that the functions at the two relays are different.

Provided the cardinality of the sources is a power of 2, we can also use the binary matrix representation of a linear mapping function. Using the same notation as before, the relay mapping functions can then be written (notice that vectors $\mathbf{b}$ are now modified to reflect the binary representation)

$$\mathbf{b}_1 = \mathbf{G}_1 \begin{bmatrix} \mathbf{b}_A \\ \mathbf{b}_B \end{bmatrix}, \quad (2.42)$$

$$\mathbf{b}_2 = \mathbf{G}_2 \begin{bmatrix} \mathbf{b}_A \\ \mathbf{b}_B \end{bmatrix}, \quad (2.43)$$

$$\mathbf{b}^r = \begin{bmatrix} \mathbf{b}_1 \\ \mathbf{b}_2 \end{bmatrix} = \begin{bmatrix} \mathbf{G}_1 \\ \mathbf{G}_2 \end{bmatrix} \begin{bmatrix} \mathbf{b}_A \\ \mathbf{b}_B \end{bmatrix} = \mathbf{G}\mathbf{b}^s. \quad (2.44)$$

In this case it is the matrix $\mathbf{G}$ that must be invertible (i.e. non-singular): again, all its columns must be linearly independent. This will not be the case if the two functions are the same.

Because there are two relays in this network, there are more options for joint mapping functions, and this means more flexibility in resolving singular fade states. For example in the QPSK case we find that the singular fade states $h = \pm 1 \pm j$ and $h = (\pm 1 \pm j)/2$ may now be resolved without necessarily compromising unambiguous decodability, though of course since the functions must be different this will not be possible if both relays are in the same singular fade state. For these singular fade states we may use the mapping matrix

$$\mathbf{G}_i = \begin{bmatrix} 1 & 1 & 1 & 0 \\ 1 & 1 & 0 & 1 \end{bmatrix}. \tag{2.45}$$

This may be combined with various mapping matrices in the second relay, provided the combination is not singular. For example, the combined matrix might be

$$\mathbf{G} = \begin{bmatrix} 1 & 1 & 1 & 0 \\ 1 & 1 & 0 & 1 \\ 1 & 0 & 0 & 1 \\ 0 & 1 & 1 & 0 \end{bmatrix}. \tag{2.46}$$

Some care is, however, needed when extending this bit-wise mapping to a coded case where the isomorphism of the hierarchical codeword is needed; see Section 6.3.4 for details.

Part II

Fundamental Principles of WPNC

3 Fundamental Principles and System Model

3.1 Introduction

This chapter is essentially all about basic definitions and classifications of various scenarios based on them. It is a bit tedious but necessary in order to develop a clear understanding of the terms. Also, the terms could have rather wide interpretations and we need to define them precisely. Proper classification of the techniques also helps to understand how they are mutually related, what they have in common, and how they differ. We attempt to present a highly modular view of the roles and functions of individual nodes in the WPNC network. This will help us later to develop various techniques (NCM design, decoding technique) that are universally usable in nodes serving a variety of roles.

We start with scenarios and models where we describe the roles of nodes, the constraints imposed by their radio interfaces, and various issues related to the topology of the network. Then we continue with the core *hierarchical principle*. It describes how data functions flowing through the network are encapsulated hierarchically. We also show how a direct neighborhood of the node affects the overall end-to-end description of the network.

Then we turn our attention back to the individual node. We show how its operation can, under very general conditions, be decomposed into *front-end processing*, *node processing*, and *back-end processing* operations. We show how the node processing operation is related to the many-to-one function of the source nodes' data, which will lead to the definition of the hierarchical symbol, and we also define a form of information measure that is used to represent it to the rest of the network. Depending on a node's predecessor path in the network graph, this form of information measure can have various forms of usefulness from the point of view of the given node. We will define *hierarchical information*, *hierarchical side-information* (friendly interference), and *classical interference*.

Previous definitions help us to classify various *strategies* of the node. The node processing operation will be classified (e.g. amplify and forward, decode and forward, soft forward, compress and forward). Depending on the properties of the hierarchical symbol and its associated hierarchical NC map we will introduce full, minimal, extended, and lossy hierarchical maps.

Then we classify back-end strategies from the source-encoding viewpoint (direct and analog hierarchical broadcast, and source-encoded NC broadcast). Even more

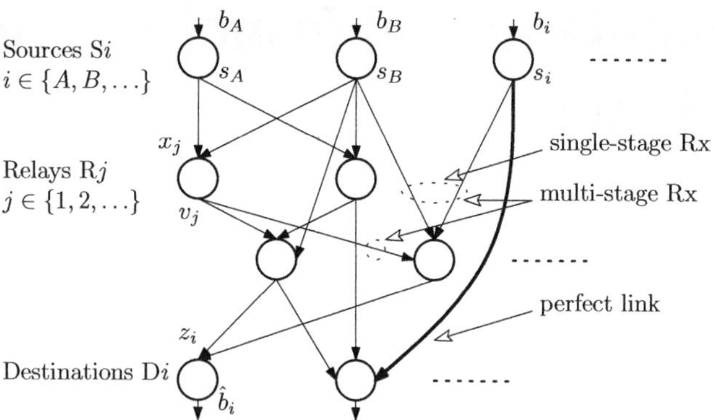

Figure 3.1 System model – multi-terminal and multi-node.

important, however, will be the network structure awareness of the coded modulation transmitted from the node. This will lead to the fundamental term – *network coded modulation*. The front-end strategies are classified from the perspective of how they incorporate the various forms of information available to the receiver and what form of the received signal preprocessing they perform.

Subsequently we define the fundamental end-to-end description of the WPNC network called a *global hierarchical network code map*, and the conditions that guarantee its solvability. This will lead to the *generalized exclusive law* theorem. The global hierarchical network code map describes fundamental properties on the level of discrete symbols and the "information flow" in the network. Its related "partner," highly specific to WPNC, on the *constellation space* level is the term *hierarchical constellation*. We will present its definition and demonstrate its close and highly WPNC-specific relation to the channel parametrization – the definition of *singular states*.

3.2   Scenarios and System Model

3.2.1   Nodes

WPNC generally applies to an arbitrary wireless network with multiple sources, multiple destinations, and multiple relays (Figure 3.1). Sometimes we refer to sources and destinations jointly as terminals. Sources and destinations are typically indexed by letters $i \in \{A, B, \ldots\}$. Relays are typically indexed with numbers $j \in \{1, 2, \ldots\}$. Data, codes, and signals corresponding to individual nodes follow the same notation. In simple cases, when no confusion is possible, we may use a simplified notation to make clearer how the individual entities are interconnected. For example, we may denote the data processed by the relay as b_{AB} to stress that they were obtained as some function of the source data b_A and b_B, i.e. $b_{AB}(b_A, b_B)$. Data, codes, and signals can be described at various levels of granularity and can have various forms: scalar discrete symbols, vectors, constellation space symbols and vectors, etc.

Each physical source or destination node is usually associated with its own individual source or target data stream. We assume that each source Si has its own information data b_i independent to other sources. Similarly, the destination Di is a target for the data estimate $\hat{b}_i$. Relay Rj has input observation x_j, which is the received signal from all radio-visible sources. Participating source node signals superpose directly at the receiver antenna and there is no way to directly distinguish between them without the help of signal processing or decoding. This is an important feature of WPNC. The situation when the received signals come from multiple transmitters sharing and interacting in a common signal subspace will be referred to as a single-stage reception. WPNC stages (more details will be given in Section 3.2.2) are defined by the transmit signal space partition, and in this case the signals share one stage. Relay can, however, also receive noninteracting signals from orthogonal subspaces of multiple stages and this situation will be referred to as multi-stage reception. The received signal processed by the relay j is transmitted as v_j. The signal transmitted by the relay broadcasts to other relay or destination nodes.

Several nodes can be sometimes collocated at one physical place, e.g. the source of one data stream can be also a destination of another data stream. The collocated nodes are assumed to perfectly share any form of the processed information. From the formal processing point of view, it is equivalent to the nodes being interconnected by perfect infinite capacity links. We will use bold lines to denote these perfect links. Perfect links can sometimes be used also for a simplified analysis where we want to focus only on a particular part of a more complex network. Collocated nodes can model a variety of situations, e.g. S/D generating or decoding multiple data streams, or relays that are source or destination for some data at the same time.

3.2.2 Radio Resource Sharing and Network Stages

Signals transmitted by nodes in WPNC network share common radio resources, spectral and temporal, inside the radio visibility area. A trivial and traditional way of the sharing uses orthogonal subspaces for each link. The orthogonality might be achieved in a number of ways – slicing in time, frequency, or using generally defined orthogonal signals (e.g. direct sequence orthogonal spreading). Orthogonal sharing is, however, suboptimal and one of the important features and benefits of the WPNC network is to use non-orthogonal sharing.

There is, however, one important technology-implied constraint limiting the possibilities of all nodes sharing one common signal space. It is specific to relay nodes that both transmit and receive the signals. Current technology does not allow a sensitive receiver to receive and process weak signals in the situation where a collocated transmitter uses a high-power output signal.[1] A solution is the separation of transmission and reception of the relay node into two orthogonal subspaces; this is typically achieved by time-sharing. This is called the half-duplex constraint.

---

[1] As the technology develops, this constraint gradually relaxes. The receiver front-ends can be constructed linear in the high dynamic range, and the dynamic range of analog-to-digital convertors and digital processing increases. A multiple antenna technique together with smart self-interference canceling algorithms also improve the transmit–receive separation ratio.

In this book, we assume that all relay nodes operate under the half-duplex constraint. Generally, it means that for each relay Rj the input and output signals are orthogonal $\langle x_j; v_j \rangle = 0$, where the inner product is defined in the space corresponding to the form of the signal representation.

The particular way that orthogonality among stages is achieved (time, frequency) does not play an essential role from the perspective of coding and processing conceptual design at a given relay node. The simplest case is the time-division sharing of stages,[2] and this will typically be used throughout the book. The simplest time-division-based network stages solution uses several time domain non-overlapping frames (slots) each dedicated to one stage. Indeed, there are many practical and theoretical implications. Time-division is the most robust one but the time-sharing is the least efficient sharing. Also the time-division half-duplex directly implies a causality between transmission and reception phases. Frequency-division (and similarly other general orthogonal divisions) is more effective from the information-theoretic perspective; however, it opens an issue of relay input–output causality.

The half-duplex constraint inevitably divides the WPNC network operation into stages. The stage is defined by a subspace of all WPNC radio resources used by some set of *transmitting* nodes (sources and relays) in such a way that the half-duplex constraint is fulfilled for all nodes in the network. Signals transmitted and received in a common stage will be graphically denoted by lines having a joint start and/or end point in block diagrams (Figure 3.1).

DEFINITION 3.1 (Network Stages with Half-Duplex Constraint)   Denote by $\mathcal{S}$ a complete signal space of the network, $\mathcal{S}_{Tx}(k)$ the signal subspace used by some transmitting node k (the source or the relay), and $\mathcal{S}_{Rx}(k)$ the subspace used by the receiving node k.

The mth network stage is defined by its associated subspace $\mathcal{S}(m) \subset \mathcal{S}, m \in \{1, \ldots, M\}$. The node k belongs to the stage if $\mathcal{S}_{Tx}(k) \cap \mathcal{S}(m) \neq \emptyset$. The set of all nodes belonging to that stage is denoted $\mathcal{K}(m)$.

The network fulfills the half-duplex constraint if the receive and transmit subspaces of any node k do not overlap $\mathcal{S}_{Rx}(k) \cap \mathcal{S}_{Tx}(k) = \emptyset$.

Stages generally do not need to create a partition of the complete network signal space defined *only* in the temporal domain. A very natural case, even in time-division stage slicing, is the one where some parts of the network are not mutually radio-visible. This naturally creates empty intersection subspaces separated in a *spatial* domain. The form of the signal space used in the definition of the stage must therefore also include the spatial domain. In any nontrivial network, there is a large number of possible settings for the stages. Typically we want to minimize the number of stages M. The stages must indeed be set in such a way that the network guarantees end-to-end connectivity and causality for all data flows.

Empty-intersection subspaces imply that the second-order moment, the inner-product, will be zero for any signal coming from two different stages. The converse, however, does not generally hold. In WPNC, all radio inputs and outputs can be represented as a

---

[2] This statement refers only to sharing among *stages*. The majority of WPNC technique is, indeed, aimed for solving sharing within one stage.

signal space point. This, together with a usual assumption of Gaussian noise, makes a *second-order* inner-product-based processing absolutely dominant. In this case we can use a relaxed definition that the signals belonging to different stages are *orthogonal* instead of strictly requiring their empty intersection.

A network where some nodes can receive the signals from several stages will be called a mixed stage network. Particular caution needs to be paid to this phenomenon when it affects the relay. In this case, several delayed versions of the signals carrying correlated information from preceding nodes might enter the relay data mapping function and this situation must be correctly addressed. A specific form of mixed stage network is the case where the only mixed stage is the stage of the source node transmission. There is only one preceding node, the source, and its HNC map has a singular form – the source data themselves, and thus it does not present any problems to cope with. If there are no mixed stages in the network, we call it a homogenous stage network. It can simply be viewed as a serial concatenation of two-stage subnetworks.

Several examples of defining half-duplex compliant stages for simple system scenarios are in Figure 3.2. The two-way relay channel is the simplest scenario with collocated sources and destinations connected by perfect links (Figure 3.2a). The minimal number of stages is two. We usually call the first one a MAC stage and the second one a BC stage due to the resemblance with traditional multi-user scenarios. Notice, however, that in the WPNC paradigm we do not try to distinguish individual multiple sources in the MAC stage and the name needs to be understood only in a relaxed form related to the topology. The MAC and BC stages, as seen in this minimalistic scenario, also appear as a more generic principle in more complex networks. A stage where the relay is receiving multiple signals coming from other nodes belonging to this stage is, from the perspective of this relay, a MAC stage (and similarly for BC stage).

A butterfly network (Figure 3.2b) allows a separation of source and destination nodes where the MAC stage signals are also overheard by the destinations and they can no longer be modeled as perfect links. The S–D side links can also be active in the second (BC) stage (Figure 3.2c). Notice that the relay cannot receive in stage 2 since it is transmitting at that time. Signals transmitted in stage 2 by sources thus can be processed only by destinations.

3.2.3 Network with Cycles

The *wireless* WPNC network can quite easily produce radio links with *cycles* in connectivity. This is a consequence of rather uncontrolled radio-wave propagation and it contrasts with the cable network where this would require a dedicated deliberate connection. The cycles might open new possibilities for WPNC but at the price of additional substantial design complications. The received signals at the node can potentially depend on its own transmission history through some complicated hierarchical data mapping functions. This can be used for advanced coding/processing to improve the performance. However, the WPNC coding and processing design for this situation is largely (apart from some very simple special cases) an open research issue. A simple, and of course suboptimal, solution is either to ignore the information flow history (i.e. to

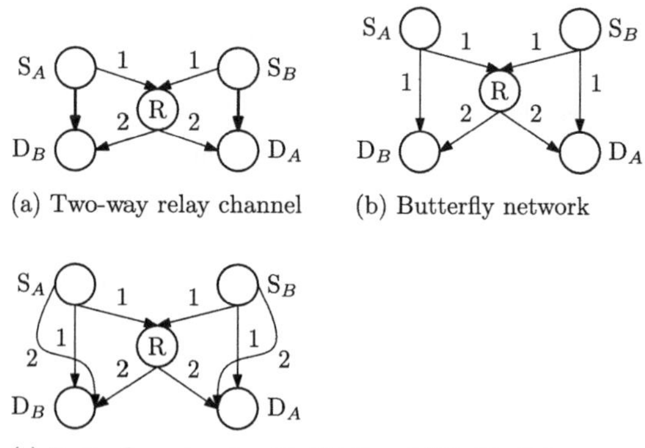

Figure 3.2 Example systems with stages fulfilling the half-duplex constraint.

ignore that hierarchical data function depends on the node transmitted data) or even to directly ignore the radio links creating the feed-back. The receiver of the particular node does not need to participate in a stage creating the cycle (even if the radio signal is available there), or we can consider the signal as a classical interference. Throughout the rest of the book, we assume cycle-free networks or the networks where the cycles are ignored or not directly utilized for the processing/coding.

3.3 Core Principles of WPNC Network

3.3.1 Hierarchical Principle

A large network consists of many nodes. Each node processes some *function* of the data (or any other entity, e.g. soft information measure) forming its inputs. The relay operation always directly depends on its neighbors and this creates a *hierarchy* of encapsulations. At each hierarchy level, we solve tasks regardless of the actual particular lower hierarchy level data value: (1) the node receives multiple signals, each carrying some form of information, and it tries to process (decode) a function of that information; (2) the information function is processed somewhat (e.g. a data decision is made); and finally (3) this processed information is sent further to other nodes. The fact that these three tasks are performed at each encapsulation direct neighbor hierarchical level *regardless* of the rest of the network is called a hierarchical principle (Figure 3.3).

The hierarchical principle allows a decomposition of an arbitrary network into individual nodes where each node fulfills its three basic tasks. Design of its receiving (front-end), processing, and transmitting (back-end) algorithms then can be treated in common. We will see that even sources and destinations are special cases in such hierarchy.

At a given hierarchical encapsulation level, a relay with arbitrary processing operation connected to an arbitrary number of input and output stages can always be described (Figure 3.4 and more details in Figure 3.5) by its front-end processing, the

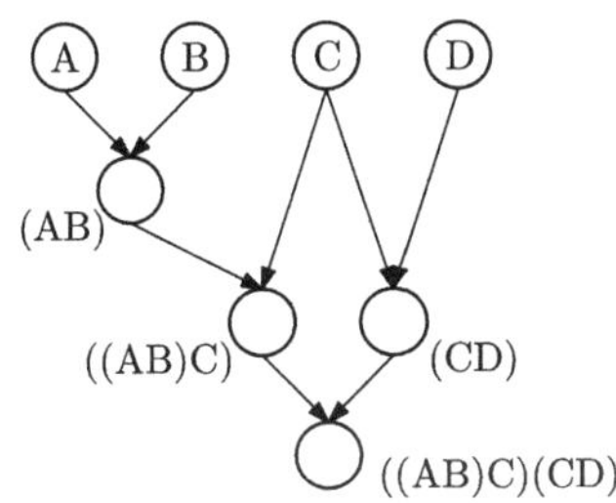

Figure 3.3 Hierarchical principle.

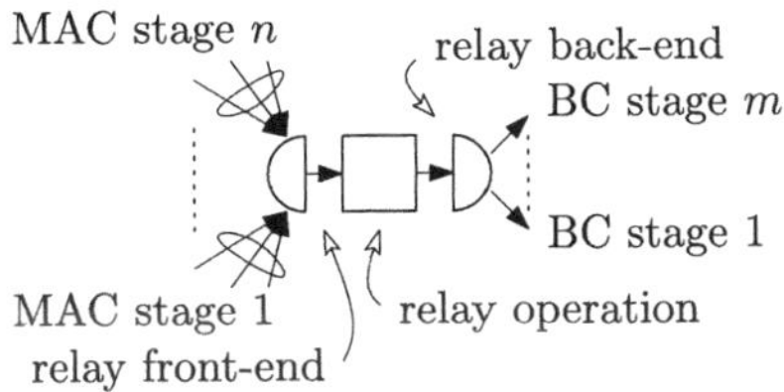

Figure 3.4 Relay processing – front-end, relay operation, back-end.

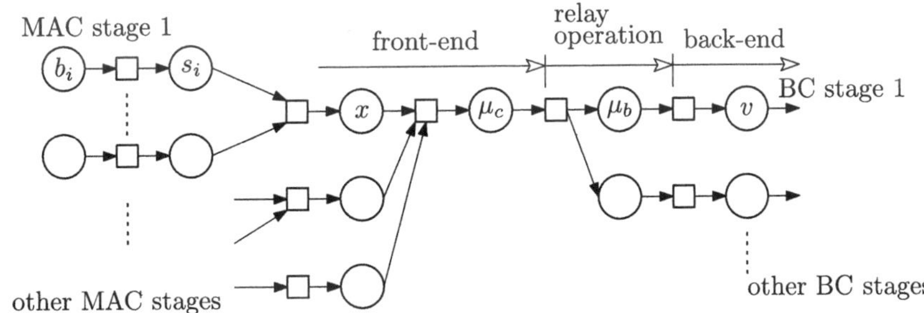

Figure 3.5 Relay processing – hierarchical MAC and BC stage, relay operation on hierarchical information.

relay operation, and the back-end processing.[3] All received front-end signals can be viewed as a MAC stage from the perspective of this particular relay (ignoring the rest of the network). If the received signals belong to multiple subspaces, we use multiple MAC stages. Similarly on the back-end side, the output can be viewed as possibly multiple BC stages. In order to stress that the stages are interpreted from the perspective of one relay possibly buried deep in the hierarchy of the network and processing hierarchical information, we frequently refer to it as a *hierarchical MAC* (H-MAC) and *hierarchical BC* (H-BC) stage.

---

[3] Compare this with a traditional generic information-theoretic way of describing the relay as a global operator $v = v[x]$. Here we intentionally introduce a chain operations $\mu = \mu[x]$, and $v = v[\mu]$ where μ is forcibly tight to some function $b(\tilde{b})$ of input data components. This enforced "key-hole" allows us to have better control of the relay procedures.

More detailed capture of H-MAC stage, relay operation, and H-BC stage is in Figure 3.5. It shows received signals x from individual H-MAC stages. All these signals can, but do not need to, be jointly used to form receive information measure (e.g. decoding metric for some hierarchical data, see details in Section 3.3.5). This is in turn used in the relay operation forming (typically by decoding) the relay processing information measure $\mu_b(x)$ (a detailed definition will come later in Section 3.3.2), which can be individual for each of BC stages. The simplest case is the one where each H-MAC stage produces information measure for a single H-BC stage independently of the other stages. The information measure is the *source* information for the back-end H-BC stage, which might be source and/or channel encoded by the back-end into its transmit variable v (details in Section 3.3.4).

The granularity of the description can vary. We can relate processing operations to individual symbols (either data or code), vectors or data/code words, signal space points, etc. We denote b_i arbitrary generic *discrete* data or code (or possibly some function of them) on one component in one H-MAC stage, s_i a corresponding *signal space* representation of the Tx output, x a *signal space* representation of the relay input, and v the *signal space* representation of the relay output. Signals at different stages are orthogonal by definition and therefore their signal space representations are separate variables and we can focus in notation on one H-MAC stage and one H-BC stage. We also denote an arbitrary set of H-MAC stage component variables $\tilde{b} = \{b_i, \ldots, b_{i'}\}$. In what follows, we formally focus only on a single H-MAC stage and single H-BC stage. The generalization using the hierarchical principle is straightforward.

3.3.2 Relay Processing Operation and Data Function

Now we turn our attention to various forms of data/symbol functions that are processed by nodes or form their inputs/outputs in the sense of the hierarchical encapsulation principle. All the following definitions are related to a *local* single hierarchical level.

Individual signal space components $s_i = s_i(b_i)$ of the H-MAC stage are generally signal space codes of the component data b_i; see Figure 3.5.[4] The receiver H-MAC stage signal space observation is a function of these components $x = x(s_i, \ldots, s_{i'})$ and in turn also $x = x(b_i, \ldots, b_{i'}) = x(\tilde{b})$. The relay processes this into the front-end H-MAC decoding metric μ_c, which is then processed by the relay operation into $\mu_b(x)$. The index b means that it is related to some (typically many-to-one) function of the H-MAC component data $b = \chi(\tilde{b})$. This is called a *hierarchical symbol* and the function $\chi(.)$ is called a *Hierarchical Network Code map (HNC map)*. The hierarchical symbol is the *only* form of information passed through the relay. This information, however, does not need to be passed in an open form as a discrete symbol b. It can be represented by an arbitrary information measure $\mu_b(x)$. This representation (function, operator) will be referred to as a relay processing operation, and the resulting quantity as relay processing information measure $\mu_b = \mu_b(x)$. It can have a variety of forms, e.g. the linear function

---

[4] As mentioned before, we now focus only on a single H-MAC and H-BC stage.

passing only scaled value of x, decoding soft likelihood metric, various forms of hard decoding decision on b, various information compression functions, etc.

The relay processing operation output is then used to form (encode) signal space representation $v = v(\mu_b)$ transmitted into the H-BC stage. In the case of multiple H-MAC stages the relay processing operation information measure μ is a joint function of all H-MAC inputs. In the case of multiple H-BC stages we generally assume different information measures used in each H-BC stage. Individual relay processing operations may, but do not need to, use the observation from multiple H-MAC stages. A causality principle may impose some constraints.

DEFINITION 3.2 (Hierarchical Symbol, HNC Map, and Relay Processing Operation) An arbitrary subset of discrete input symbols of the nodes participating in the H-MAC stage is denoted $\tilde{b} = \{b_i, \ldots, b_{i'}\}$. Generally, a many-to-one discrete function $\chi(\tilde{b})$ is called a *Hierarchical Network Code map* and the resulting symbol $b = \chi(\tilde{b})$ is called the *Hierarchical Symbol* and it belongs to the HNC map alphabet $\mathcal{A}_b$. A hierarchical symbol is provided to the relay output only in the form of the *relay processing information measure* μ_b obtained from the input by the *relay processing operation* $\mu_b = \mu_b(x)$.

We may say that $b = \chi(\tilde{b})$ describes the *contents* of the information flow through the network, while $\mu_b(x)$ describes its *form*. For example, the contents could be the data of one source b_A or it could be a linear function over the GF $b = b_A + b_B$. The form of representing this data symbol can be, for example, directly its discrete value (i.e. the hard decision) or it could be any form of the decision metric (e.g. likelihood). Particular practically important cases will be discussed and classified later.

Sources and destinations are special cases of the above-stated generic node processing chain. The source has only the back-end and the node operation μ_b is a representation (code) of the source data where the HNC map is $b = b_i$, i.e. directly the data source. The destination has only a front-end and the node operation $\mu_b(x)$ produces desired data estimates where the HNC map is $b = b_{i'}$ and where $b_{i'}$ are the target data of the destination node.

Signals in the WPNC network can generally depend on many preceding ones in a network graph, possibly even those coming from multiple stages. Each of these signals can be formed from a different subset of input variables. This raises a natural question of how to classify these signals from the perspective of the given node processing aimed for a given target hierarchical symbol. We want to find out how a particular signal helps the node processing (decoding). For that purpose, we define two arbitrary subsets of input symbol variables $\tilde{b}_\alpha = \{b_{\alpha i}, \ldots, b_{\alpha i'}\}$ and its complement to the complete set $\tilde{b} = \{b_i, \ldots, b_{i'}\}$, $\tilde{b}_{\bar{\alpha}} = \tilde{b} \setminus \tilde{b}_\alpha$. The complete set $\tilde{b}$ contains all the variables that are graph predecessors for the node of interest found on some network cut possibly several stages earlier (Figure 3.6). The components $\{b_{\alpha i}, \ldots, b_{\alpha i'}\}$ can be either directly the source node data (symbols) or any other symbol encapsulation at arbitrary hierarchical level. The desired hierarchical symbol is $b = \chi(\tilde{b}_\alpha)$ and this depends only on the subset $\tilde{b}_\alpha$. See Figure 3.6 for an example scenario.

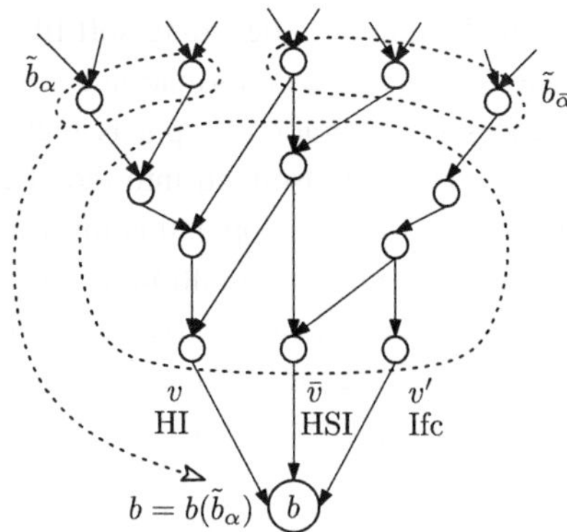

Figure 3.6 Hierarchical information (HI), hierarchical side-information (HSI), and interference (Ifc).

DEFINITION 3.3 (Hierarchical Information (HI))   We say that the signal v carries hierarchical information relative (w.r.t.) to the target hierarchical symbol $b = \chi(\tilde{b}_\alpha)$ being a function of a subset $\tilde{b}_\alpha$ if and only if $I(v; \tilde{b}_\alpha | \tilde{b}_{\bar{\alpha}}) > 0$.

The HI definition means that v carries some useful information about the variables forming the desired hierarchical symbol b provided that we remove all potentially "interfering" influences[5] of the variables that are not part of b (it is helped by others). HI carries the "information contents" related to b. From the graph theory point of view, all $\tilde{b}_\alpha$ are oriented graph predecessors of v.

DEFINITION 3.4 (Hierarchical Side-Information (HSI))   We say that the signal $\bar{v}$ carries hierarchical side-information relative (w.r.t.) to the target hierarchical symbol $b = \chi(\tilde{b}_\alpha)$ being a function of a subset $\tilde{b}_\alpha$ if and only if $I(\bar{v}; \tilde{b}_\alpha | \tilde{b}_{\bar{\alpha}}) = 0$ and $I(\bar{v}; v | \tilde{b}_\alpha) > 0$ where v is HI w.r.t. b.

The HSI is *not* an HI w.r.t. b but it *affects* its HI v through complementary set $\tilde{b}_{\bar{\alpha}}$. HSI carries helping information (complementary) that helps resolving ambiguity in HI caused *only*[6] by a complementary set of component variables $\tilde{b}_{\bar{\alpha}}$. We can see HSI as a *friendly interference*. From the perspective of graph theory, HSI must not have $\tilde{b}_\alpha$ in the oriented graph predecessors and HI and HSI must have some common predecessor, which in turn has a predecessor in the complementary set $\tilde{b}_{\bar{\alpha}}$.

DEFINITION 3.5 (Interference (Ifc))   We say that the signal v' carries interference relative (w.r.t.) to the target hierarchical symbol $b = \chi(\tilde{b}_\alpha)$ being a function of a subset $\tilde{b}_\alpha$ if and only if $I(v'; \tilde{b}_\alpha | \tilde{b}_{\bar{\alpha}}) = 0$ and $I(v'; v | \tilde{b}_\alpha) = 0$ where v is HI w.r.t. b.

Interference, in a *classical harmful sense*, is the signal that is neither HI nor HSI. It neither carries any useful information nor can it help in decoding of the hierarchical symbol (Figure 3.6).

---

[5] Notice the similarity with cut-set bound mutual information definition.

[6] Notice the conditioning by $\tilde{b}_\alpha$ in the mutual information. It leaves the only degree of freedom in complementary set $\tilde{b}_{\bar{\alpha}}$ when evaluating the stochastic connection between v and $\bar{v}$.

Example 3.1   Two-Way Relay Channel (Figure 3.2a)

Relay R_1 has an HNC map $b_1 = \chi_1(b_A, b_B)$ and the relay output signal is $v_1 = v_1(\mu_{b_1})$. At the destination D_A, our target map is $\chi_A(b_A, b_B) = b_A$, i.e. the pure source data b_A. The destination receives signals from the relay and has available perfect observation of its own symbol $\bar{v}_B = b_B$. Signal v_1 is HI for b_A and signal $\bar{v}_B$ is perfect HSI for b_A. There is no classical harmful interference.

Example 3.2   Butterfly Network (Figure 3.2b)

The situation is exactly the same as for two-way relay channel with the only difference that the HSI information is no longer perfect. It is only overheard over the radio channel.

3.3.3   Classification of Node Processing Operation Strategies

Node operation strategies can be viewed as a high-level classification of (1) the node (relay) processing operation $\mu_b(x)$ form, and (2) properties of the hierarchical symbol alphabet and the associated HNC map. The classification of these properties can be, to a large extent, completely decoupled. The node output v is a function of the information measure $\mu_b = \mu_b(x)$, which is related to a particular HNC map $b = \chi(\tilde{b})$. We can classify the strategies as follows.[7]

An important aspect that needs to be stressed is the fact that all statements discussed here apply to the mapping between constellation space observation x and the information measure μ_b, i.e. to the relay operation $\mu_b(x)$. The relay back-end processing $v = v(\mu_b)$ might introduce additional functions and operations on top of it. But the processing chain is $x \mapsto \mu_b \mapsto v$ and the "bottleneck" discussed here is μ_b.

Node Processing Operation

Node processing operations are classified according to the form of the $\mu_b(x)$ operation *regardless* of the particular form of the HNC map $b = \chi(\tilde{b})$.

AF (Amplify and Forward) is defined as operation $\mu_b(x)$ *linear* in x and completely ignoring map b. It means

$$\mu_b(x) = \alpha x \tag{3.1}$$

where α is a scaling coefficient (or a matrix) with a proper dimension corresponding to the interpretation of the dimensions of the input x (temporal, spatial, frequency domain). An obvious advantage of AF is its independence on the HNC map and even on the alphabet $\mathcal{A}_b$ itself.

SF (Soft Forward) (also called by some authors Estimate and Forward, or Soft-Information Forward, or other variants) means that the processing operation provides

---

[7] The terminology, particularly in the context of WPNC, does not seem to be strict and some authors slightly differ.

a soft-information measure (decoding metric). The measure is a set of values for each particular alphabet $\mathcal{A}_b = \{b^{(1)}, b^{(2)}, \ldots\}$ member

$$\mu_b(x) = \left\{ \mu(x, b^{(1)}), \mu(x, b^{(2)}), \ldots \right\}. \tag{3.2}$$

An example is the soft-output demodulator set of likelihoods $\mu_b(x) = \left\{ p(x|b^{(i)}) \right\}_i$. The form of soft metric can vary. It might, but does not necessarily need to, be optimized w.r.t. some performance utility target, e.g. bit error rate at the target destination. However, the values of the metric are *continuous valued*.

QF (Quantize and Forward) is a variant of SF where instead of a continuous valued soft metric a quantized *discrete* valued metric is used.

DF (Decode and Forward) is the strategy where the node processing operation makes a *decision* on the b symbol, which can be a constellation symbol/vector, or a code symbol/vector, or a data symbol/vector. The alphabet of the processing information measure $\mathcal{A}_{\mu_b}$ is the same as the symbol b alphabet $\mathcal{A}_b$, $\mathcal{A}_{\mu_b} = \mathcal{A}_b$. A typical example is the ML estimate $\mu_b = \hat{b} = \arg\max_b p(x|b)$. Also note that the DF strategy is in fact a special case of QF (hard quantization) where the quantization levels are 0 and 1. All alphabet members are assigned 0 except one assigned 1, which denotes the decision. In the DF case we simply name the resulting value. The technique called Denoise and Forward (used in the context of WPNC) can be considered as decision making (or hard quantization) at the uncoded constellation symbol level.

CpsF (Compress and Forward) provides (possibly lossy) compressed discrete symbols (codewords). The size of processing information measure alphabet (codebook) is smaller than the size of the symbol alphabet $|\mathcal{A}_{\mu_b}| < |\mathcal{A}_b|$. Notice that CpsF compresses the symbol alphabet (codebook) while QF compresses the soft-information measure.[8]

Hierarchical Symbol and HNC Map

The hierarchical symbol alphabet properties and corresponding HNC map classify the node according to the processed information flow content. Depending on the cardinality of the HNC map alphabet, we distinguish the following cases. All statements (e.g. the name *source node* symbols) are related to direct radio neighbors of the node. Their transmitted symbols can already be, in the sense of the hierarchical principle, functions of the symbols from previous encapsulation levels.

Full HNC map The cardinality of the HNC map alphabet is such that all combinations of $\tilde{b} = \{b_i, \ldots, b_{i'}\}$ are one-to-one mapped to the symbol b, i.e. $|\mathcal{A}_b| = |\mathcal{A}_{\tilde{b}}|$ where $\mathcal{A}_{\tilde{b}}$ is Cartesian product alphabet $\mathcal{A}_{\tilde{b}} = \mathcal{A}_{b_i} \times \cdots \times \mathcal{A}_{b_{i'}}$ and clearly $|\mathcal{A}_{\tilde{b}}| = \prod_{k=i}^{i'} |\mathcal{A}_k|$. This case corresponds to a classical multi-user decoding where all neighbors are fully decodable at the node and no other side-information is required. In this case, using a word "hierarchical" is not needed, or has a singular meaning, since we simply *jointly* decode everything at all stages.

---

[8] Some authors use CpsF term in a relaxed manner comprising both codebook and/or soft-information compression.

Minimal HNC map The minimal map has a *minimum* cardinality that allows solvability (necessary but not generally sufficient condition) for the arbitrary source node symbol from the set $\{b_i, \ldots, b_{i'}\}$ provided that all others are *perfectly* known. The HNC map alphabet must be just capable of distinguishing the largest source size, i.e. $|\mathcal{A}_b| = \max\left(|\mathcal{A}_{b_i}|, \ldots, |\mathcal{A}_{b_{i'}}|\right)$.

Extended HNC map The cardinality of HNC map alphabet is in between the size of the minimal and full ones $\max\left(|\mathcal{A}_{b_i}|, \ldots, |\mathcal{A}_{b_{i'}}|\right) < |\mathcal{A}_b| < |\mathcal{A}_{\tilde{b}}|$. A full decoding without additional information is not possible. Some independent information (either HSI or other HI) is required but it does not need to be a perfect one. Only partial HSI/HI is needed.

Lossy HNC map The HNC map alphabet is smaller in cardinality than required by the minimal map $|\mathcal{A}_b| < \max\left(|\mathcal{A}_{b_i}|, \ldots, |\mathcal{A}_{b_{i'}}|\right)$. In this case, even a perfect HSI does not suffice on its own to allow solvability and it must be complemented by other HI.

Hierarchical Decoding

The core task of the relay operation is to exploit the codebook structure of the received signals. Depending on the target data HNC map form and the form of the decoder input, we distinguish the following situations.

Single Source Decoding This is a classical case when the front-end received signal comes from a single stage and there is only one transmitting source node in that stage and the target is the single-user data. The traditional single-user decoding technique applies here.

Classical Multi-user Decoding This is the case where the decoding target HNC map has *full cardinality* and all component symbol combinations have their representation in the metric.

Hierarchical Decoding We talk about the hierarchical decoding in the situation when the front-end receives the signal only in *one stage* and there are *multiple* transmitters active (non-orthogonal superposed signals) in that stage and the task is to obtain the *hierarchical* information measure μ_b. Alternatively, we may talk about hierarchical soft-decoding (or processing) if the target measure μ_b has a form of a soft-information measure.

Hierarchical Side-Information Decoding Relay processing performs hierarchical side-information decoding if, apart from the single-stage received signal that contains multiple non-orthogonal superposed contributions, we also have available an *orthogonal* observation from a different stage, and it carries either other hierarchical information or hierarchical side-information. The decoder then can properly combine all this when decoding its target hierarchical information measure μ_b.

3.3.4 Classification of Back-End Strategies

The hierarchical information measure μ_b, which is the input to the back-end, is in fact *source* data for the back-end channel encoder and modulator. The first aspect that distinguishes the back-end strategies is whether we apply any additional *source* encoding before channel encoding and modulation.

Direct H-BC (Direct Hierarchical Broadcast)   This is the simplest case. The discrete finite cardinality hierarchical measure μ_b is encoded and transmitted by the back-end as it is, with no additional source coding.

Analog H-BC (Analog Hierarchical Broadcast)   In the case of *continuous* valued μ_b the broadcast stage uses an analog modulation technique. Notice that the quantized μ_b case falls into the Direct H-BC category.

NC-H-BC (Network Coding in Hierarchical Broadcast)   This defines the case where we have reliably decoded individual (full HNC map) sources represented in μ_b, and we apply *standard* NC on them. The resulting NC (source encoded) symbol is then transmitted into H-BC.

The second classification criteria relate to the *channel encoding and modulation* for the H-BC stages. They distinguish whether the broadcast signal is expected to be received by the direct neighbors as the *only* signal or whether we should count on the fact that our node's signal will be received in a superposition with other nodes' signals. In other words, whether the node under the consideration is the only transmit source of the given stage or if there are other transmit sources in that stage. This reflects the node's awareness of the network structure.

NCM (Network Coded Modulation) NCM is the channel encoding and modulation *aware* of (and specifically designed for) network structure and hierarchical information flows – *hierarchical network-aware coded modulation.*
  (1)   The signal will be received in a superposition with other NCM signals from other sources, i.e. sharing a common stage with other transmitters.
  (2)   The performance optimization target is related to the *HNC map* of the source node data rather than source data themselves, i.e. optimization is done with respect to the hierarchical information measures.
  (3)   NCM is a constellation space channel code jointly designing coding and modulation (constellation space), i.e. it is a coded modulation.
Standard Coded Modulation   If the node back-end transmitted signal is the only transmission source of the given stage, we use a standard channel encoded modulation.

3.3.5   Classification of Front-End Strategies

The front-end processes received signals, possibly being transmitted in multiple stages, into the decoding metric μ_c. The decoding metric serves as the input for the relay processing operation, which is some form of the decoding. The decoding metric can be any form of the information preprocessed from the received signal. It varies in (1) the form of the metric, and (2) the variable(s) it describes. The formal definition is quite similar to the definition of relay processing measure. But because the variable that is described by the metric is typically (except for some singular cases – uncoded case, AF case, etc.) different from the target variable b of the relay processing, we use *channel* symbols $\tilde{c}$ in the notation. Typically these are encoded discrete symbols.

DEFINITION 3.6 (Front-End Decoding Metric)   An arbitrary subset of discrete *channel* symbols of the nodes participating in the H-MAC stage is denoted $\tilde{c} = \{c_i, \ldots, c_{i'}\}$

and the HNC map associated with them is $c = \chi_c(\tilde{c})$. The front-end decoding metric is a preprocessing operation on the received signal $\mu_c = \mu_c(x)$. In the case of discrete-valued symbols it has the form of the metric value set

$$\mu_c(x) = \left\{ \mu(x, c^{(1)}), \mu(x, c^{(2)}), \ldots \right\} \tag{3.3}$$

or any other equivalent form. The metric is called *hierarchical* if the associated map is *not a full* cardinality map.

The front-end can be interpreted as a generalization of the classical soft-output demodulator[9] and the decoding metric is then its output. The form of the metric in the simplest case is no preprocessing at all, $\mu_c = x$. A second, very common, case is a set of channel symbol likelihoods or its equivalent form, obtained by using a standard matched filter demodulator output. It is nothing else than the sufficient observation statistic for the given channel symbols. More complicated cases can include some nonlinear preprocessing, e.g. modulo lattice operation.

The variable(s) described by the metric can also be of various forms. The simplest case is no association at all, i.e. $\mu_c = x$. Second, a classical case is an associated *full* cardinality HNC map. It corresponds to the classical multi-user decoding where each component variable c_i is distinguished in the metric. From the WPNC perspective, the most interesting one is the case where the χ_c HNC map is *not* the full map. This case will be called the *hierarchical* decoding metric.

A properly chosen decoding metric with less than full map cardinality can significantly simplify processing/decoding at the relay. On the other hand, the processing chain $x \mapsto \mu_c \mapsto \mu_b$ should not impose any constraints or additional bottlenecks from the perspective of target information measure μ_b and, as a consequence, also on the end-to-end WPNC performance. In other words, the metric μ_c should be a *sufficient statistic* for decoding μ_b. A trade-off between these two requirements is an important aspect of the WPNC design.

A high-level classification of the front-end processing strategies characterizes the associated variable that is described by the decoding metric. This is applicable on each reception stage signal.

Joint-metric demodulator (front-end) This is the classical multi-user (soft-output) demodulator. The channel symbol map is a full cardinality map.

Hierarchical demodulator (front-end) The decoding metric provides values $\mu_c(x)$ for the HNC channel symbol map with *less* than full cardinality (see Section 4.4 for more details).

3.3.6   Classification of Relay Node Strategy

An overall input–output strategy of the (relay) node can be viewed as a combination of the previously defined front-end, processing, and back-end strategies. The number of options is high and sometimes not rigorously defined, but here we name a few important cases.

---

[9] Sometimes, we stress this fact by using explicitly the SODEM term.

HDF (Hierarchical Decode and Forward) is defined by the node processing operation being DF, and the HNC map for the associated measure μ_b being anything except a full map. HDF is a strategy that makes a *decision* on some *many-to-one hierarchical data* function. The input decoding metric of the HDF decoder is directly obtained as the constellation space metric (compare this to NC-JDF case). The input metric of the decoder (which is the output of the front-end) can have a variety of forms, e.g. joint metric or hierarchical metric (see Sections 4.3 and 5.7 for more details). Decoding may or may not also exploit the NCM codebook structure.

JDF (Joint Decode and Forward) uses a DF processing operation on the *full* HNC map, i.e. decoding all source nodes individually (classical multi-user decoder).

NC-JDF (Network Coding over JDF) first performs JDF and then the back-end applies classical NC source coding on the previously reliably decoded individual source node data.

3.4   Global HNC Map and Generalized Exclusive Law

Now we turn our attention to the global end-to-end connectivity provided by the WPNC network. The network itself might transfer various forms of hierarchical many-to-one functions of the source node data. The hierarchical symbols can be represented by a variety of information measures. But at the end, regardless of how many ways and forms of the information we collect at the destination node, we must ensure the solvability for the target data. We may visualize this problem as collecting a set of equations at the final destination. One of the unknowns is our target data and we must make sure that we can solve it out of the provided equations.

We will now develop the necessary conditions for this task. It would be relatively easy, if all the involved data hierarchical functions performed by the nodes were *linear* over a *common discrete alphabet* and the hierarchical information measure μ was a *reliable* decision. Then the overall network map could be described as a linear set of equations over GF and we could use all the classical results. This is a common situation in traditional NC.

Unfortunately, it has only a very limited applicability to WPNC networks. The most important constraint is that, owing to WPNC properties being strongly connected to the constellation space and channel fading parametrization, we are frequently forced to use different alphabets at each node and also quite frequently these alphabets might not form extended GF. Second, in many situations it is useful to use nonlinear or adaptive HNC maps. Also of importance is the fact that owing to the noisy constellation space channel parametrized observations and potentially unreliable decisions (or other, e.g. soft-information forms of hierarchical information measures), the solvability conditions can typically be constructed only as the necessary ones, not generally sufficient.[10] The problem of analyzing the end-to-end solvability can be split into two parts.

---

[10] Compare this with a traditional NC coded network over reliably decoded source symbols using a common GF linear map at all nodes.

We first model the "information flows" and then we state the conditions of solvability. The information flow in the network is described by a global HNC map associated with each destination node for its target data. Only then in the second step, after making sure that the information flows as desired, can we analyze the reliability and the achievable rates.

DEFINITION 3.7 (Global HNC Map – Cycle-Free Network)   Assume a cycle-free network with source node symbol set $\tilde{b} = \{b_A, b_B, \ldots\}$. Relay R$j$ at its stage m uses the HNC map $b_{j,m} = \chi_{j,m}(\tilde{b}_j)$ associated with its relay processing information measure $\mu_{b_{j,m}}$, where $\tilde{b}_j$ is a set of hierarchical symbols from direct neighbor predecessors. Destination Di for target symbol b_i, $i \in \{A, B, \ldots\}$ uses at its mth stage HNC map $b_{\hat{i},m} = \chi_{\hat{i},m}(\tilde{b}_{\hat{i}})$, where $\tilde{b}_{\hat{i}}$ is the set of its neighbor predecessor hierarchical symbols.

We define the *global HNC map* vector at Di as a compound hierarchically *nested* HNC map of all network graph predecessors from the destination point of view

$$\vec{b}_{\hat{i}} = \vec{\chi}_{\hat{i}}(\tilde{b}) = \left[\chi_{\hat{i},1}(\tilde{b}), \chi_{\hat{i},2}(\tilde{b}), \ldots \right]^{\mathrm{T}} \tag{3.4}$$

where each vector component corresponds to one stage observed at Di. The fact that the individual stages go into separate vector components is a consequence of the orthogonality of the stages.

The global HNC map describes "the flow of the information" in the WPNC network as if all involved nodes performed the *DF strategy* in their processing operations $\mu(.)$ and the decisions were *perfect (reliable)*. This model completely ignores any information rate constraints imposed by noisy and parametrized radio channels, or any suboptimality associated with the form of node processing metric used in the network. We can view it as an idealized deterministic input–output model. Notice also that even the destination node can receive multiple signals in one stage and then its front-end and node operation processing becomes the same as for any other relay node, i.e. it evaluates its own hierarchical HNC map.

Example 3.3   Consider the example network in Figure 3.7. For simplicity, we assume that the nodes are active only in one stage – SA, SB in stage 1; R1, SC, SD in stage

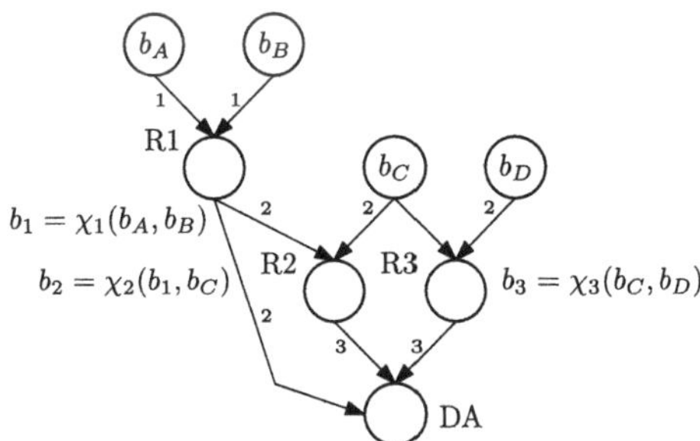

Figure 3.7 Global HNC map example.

2; and R2, R3 in stage 3. We also show the situation only for one destination DA. The construction of the global map for multiple destinations is a straightforward extension. The global HNC map for the destination DA has two components, since R1 and R2 must be inevitably in the two different stages due to the half-duplex constraint on R2. The first component is the only signal received by DA at that stage and thus the HNC map is simply an identity $\chi_{\hat{A},1}(b_1) = b_1$. The second component is, however, a superposition of signal from R2 and R3 and the destination DA evaluates its own map $\chi_{\hat{A},2}(b_2, b_3)$. The resulting global HNC map is

$$\vec{b}_{\hat{A}} = \vec{\chi}_{\hat{A}}(b_A, b_B, b_C, b_D) = \begin{bmatrix} \chi_1(b_A, b_B) \\ \chi_{\hat{A},2}\left(\chi_2\left(\chi_1(b_A, b_B), b_C\right), \chi_3(b_C, b_D)\right) \end{bmatrix}. \tag{3.5}$$

The example also nicely illustrates the hierarchical encapsulation principle.

Now we state a theorem that guarantees solvability for a given source symbol at a given destination node in the sense of a "deterministic" model of a global HNC map.

THEOREM 3.8 (Generalized Exclusive Law) *We denote $\tilde{b} = \{b_A, b_B, \ldots\}$ as a set of all source node symbols and $\vec{\chi}_k(\tilde{b})$ as a global HNC map at the destination node Dk, and $\vec{b}_k = \vec{\chi}_k(\tilde{b})$ the available global hierarchical symbol vector at Dk. We also denote $\tilde{b}(b_i) = \{\tilde{b} : b_i\}$ as a set of all $\tilde{b}$ consistent with b_i, i.e. having a given b_i at its particular position.*

The global HNC map at the destination Dk is solvable for the symbol b_i (i.e. for a given $\vec{b}_k$ we can uniquely find b_i) if and only if the following holds

$$\forall \tilde{b}(b_i), \tilde{b}(b_i') : b_i \neq b_i' \Rightarrow \vec{\chi}_k\left(\tilde{b}(b_i)\right) \neq \vec{\chi}_k\left(\tilde{b}(b_i')\right). \tag{3.6}$$

Proof The key is to realize that the condition (3.6) defines a partition of the set of all possible $\vec{b}_k$ values w.r.t. values b_i (Figure 3.8) with one-to-one mapping to b_i values. The solvability then simply follows. $\square$

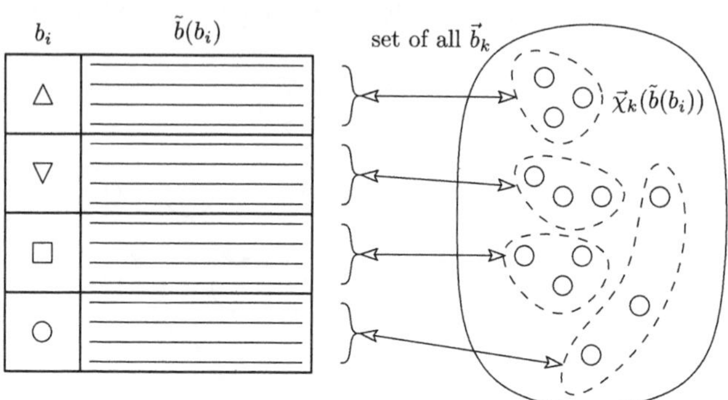

Figure 3.8 Generalized exclusive law – set partition.

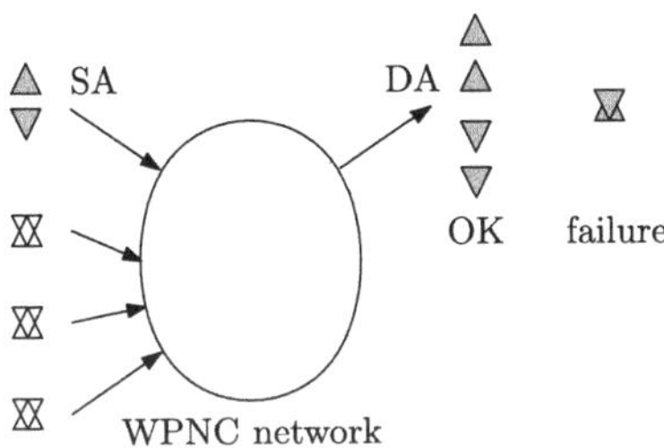

Figure 3.9 Visualization of the generalized exclusive law. Two different symbols (upward/downward facing triangles) on the source SA must be mapped into distinct (possibly multiple) symbols at the destination DA regardless of the value of the symbols at all other sources.

Notice that the inequality condition in (3.6) allows multiple global HNC map values $\vec{\chi}_k$ for one value b_i depending on the other nuisance source symbol values. It is absolutely *not* required to be a one-to-one mapping. The condition (3.6) has a form of implication only. The cardinality of all $\{\vec{b}_k\}$ might be higher than the cardinality of all $\{\tilde{b}\}$ (Figure 3.8). It allows, for example, a random HNC map where several distinct values of $\vec{b}_k$ correspond to one $\tilde{b}(b_i)$. Also notice that it works for arbitrary HNC maps – for example nonlinear ones, the ones with mixed alphabet cardinalities, etc. The theorem generally allows $i \neq k$ (e.g. when one destination decodes multiple sources) but in the majority of cases these two are equal. In the case when (3.6) does not hold, we talk about generalized exclusive law failure (Figure 3.9).

3.5 Hierarchical Constellation

3.5.1 Hierarchical Constellation and Hierarchical Codebook

The hierarchical constellation describes how the observed received superposed *constellation space* points are connected to the *HNC symbol map*. The symbol can have again various meanings, including uncoded channel symbol, codeword, or data symbol. The received constellation is viewed as a function of the hierarchical symbol.

There are two important aspects that are characteristic for WPNC. First, owing to the channel parametrization, the *shape* of the hierarchical constellation strongly depends on the actual channel state, particularly on a relative fading (see Section 3.5.2 for a detailed discussion). Second, since the cardinality of the HNC map alphabet is typically smaller than the product of source symbol cardinalities (for any except the full HNC map), the hierarchical constellation is a *set of multiple* points for one particular hierarchical symbol b value.

The principle is demonstrated for two binary sources and a minimal HNC map in Figure 3.10. We can clearly see that there are two independent relationships between sources and the relay observation. The first is on the discrete symbol level and it is given by the HNC map. The second is at the constellation level and it is dictated by the radio-wave superposition and by the channel model characteristic. Notice that the former is

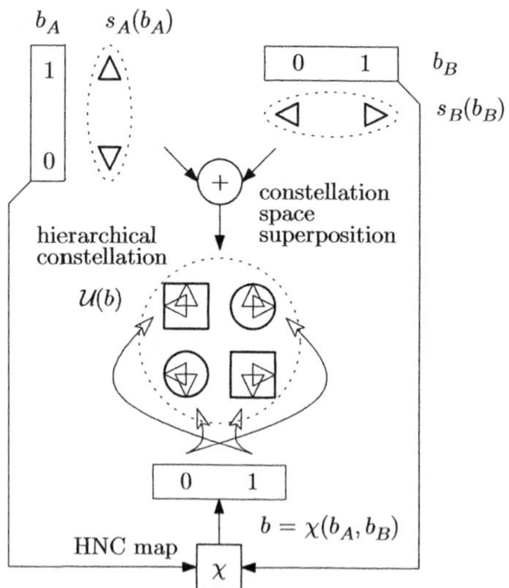

Figure 3.10 Hierarchical constellation example for two binary sources and a minimal HNC map.

mathematically modeled by discrete number space functions (e.g. over a GF); the latter is mathematically modeled in a very different constellation space. The coexistence of these two very differently modeled relationships is a distinctive feature of WPNC that distinguishes it from the classical NC (where only the discrete model counts) and from the classical multi-user PHY layer (where only the constellation space part counts).

Assume that the signal space representation of the source neighbor node is $s_i(b_i)$, which is a function of the source symbol b_i. Assume that the channel model between the node Si and the relay R is parametrized by h_i (e.g. phase shift, amplitude, multipath fading, etc.), and the set of sources received by the relay in a given stage is $\mathcal{S}$. Owing to the interaction of the transmitted signals in the channel (e.g. by a superposition), the resulting "channel combined" constellation points differ from the individual transmitted component source constellations. It will be termed hierarchical channel-combined constellation space symbol $u \in \mathcal{A}_u$ and the corresponding set $\mathcal{A}_u$ will be called the hierarchical alphabet.

Also assume that the channel stochastic input–output description is a Markov chain

$$\left\{ \{s_i(b_i)\}_{i\in\mathcal{S}}, \tilde{h} \right\} \mapsto u \mapsto x \tag{3.7}$$

where x is the received signal at the relay with the observation likelihood

$$p\left(x | u\left(\{s_i(b_i)\}_{i\in\mathcal{S}}, \tilde{h} \right) \right) \tag{3.8}$$

where $\tilde{h} = \{h_i\}_{i\in\mathcal{S}}$ is the set of all channel parametrizations. The relay HNC map is $\chi(\tilde{b})$, where $\tilde{b} = \{b_i\}_{i\in\mathcal{S}}$.

DEFINITION 3.9 (Hierarchical Constellation, Hierarchical Alphabet) The hierarchical constellation (H-constellation) is defined to be a signal space representation of the hierarchical channel-combined constellation symbol $u \in \mathcal{A}_u$ *viewed as a function* of the hierarchical symbol $b = \chi(\tilde{b})$. A particular subset corresponding to a particular b is

$$\mathcal{U}(b) = \left\{ u : u = u\left(\{s_i(b_i)\}_{i \in \mathcal{S}}, \tilde{h} \right) \mid b = \chi(\tilde{b}) \right\}. \tag{3.9}$$

A complete hierarchical constellation for all values b is denoted by $\{\mathcal{U}(b)\}_b = \{\mathcal{U}(b)\}$ or simply $\mathcal{U}$. The set of all hierarchical channel-combined symbols $\mathcal{A}_u$ will be called the hierarchical alphabet (H-alphabet).

Definition (3.9) identifies individual constellation points $u \in \mathcal{A}_u$ such that they correspond to one given hierarchical symbol b related by HNC map $\chi(\tilde{b})$ to the component symbols $\tilde{b}$. A constellation $\mathcal{U}(b)$ is not a single point but is generally a set of points corresponding to one hierarchical symbol b. It is important to distinguish between H-alphabet and H-constellation. H-alphabet is the set of all channel-combined symbols ignoring any relationship to the HNC map. On the other hand, H-constellation *groups* the points from the H-alphabet according to the corresponding HNC map value. $\mathcal{U}(b)$ depends on (1) *HNC map* $\chi(.)$, (2) *source node constellation mappings* $s_i(b_i)$, and (3) *channel parametrization* h_i. Notice that, apart from (1), these characteristics are very specific to WPNC, namely its dependence on channel parametrization. As we will describe in the following, the parametrization can even completely change the constellation shape.

Sometimes, when we want to stress that the points u are multidimensional and have some *codebook structure* we use the term hierarchical codebook (H-codebook). Definition 3.9 changes only in the notation.

DEFINITION 3.10 (Hierarchical Codebook)   The hierarchical codebook (H-codebook) is defined to be a signal space representation of channel-combined signal u viewed as a function of the hierarchical symbol $b = \chi(\tilde{b})$. A particular subset corresponding to a particular b is

$$\mathcal{C}(b) = \left\{ u : u = u\left(\{s_i(b_i)\}_{i \in \mathcal{S}}, \tilde{h} \right) \mid b = \chi(\tilde{b}) \right\}. \tag{3.10}$$

A complete hierarchical codebook for all values b is denoted by $\{\mathcal{C}(b)\}_b = \{\mathcal{C}(b)\}$ or simply $\mathcal{C}$.

3.5.2   Common and Relative Channel Parametrization

The hierarchical constellation strongly depends on the channel parametrization. Even under a very simplistic channel model, the hierarchical constellation completely changes its shape[11] and it has important consequences on the system performance. A usual common-sense feeling that the strong received signal and perfect channel state knowledge guarantee high performance does not necessarily hold. The fact that this phenomenon is caused by mutual interaction of all channel states of *all* involved source nodes makes the situation even worse. An individual transmitting source thus has very limited possibilities for preventing this phenomenon.

---

[11] Compare this with classical point-to-point communications over a parametric channel where the *shape* of the constellation viewed at the receiver is purely given by its shape at the transmitter.

The dependence of the hierarchical constellation (*the shape in association with the HNC map*) on the relative (nonlinear) channel parametrization is undoubtedly one of the most specific aspects of WPNC and represents a major target of the NCM design and performance analysis. Also it is practically unavoidable and demonstrates itself even in very simplistic scenarios. Because of its importance, we must precisely define and classify the conditions that affect this behavior.

We will define two forms of the channel parametrization (fading) according to their influence on the *shape* of the hierarchical constellation. The first one, a *common (linear)* fading, purely linearly scales (complex valued) the hierarchical constellation. It is relatively easy to cope with in the sense that it is quite similar to what we used to do for a classical point-to-point single-user channel. The second one, *relative (nonlinear)*, completely changes the shape of the hierarchical constellation with the inherently associated HNC map. The cause for this behavior comes from the relative difference in the parametrization influence on one and the other link. The parametrization effects do not go "hand-in-hand" but they relatively depart, hence the name relative parametrization. And because the resulting H-constellation shape changes, we also call it nonlinear. Notice that even a simple linear channel model can produce a severe nonlinear H-constellation shape change. The relative fading fundamentally affects the design of the NCM and its performance. The performance can be heavily affected even for very basic channel models and perfect channel state information available at the receiver.

DEFINITION 3.11 (Common Channel Parametrization) The common channel parametrization (linear scaling) is such that it causes only (generally complex multidimensional) a scaling of the constellation (it keeps its shape), i.e.

$$\forall \tilde{h}, \tilde{h}' \exists \alpha \in \mathbb{C}^{n \times n} : \{\mathcal{U}(b, \tilde{h})\}_b = \alpha \{\mathcal{U}(b, \tilde{h}')\}_b \tag{3.11}$$

where $\{\mathcal{U}(b, \tilde{h})\}_b = \{\mathcal{U}(b, \tilde{h})\}_{b \in \mathcal{A}_b}$ is an n-dimensional H-constellation (3.9) with explicit channel parametrization argument.

The meaning of the common parametrization is such that we can find a proper linear transformation transferring the H-constellation for two arbitrary channel coefficients.

DEFINITION 3.12 (Relative Channel Parametrization) The channel parametrization is called relative (nonlinear) if there exist some channel parametrizations where the linear transformation of the H-constellation is impossible, i.e.

$$\exists \tilde{h} \neq \tilde{h}' \forall \alpha \in \mathbb{C}^{n \times n} : \{\mathcal{U}(b, \tilde{h})\}_b \neq \alpha \{\mathcal{U}(b, \tilde{h}')\}_b \tag{3.12}$$

where $\{\mathcal{U}(b, \tilde{h})\}_b = \{\mathcal{U}(b, \tilde{h})\}_{b \in \mathcal{A}_b}$ is an n-dimensional H-constellation (3.9) with explicit channel parametrization argument.

The meaning of the definition is that there are (at least) some "unfavorable" channel parametrizations such that we cannot get one from the other one by a linear transformation.

Figure 3.11 demonstrates the common (linear) and relative (nonlinear) channel parametrization. The real system with real physically existing channel parametrizations h_A, h_B (Figure 3.11a) can be also viewed as a virtual multi-input system with "global"

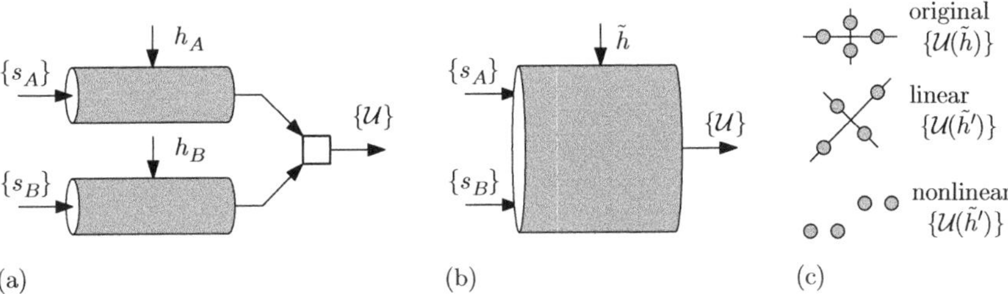

Figure 3.11 Common (linear) and relative (nonlinear) channel parametrization.

parametrization $\tilde{h}$. Their components do not necessarily need to be exactly the real channel component values. We can create some form of virtual (existing only in the model) components of $\tilde{h}$. For example, the relative channel gain (see Example 3.4). The linear and nonlinear names for the channel parametrization are motivated by the property of the set $\{\mathcal{U}(b, \tilde{h})\}_b$ viewed as a linear or nonlinear function of the parameter $\tilde{h}$.

Example 3.4 Consider a simple flat fading block-constant two-source channel with the channel-combined signal at the relay $u = h_A s_A + h_B s_B$ with all signals from a single-dimensional complex constellation space $\mathbb{C}^1$. We can rearrange the fading coefficients into

$$u = h_A \left(s_A + h s_B\right) \tag{3.13}$$

where $h = h_B/h_A$.

The parameter h_A is a common (linear) fading and h is the relative fading. Notice also that we can swap the factorized coefficients and we get $u = h_B \left(h' s_A + s_B\right)$ with $h' = h_A/h_B$.

Example 3.5 Assume again a two-source system with the channel model as in Example 3.4 where we set the common fading to unity $h_A = 1$ and we keep only the relative fading $u = s_A + h s_B$.

Figures 3.12 and 3.13 show the source constellation with the discrete symbol mapping $s_i(b_i)$ and the H-constellations for two different values of the relative fading. The notation $b(b_A, b_B)$ describes which particular HNC map symbol b is mapped to a corresponding pair of the source symbols (b_A, b_B). In the case of BPSK, the discrete symbols are $\{0, 1\}$. In the case of QPSK, we used binary pairs $\{00, 01, 10, 11\}$. You may notice that the HNC map is a bit-wise XOR function.

We can observe several phenomena. The H-constellation changes its shape as a function of the relative fading. At some relative fading values, several H-constellation points can overlap in one place. They can, but do not need to (see Section 3.5.3), correspond to the same HNC map symbol b – for example in cases (b). The HNC map is clearly the minimum one and, for example, in the QPSK case each hierarchical symbol corresponds to four combinations of source symbols.

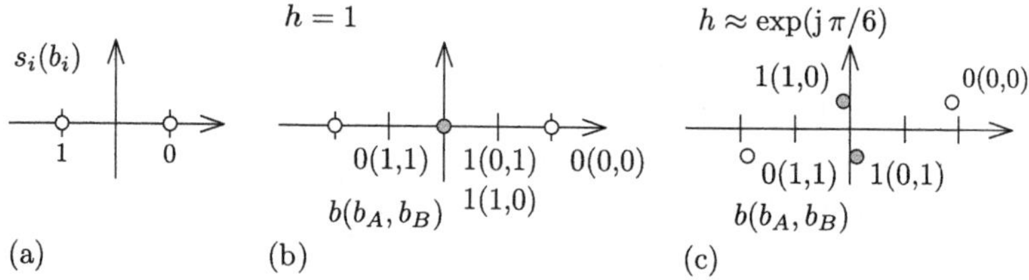

Figure 3.12 Hierarchical constellation and relative fading – BPSK example. (a) Source constellation; (b) and (c) hierarchical constellation for two different relative fading values.

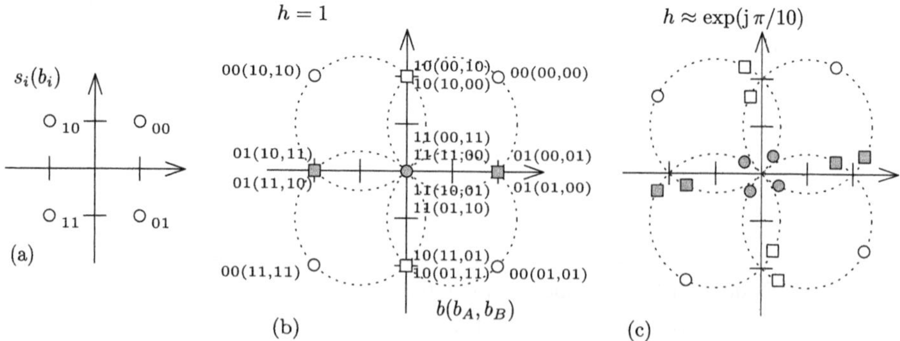

Figure 3.13 Hierarchical constellation and relative fading – QPSK example. (a) Source constellation; (b) and (c) hierarchical constellation for two different relative fading values.

Since the only critical impact on the design and performance is the one caused by the relative fading, we frequently consider system models dropping the linear common scaling completely (or setting it to some default, easy to handle, value). It is of course present in a real system; however, the simplified notation of the model allows us to concentrate on the important, WPNC specific, aspects. The common linear scaling can usually be handled by standard synchronization approaches as they are used in classical single-user systems.

3.5.3   Singular Fading

Singular fading is a phenomenon very specific to WPNC and it strongly affects the overall performance of the system. As such, it is a major H-constellation and NCM design concern. In simple words, it describes the situation when some value of the relative fading causes two or more constellation points of the superposed signals to fall into the same point for different combinations of the source node symbols. If these source node symbols correspond to *different* hierarchical symbols, we call this *unresolved* singular fading (or *HNC mapping failure* or *conflict*). In the case when these multiple overlapping constellation points correspond to a common hierarchical symbol, there is no problem and we call it *resolved* singular fading.

Having an unresolved singular fading state means that multiple hierarchical symbols that need to be distinguished (decoded) at the relay thus have indistinguishable constellation space representation. This ambiguity, in some sense similar to the classical erasure channel model, has a strong impact on the achievable rates or bit error rate performance. Even for relative fading values approaching the singular fading value, the performance is significantly influenced. It is important to note that it can happen only for some relative fading values. So, even if the H-constellation and its corresponding HNC map were properly designed for some particular relative fading value, it does not mean that unresolved singular fading cannot happen at some other value.

DEFINITION 3.13 (Singular Fading)   Denote two sets of particular source symbol values $\tilde{b} = \{b_i\}_{i\in\mathcal{S}}$, $\tilde{b}' = \{b'_i\}_{i\in\mathcal{S}}$, and source node constellation mapping $s_i(b_i)$. We say that the fading coefficient set $\tilde{h}$ is a singular fading iff

$$\exists \tilde{b} \neq \tilde{b}' : u\left(\{s_i(b_i)\}_{i\in\mathcal{S}}, \tilde{h}\right) = u\left(\{s_i(b'_i)\}_{i\in\mathcal{S}}, \tilde{h}\right). \tag{3.14}$$

The singular fading is called *unresolved (or fading with HNC mapping failure)* iff

$$\exists \tilde{b} \neq \tilde{b}' : \chi(\tilde{b}) \neq \chi(\tilde{b}') \wedge u\left(\{s_i(b_i)\}_{i\in\mathcal{S}}, \tilde{h}\right) = u\left(\{s_i(b'_i)\}_{i\in\mathcal{S}}, \tilde{h}\right). \tag{3.15}$$

The singular fading is called *resolved* iff

$$\exists \tilde{b} \neq \tilde{b}' : \chi(\tilde{b}) = \chi(\tilde{b}') \wedge u\left(\{s_i(b_i)\}_{i\in\mathcal{S}}, \tilde{h}\right) = u\left(\{s_i(b'_i)\}_{i\in\mathcal{S}}, \tilde{h}\right). \tag{3.16}$$

The meaning of unresolved singular fading is that there is at least one pair of source symbol sets that each has different value of the HNC map although their channel-combined superposed constellation falls into one point. As we see, the validity (e.g. for global end-to-end solvability) of the HNC map itself is *not enough* to guarantee a proper NCM design. The consequences of the unresolved singular fading depend on what the symbols $\tilde{b}$ represent. If it represents the information carrying data, then it means actual erasure of the information. If it represents the channel encoded symbols, some of the detrimental effect can be recovered by a proper NCM codebook design, and we can expect only some drop of the achievable rate (see Chapter 6 for some quantitative results).

Notice also that the split definition of singular fading and its resolvability allow us to simplify the design by separating the properties of the constellation and HNC map. The

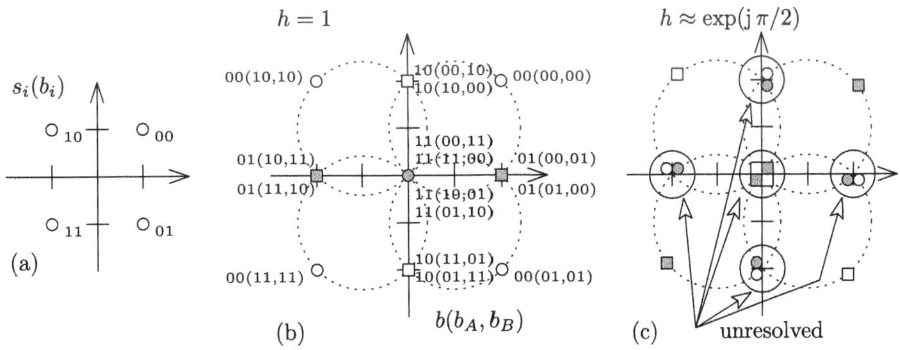

Figure 3.14 Singular fading – QPSK example.

singular fading depends only on the source node constellations and the channel. The HNC map enters into the design only when deciding about the resolvability.

Example 3.6   We now continue with Example 3.5, particularly the QPSK case. Figure 3.14b shows the example of *resolved* singular fading at $h = 1$. There are multiple superposed constellation points at various positions, for example at the origin, but they all correspond to the same hierarchical symbol. In contrast, Figure 3.14c shows the *unresolved* case with the relative fading value close to the singular fading value $h = \exp(j\,\pi/2)$. Again there are multiple unresolved HNC map conflicts, for example at the origin. There are four points but two of them correspond to the hierarchical symbol 01 and two of them to the hierarchical symbol 10.

---

4   Components of WPNC

4.1   Introduction

This chapter still contains general material on the processes carried out at each node, especially the relays, but it goes deeper into the details of each WPNC component, as previously defined in Chapter 3. It defines design criteria and models, expanding on Chapter 3, which gives only definitions and classifications. On the other hand, more advanced material and particular methods of constructing these components will be described later, in Part III.

The topics are structured according to the division: (a) MAC front-end Rx processing, (b) relay HNC map and processing operation, and (c) BC back-end Tx processing. Note that source and destination nodes are special cases with BC back-end only and MAC front-end only processing and a singular form of HNC map and processing operation. In the sense of the *hierarchical* principle explained in Chapter 3, we focus here on *direct local neighborhood* of the node. Global whole WPNC cloud aspects are treated in Chapter 5.

We start with the description of the back-end processing, since it determines the transmitted signals. Then we discuss demodulation and decoding techniques associated with the front-end processing. Both front-end and back-end processing assume that the HNC map is given. Particular aspects related to the HNC map are discussed at the end.

4.2   Network Coded Modulation

4.2.1   Multi-source Network Structure Aware Constellation Space Codebook

Network Coded Modulation (NCM) is the name for a multi-source network structure aware signal space code. NCM is an implementation of the back-end node strategy that is suited for maximal a priori knowledge utilization of WPNC network structure. Such back-end strategy should fulfill the following.

(1) WPNC operates in the wireless domain and the information must be encoded by the channel coding that fully respects that the *encoded* transmitted symbols are the *signal space constellation symbols*. NCM is a channel coding that must provide forward error correction/protection capabilities. This is reflected by "coded

modulation" in the name, and it is quite obvious and not much different from the classical single-user systems.

(2) NCM must be *network structure aware* and must respect it, hence the name "network coded modulation." NCM must "see around the corner" in the network. Clearly, decoding individual source component data at the node that receives NCM multiple signals might not be the optimal strategy. Instead, the receive node processing information measure μ_b related to some more complicated HNC map might be a better option from the global point of view. However, in order to take this advantage, all participating back-end generated NCM signals must be aware of

  (a) the fact that they are *not alone* in the given stage, and

  (b) that the performance optimization target and quality of NCM design are related to the measure μ_b related to some given HNC map $b = \chi(\tilde{b})$ (and *not* to the individual source components $\tilde{b}$), and

  (c) that the H-constellation depends on *all* involved component channels (not just the channel associated with the given NCM transmitting node).

(3) The NCM codebook is a multi-source and distributed one. Several component sources are contributing their signals; however, the final codeword, as visible to the receiver, is actually created by the radio-wave superposition in the channel additionally influenced by channel parameters. Each transmission node independently creates its component codeword but the final received composite codeword is combined by the channel without transmission nodes having a chance to influence it. Other participating nodes do not have available the data of remaining NCMs and typically also not the channel state of the others. This is particularly problematic, since the H-constellation and the associated performance metric related to μ_b is affected by all participating channel states; however, the given transmitting node does not have this information available. It must use its own NCM component codeword in a *non-cooperative* way with respect to other nodes' data and channels. The component transmission nodes generally do not have a link allowing their on-line cooperation and mutual adjustment.

(4) NCM is only network structure *aware*, not blind[1] against it. The information about the network structure (items (a), (b), (c) above) must be known a priori to the NCM design (see Section 10.2 for a relaxation of this constraint). It also includes the knowledge of other nodes' NCM component codebooks or channel symbol alphabets. The knowledge of the codebooks or alphabets must not be confused with the knowledge of the particular transmitted codeword or symbol. Some global authority is responsible[2] for making sure that the information about the network provided to individual nodes is correct and that the choice of HNC maps and associated μ_b fulfills the global performance targets. The global performance target depends also on all other relay processing operations of all involved WPNC nodes.

---

[1] Blind (oblivious) processing is understood in the sense that the algorithm does not need to have that particular information and works well without it. Similarly as we use e.g. (data) blind equalizer algorithm.

[2] This aspect is, however, not solved in this chapter.

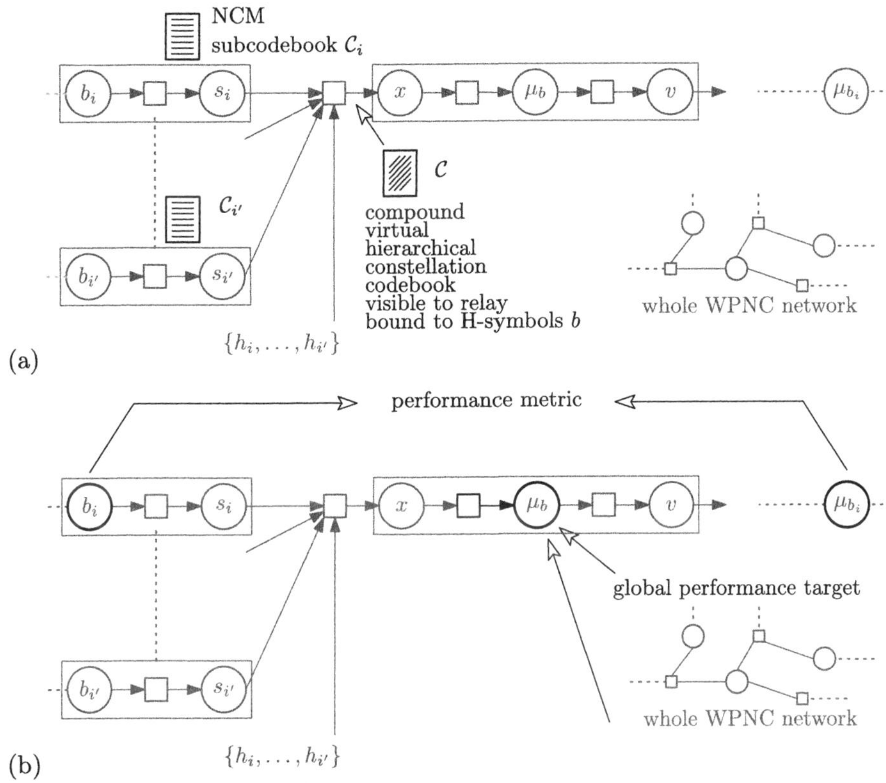

Figure 4.1 Network coded modulation design goals.

Mutual interactions of these principles are visualized in Figure 4.1.[3] NCM is the set of component codebooks (subcodebooks) $\mathcal{C}_i$ (Figure 4.1a). After their codewords pass through the channel parametrized by $\{h_i, \ldots, h_{i'}\}$ and superpose at the receiver antenna they form a virtual hierarchical codebook. The hierarchical codebook $\mathcal{C}$ is indexed by hierarchical symbols b. Essentially it is a hierarchical constellation over multidimensional code space formally following Definition 3.9. The important consequence is the fact that one particular b symbol corresponds generally to a *multipoint* set $\mathcal{C}(b)$. The relay uses the structure of H-codebook to obtain its processing information measure μ_b related to the hierarchical symbol b and its HNC map χ. Of course, a major question is whether there exists some easily interpretable relation between component codebooks and H-codebook and between the component messages and H-message. This will be solved later in Section 4.2.4.

The measure μ_b and the corresponding HNC map are set to match the global performance target (Figure 4.1b), which must use a global knowledge of the whole network structure and operations. The quality of NCM design is not directly evaluated by the

---

[3] In this section, we use a *generic* notation (e.g. b, c, s) to denote messages, codewords, constellation codewords regardless of how they would be practically represented in a real system (e.g. scalars, vectors on GF, etc.).

quality of individual component codebooks C_i or directly by H-codebook quality. It is measured by a chosen performance metric (typically the error rate) related to b_i and its information measure μ_{b_i} somewhere further down in the network. Most typically, b_i are the source node data and $\mu_{b_i} = \hat{b}_i$ is the final hard decision on the target destination node. However, these two reference points can be anywhere inside the network and they can be some hierarchical functions themselves. This is most frequently used for a simplification of the complex network WPNC design where we split the network into some smaller hierarchical subnetworks and each subnetwork is optimized individually.

DEFINITION 4.1 (Network Coded Modulation)   Network coded modulation (NCM) is a set of component codebooks (subcodebooks) $\{C_i, \ldots, C_{i'}\}$ of all nodes participating in one common H-MAC stage received by the given relay node. Component codebooks encode component messages b_i, $s_i = C_i(b_i)$. Component codebooks are designed for a *given* receive-node HNC map $\chi(\tilde{b})$, $\tilde{b} = \{b_i, \ldots, b_{i'}\}$, and associated information measure μ_b that is the only value passed to the node back-end. The performance metric of NCM is a set of performance metric functions $\{P_{e,i}(b_i, \mu_{b_i})\}_i$, where μ_{b_i} is the processing information measure of the given component symbol b_i at an arbitrary node in the network at which it is evaluated. The hierarchical constellation codebook corresponding to the given NCM and associated to the given relay node HNC map $b = \chi(\tilde{b})$ is

$$C(b) = \left\{ u : u = u\left(\{C_i(b_i)\}_{i\in\mathcal{S}}, \tilde{h}\right) \mid b = \chi(\tilde{b}) \right\}.$$

In simple words, NCM is a set of constellation space subcodebooks that are received in a superposition over parametric channels and are designed for a given receive node processing ($\chi(\tilde{b})$ and μ_b) and with the performance measured from the end-to-end single component symbol (b_i, μ_{b_i}) perspective. The most common performance metric is the error rate but it can be the outage probability, achievable rate, or any other. One component coder cannot see the data of others (and cannot adjust its behavior); however, the performance is affected through the common map $b = \chi(\tilde{b})$ presented through μ_b. The design target for NCM is (1) to provide channel error protection coding, while (2) respecting that the relay target is the information measure μ_b related to HNC map $b = \chi(\tilde{b})$.

Example 4.1   NCM for two-way relay channel (Figure 3.2a)

There is only one H-MAC stage for which we can use the NCM design – stage 1. Stage 2 has only one transmitter, the relay, and there are no interfering signals and thus a classical coding and modulation can be used. NCM for stage 1 consists of two subcodebooks $s_A = C_A(b_A)$, $s_b = C_B(b_B)$ used at sources SA and SB. Data $b_A, b_B \in \mathbb{F}_{2^m}$ are binary vectors of equal size. The global performance target is to minimize the H-BC stage information rate, so we choose a linear minimal HNC map over $\mathbb{F}_{2^m}$, $b = b_A + b_B$. The quality of the NCM design is measured by the end-to-end (SA-DA and SB-DB) error rate performance $P_{e,A} = \Pr\{b_A \neq \hat{b}_A\}$ and $P_{e,B} = \Pr\{b_B \neq \hat{b}_B\}$. In this *very special case*, because of perfect HSI links and the use of a minimal map, the end-to-end performance will be given by a simple serial concatenation of H-MAC and H-BC stage performance. The error rate of the H-MAC stage is given by the error rate of hierarchical symbols b.

Example 4.2 NCM for a butterfly network (Figure 3.2b)

Again, there is only one applicable H-MAC stage for NCM design – stage 1. NCM consists of two subcodebooks the same as in the previous example. The global performance target is again to minimize the H-BC stage information rate. The HSI links SA-DB and SB-DA are wireless and we assume that they cannot support reliably the full source SA and SB rates. The relay HNC map can no longer be minimal. The quality of the NCM design is measured by the end-to-end (SA-DA and SB-DB) error rate performance $P_{e,A} = \mathrm{Pr}\{b_A \neq \hat{b}_A\}$ and $P_{e,B} = \mathrm{Pr}\{b_B \neq \hat{b}_B\}$. However, in this case, the overall performance will not be given by a simple H-MAC and H-BC stage concatenation. It will be a complicated function depending on the particular relay HNC map (including associated hierarchical symbol error rate) and the quality and the rate of HSI links and also H-BC stage links.

4.2.2 NCM with Hierarchical Performance Target

Example 4.2 above nicely shows that the overall end-to-end utility target depends, apart from the NCM properties in the given stage and the given hierarchical encapsulation subnetwork level, also on the rest of the WPNC network. In order to simplify the NCM design, we may, instead of aiming for a global end-to-end performance target, use a hierarchical symbol associated with the relay node as a design goal. This factorizes the large network into smaller pieces defined by individual hierarchical encapsulation levels.

The price paid on the global performance depends on a number of factors including namely: the form of the performance metric, the overall structure of the network and the form of all HNC maps and associated information measures. On the other hand, this separation of the global and the local target allows us to build a more complex network using smaller building blocks (see also Section 10.3). Global aspects of WPNC cloud structuring, performance limits, and solvability are discussed in Chapter 5.

DEFINITION 4.2 (NCM with Hierarchical Performance Target) NCM with hierarchical performance target is the NCM where, instead of the global end-to-end performance metric for all participating component symbols, we use the hierarchical performance metric $P_e(b, \mu_b)$ for one given hierarchical symbol processed by the receiving node.

4.2.3 Layered NCM

A design of NCM that would *jointly* address both the error protection capabilities and the fact that the relay target is the information measure μ_b bound to the HNC map $b = \chi(\tilde{b})$, is a complicated task. A substantial simplification is reached when we *decouple* the error correcting capability and the capabilities allowing to obtain the hierarchical symbol measure μ_b. This form of NCM design will be called a *layered* design and the resulting NCM will be called layered NCM.

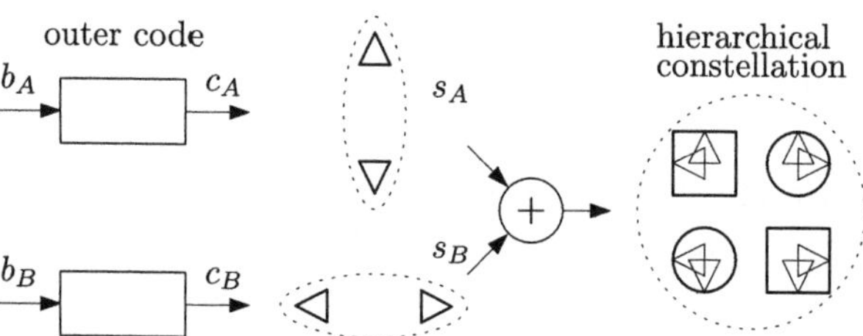

Figure 4.2 Layered NCM principle.

Layered NCM works in two layers (Figure 4.2). The *outer* layer is responsible for the error correcting capabilities and it is formed by a standard single-user error correcting code producing *discrete* encoded symbols $c_i = C_i(b_i)$ for each component node. It does not need to know any details of the network structure.

The second, *inner*, layer is responsible for the hierarchical properties and, in turn, it does not have to have error correcting capabilities. But since it operates without the cooperation with the outer layer, the only symbols over which it can define the hierarchical map are the *encoded outer layer* symbols c_i, i.e. $c = \chi_c(\tilde{c})$. But we are ultimately interested in the *message data* hierarchical measure μ_b related to the *message data* HNC map $b = \chi(\tilde{b})$, which forms the information contents for the relay back-end. This imposes some additional constraints on the *mutual consistency* among both HNC maps χ and χ_c and all outer layer encoders C_i if we want to utilize this in a straightforward way. It will be addressed later (Section 4.2.4) by defining isomorphic layered NCM.

DEFINITION 4.3 (Layered NCM)   Layered NCM consists of (1) outer codebooks C_i with discrete encoded symbols $c_i = C_i(b_i)$, and (2) inner constellation space symbol one-to-one mappers $s_i = A_i(c_i)$. We define HNC mapping on *both* data $\tilde{b} = \{b_i\}_i$ and outer layer encoded symbols $\tilde{c} = \{c_i\}_i$ levels, i.e. $b = \chi(\tilde{b})$ and $c = \chi_c(\tilde{c})$. The H-constellation associated with layered NCM is the one related to the outer layer encoded symbols

$$\mathcal{U}(c) = \left\{ u : u = u\left(\{s_i(c_i)\}_{i \in \mathcal{S}}, \tilde{h} \right) \big| c = \chi_c(\tilde{c}) \right\}. \tag{4.1}$$

We also define a product component code $\tilde{C} = C_i \times \cdots \times C_{i'}$, $\tilde{c} = \tilde{C}(\tilde{b})$.

Notice that defining the code symbol map on $\tilde{c}$ is equivalent to defining it on constellation space symbols $\tilde{s}$ ($s_i = A_i(c_i)$ is one-to-one mapping). However, we prefer the former because it is easier to handle the discrete code alphabet (we can use GF arithmetics) than the constellation space points. This way we also decouple χ_c HNC map from particular component constellation maps and this gives us a bit more freedom in defining $\mathcal{U}(c)$ H-constellation.

The advantage of layered NCM becomes obvious when complemented with *layered hierarchical decoding* at the relay. The layered decoding is applicable even for some non-layered NCMs (see Sections 4.3 and 4.4) but layered NCM makes its application straightforward.

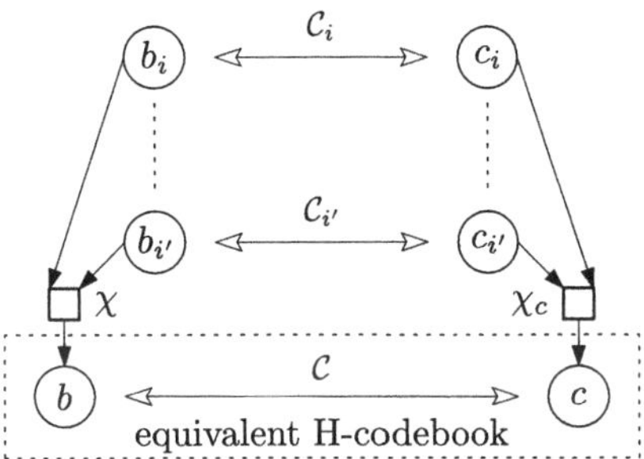

Figure 4.3 Isomorphic layered HCM and equivalent H-codebook.

4.2.4 Isomorphic Layered NCM

The layered NCM will be called *isomorphic* if the given message data $b = \chi(\tilde{b})$ and code level $c = \chi_c(\tilde{c})$ HNC maps and component outer encoders C_i are such that the virtual equivalent hierarchical code $c(b)$ forms a valid *one-to-one* codebook. If this holds, we can utilize it in layered hierarchical decoding and we will be able to turn hierarchical decoding into the case as if there were only one user with hierarchical data encoded by the hierarchical equivalent encoder. The name "isomorphic" is, indeed, used since we can find isomorphism (bijective morphism) between the data and code of component NCM encoders and also their HNC maps and it makes the map between b and c a *valid codebook* (Figure 4.3).

It is also worth to compare the isomorphic layered NCM with a general NCM (Definition 4.1) and a general layered NCM (Definition 4.3). The general NCM case defines the hierarchical codebook as a mapping between b and H-constellation $\mathcal{U}(b)$, which is generally a *multipoint set* for each b and its structure can be complicated (generally given by a product of component codes). The general layered NCM defines HNC maps for both b and c and it also defines H-constellation $\mathcal{U}(c)$, which defines hierarchical structuring at the *code* symbol level. However, it does not say anything about the relationship between b and c. The isomorphic layered NCM adds this relationship (equivalent *isomorphic* hierarchical codebook) and guarantees that it has a form of ordinary (single-user) one-to-one code and thus we can avoid the complexity of the component product code. Also, the isomorphic layered property becomes a critical condition for setting coding theorems in Section 5.7.4.

DEFINITION 4.4 (Isomorphic Layered NCM) Layered NCM consisting of outer codes C_i, $c_i = C_i(b_i)$, and HNC data and code symbol maps $b = \chi(\tilde{b})$, $c = \chi_c(\tilde{c})$ is called *isomorphic* layered NCM if there exists a valid one-to-one equivalent *isomorphic* hierarchical codebook (IH-codebook) C, such that $c = C(b)$, i.e.

$$\forall \tilde{b} : c = \chi_c\left(\tilde{C}(\tilde{b})\right) = C\left(\chi(\tilde{b})\right). \tag{4.2}$$

REMARK 4.1 (Generalized Isomorphism for Non-Identical Sets) The term isomorphic is used in a slightly more generalized form than in standard definitions. The standard form of the isomorphism definition defines a bijective function $f : \mathcal{A} \mapsto \mathcal{B}$ such that $f(x(a_1,\ldots,a_n)) = y(f(a_1),\ldots,f(a_n))$ where $x : \mathcal{A}^n \mapsto \mathcal{A}$ and $y : \mathcal{B}^n \mapsto \mathcal{B}$. In our generalized form, the domains and ranges (codomains) do not need to be the same sets, i.e. $f : \mathcal{A} \mapsto \mathcal{B}$ such that $f(x(a_1,\ldots,a_n)) = y(f_1(a_1),\ldots,f_n(a_n))$ where $x : \mathcal{A}_1 \times \cdots \times \mathcal{A}_n \mapsto \mathcal{A}$, $y : \mathcal{B}_1 \times \cdots \times \mathcal{B}_n \mapsto \mathcal{B}$, and $f_i : \mathcal{A}_i \mapsto \mathcal{B}_i$. This allows for arbitrary component encoders that do not even need to be with the common input–output alphabet.

REMARK 4.2 (Vector-Wise and Symbol-Wise HNC Maps)   Provided that the NCM maps χ and χ_c have equal sizes of the output codomains and the entities b and c are defined vector-wise, i.e. for *whole* messages or codewords, we can always find an isomorphic relationship by using tables. However, it has only a limited practical utility, since all relationships would have to be described by tables with exponential complexity w.r.t. vector length. Sometimes we can use this form in information-theoretic statements but for a practical code construction the functions should have some structure. The most useful is to define the HNC maps *symbol-wise*. But then, finding the *structured* equivalent isomorphic codebook for given component codebooks is less obvious apart from some trivial linear cases (see Chapter 6 for more details).

4.3   Hierarchical Decoder

4.3.1   Relay Operation for Decoding Hierarchical Information Measure

The relay operation should produce a hierarchical *data* measure $\mu_b(x)$ based on the observation x. In a general case (Figure 3.5) there may be multiple observations from multiple H-MAC stages. In this section, we formally assume a single-stage only front-end observation. The extension to multiple observations is treated in Section 4.6.

The decoding operation is called hierarchical decoding (H-decoding)[4] to reflect that the decoding goal is the hierarchical measure. The front-end is assumed to provide *some* suitable (depending on the performance optimality goal) form of the decoding metric $\mu_c(x)$ as the input to the H-decoder. The cardinality of the associated code symbol HNC map $c = \chi_c(\tilde{c})$, which is provided by the front-end, may be either full or less than full, which corresponds to a joint-metric or hierarchical demodulator, respectively (see Section 3.3.5). In a correspondence with these two possible input decoding metrics, we distinguish two basic forms of the H-decoder: (1) joint-metric, and (2) layered H-decoder. The term "layered" H-decoder is motivated by the fact that before the decoding itself, there is a layer performing the evaluation of the hierarchical input metric w.r.t. code symbols c. In this chapter, we set only a basic system model and evaluate the very basic properties (as e.g. the sufficient statistics).

The information-theoretic performance is further investigated in Section 5.3. On one hand, the joint-metric H-decoder may have on its input a metric that is a sufficient statistic under some conditions. However, as we will see in Section 5.7.3, the joint-metric

---

[4] If the output μ_b is some continuous valued measure, we can also use the term "soft-output" decoder to stress it. But in any case the decoding utilizes the inner structure of the received superposed signals encoded by NCM.

hierarchical decoding *directly* using the product codebook does not provide the highest achievable hierarchical rate. There are better performing options (Section 5.7.4) using the layered H-decoding (Section 4.3.3); however, these require a *layered isomorphic* NCM. The sufficient statistics on its own is not a guarantee of the best performance. The layered isomorphic property that internally couples the component codes appears to be more substantial. This is in contrast with the case when the product codebook is only coupled on its *output* HNC map and *no* H-codebook is revealed to the decoder (more details will come in Section 5.7.3).

4.3.2 Joint-Metric Hierarchical Decoder

The joint-metric hierarchical decoder is the H-decoder using a front-end *full cardinality* channel symbol metric. The advantage is that the H-decoder has available the best available observation, e.g. the likelihoods[5] $p(x|\tilde{c}) = p(x|c_i, \ldots, c_{i'})$. This clearly provides a sufficient statistic for estimating any function $b = \chi(\tilde{b})$. A disadvantage of the joint metric is the complexity. First, the cardinality of $\tilde{c}$ is greater than it would be for the hierarchical symbol and thus the metric has a larger size. Second, the joint metric does not turn the decoding to the virtual "single user" decoding as the hierarchical metric (together with isomorphic layered NCM) has a potential to do (see Section 4.3.3). As a consequence, the decoder has generally a complexity equal to the product of the component NCM code complexities.

In the following development, there will be some differences depending on whether we refer to complete data/codewords or individual symbols. For that purpose, we now use a more precise notation. Bold variables $(\mathbf{b}, \mathbf{c})$ denote whole data/codeword/vectors; variables indexed by the sequence number (typically n or k) possibly together with the index denoting the variable origin (e.g. source node) $(b_{i,k}, c_{i,n})$ denote individual data/codesymbols. Ordinary variables (e.g. b_i, c_i), as they have been used up to now, denote a universal variable in the situations where we do not need to explicitly distinguish the cases.

THEOREM 4.5 (Joint-Metric Demodulator Sufficient Statistic) *Assume NCM with one-to-one component codes* $\mathbf{c}_i(\mathbf{b}_i)$ *and one-to-one* $\mathbf{s}_i(\mathbf{c}_i)$ *constellation mappers. Also denote* $\tilde{\mathbf{c}} = \{\mathbf{c}_i, \ldots, \mathbf{c}_{i'}\}$, $\tilde{\mathbf{b}} = \{\mathbf{b}_i \ldots, \mathbf{b}_{i'}\}$. *The joint-metric demodulator joint likelihood* $\{p(\mathbf{x}|\tilde{\mathbf{c}})\}$ *is a sufficient statistic for decoding any hierarchical* data *map* $\mathbf{b} = \chi(\tilde{\mathbf{b}})$.

Proof First of all, $\mathbf{s}_i(\mathbf{c}_i)$ are one-to-one mappers, and thus all statements can be equivalently stated against either $\mathbf{s}_i$ or $\mathbf{c}_i$. We choose the latter. Second, we assume that all $\mathbf{b}_i$ and $\mathbf{c}_i$ are discrete words in arbitrary extended GF and represent complete data/codewords.

The target likelihood that forms the hierarchical symbol information measure is $\mu_{\mathbf{b}} = p(\mathbf{x}|\mathbf{b})$. It can be used as soft information passed to the relay back-end, or it can, together with a priori PDF, be used for MAP data decisions. The target measure is evaluated by the conditioning over all $\tilde{\mathbf{b}}$ consistent with a given $\mathbf{b}$

---

[5] We assume one-to-one $s_i(c_i)$ constellation space mappers.

$$p(\mathbf{x}|\mathbf{b}) = p\left(\mathbf{x}\,\middle|\,\bigcup_{\tilde{\mathbf{b}}:\chi(\tilde{\mathbf{b}})=\mathbf{b}} \tilde{\mathbf{b}}\right). \tag{4.3}$$

Individual events of $\tilde{\mathbf{b}}$ realizations are disjoint and thus we get

$$p(\mathbf{x}|\mathbf{b}) = \frac{\sum_{\tilde{\mathbf{b}}:\chi(\tilde{\mathbf{b}})=\mathbf{b}} p(\mathbf{x}|\tilde{\mathbf{b}})p(\tilde{\mathbf{b}})}{\sum_{\tilde{\mathbf{b}}:\chi(\tilde{\mathbf{b}})=\mathbf{b}} p(\tilde{\mathbf{b}})}. \tag{4.4}$$

Using Neyman–Fisher factorization theorem [25] (see Section A.3) and realizing that the set of all $\{\tilde{\mathbf{b}}\}$ is known a priori, we easily see that the set of all likelihoods $\{p(\mathbf{x}|\tilde{\mathbf{b}})\}_{\tilde{\mathbf{b}}}$ forms a trivial form of the sufficient statistic. But since the individual component codes $c_i(b_i)$ are one-to-one mappings, $\tilde{\mathbf{c}}(\tilde{\mathbf{b}})$ is also a one-to-one mapping and it holds that

$$\{p(\mathbf{x}|\tilde{\mathbf{b}})\}_{\tilde{\mathbf{b}}} = \{p(\mathbf{x}|\tilde{\mathbf{c}}(\tilde{\mathbf{b}}))\}_{\tilde{\mathbf{b}}} = \{p(\mathbf{x}|\tilde{\mathbf{c}}\}_{\tilde{\mathbf{c}}}. \tag{4.5}$$

The sets $\{p(\mathbf{x}|\tilde{\mathbf{b}})\}_{\tilde{\mathbf{b}}}$ and $\{p(\mathbf{x}|\tilde{\mathbf{c}}\}_{\tilde{\mathbf{c}}}$ are equal and they are related to one another only by reindexing, i.e. by one-to-one mapping. Thus, $\{p(\mathbf{x}|\tilde{\mathbf{c}}\}_{\tilde{\mathbf{c}}}$ forms a sufficient statistic. $\quad\square$

Notice that the situation at *individual data/codesymbol* level is generally not the same. It still holds that we can get data symbol likelihoods $p(\mathbf{x}|b_k)$ from $\{p(\mathbf{x}|\tilde{\mathbf{b}})\}_{\tilde{\mathbf{b}}}$

$$p(\mathbf{x}|b_k) = \frac{\sum_{\tilde{\mathbf{b}}:\chi(\tilde{\mathbf{b}})=b_k} p(\mathbf{x}|\tilde{\mathbf{b}})p(\tilde{\mathbf{b}})}{\sum_{\tilde{\mathbf{b}}:\chi(\tilde{\mathbf{b}})=b_k} p(\tilde{\mathbf{b}})}. \tag{4.6}$$

Assuming that the symbol level map is a function of component symbols at the same sequence number $b_k = \chi(\tilde{b}_k)$ where $\tilde{b}_k = \{b_{i,k}, \ldots, b_{i',k}\}$ we get

$$\begin{aligned}
p(\mathbf{x}|b_k) &= \frac{\sum_{\tilde{b}_k:\chi(\tilde{b}_k)=b_k} \sum_{\tilde{\mathbf{b}}:\tilde{b}_k} p(\mathbf{x}|\tilde{\mathbf{b}})p(\tilde{\mathbf{b}})}{\sum_{\tilde{b}_k:\chi(\tilde{b}_k)=b_k} \sum_{\tilde{\mathbf{b}}:\tilde{b}_k} p(\tilde{\mathbf{b}})} \\
&= \frac{\sum_{\tilde{b}_k:\chi(\tilde{b}_k)=b_k} p(\mathbf{x}|\tilde{b}_k)p(\tilde{b}_k)}{\sum_{\tilde{b}_k:\chi(\tilde{b}_k)=b_k} p(\tilde{b}_k)}.
\end{aligned} \tag{4.7}$$

Per-symbol data likelihoods $\{p(\mathbf{x}|\tilde{b}_k)\}_{\tilde{b}_k}$ form the sufficient statistic; however, there is no clear one-to-one mapping to per-symbol code likelihoods $\{p(\mathbf{x}|\tilde{c}_n)\}_{\tilde{c}_n}$, and even the length of the vectors does not need to be the same (hence we use different indices k, n). The only known relationship is between $\tilde{\mathbf{b}}$ and $\tilde{\mathbf{c}}$ and it is described by a product codebook including all NCM components. In order to exploit its structure, we would have to have $p(\mathbf{x}|\tilde{\mathbf{c}})$ available. This is possible only under specific additional constraints, particularly the memoryless channel.

THEOREM 4.6 (Symbol-Wise Joint-Metric Demodulator Sufficient Statistic) *Assume NCM with one-to-one component codes $c_i(\mathbf{b}_i)$ and one-to-one $s_i(\mathbf{c}_i)$ constellation mappers. Also denote $\tilde{\mathbf{c}} = \{\mathbf{c}_i, \ldots, \mathbf{c}_{i'}\}$, $\tilde{\mathbf{b}} = \{\mathbf{b}_i \ldots, \mathbf{b}_{i'}\}$. Symbol-wise joint-metric* *demodulator joint likelihood* $\{p(x_n|\tilde{c}_n)\}_{\tilde{c}_n,n}$ *is a sufficient statistic for decoding any hierarchical* data symbol *map* $b_k = \chi(\tilde{b}_k)$ *if the channel is memoryless,*[6] *i.e.* $p(\mathbf{x}|\tilde{\mathbf{c}}) = \prod_n p(x_n|\tilde{c}_n)$.

Proof The symbol-wise hierarchical data symbol measure is

$$p(\mathbf{x}|b_k) = \frac{\sum_{\tilde{b}_k:\chi(\tilde{b}_k)=b_k} \sum_{\tilde{\mathbf{b}}:\tilde{b}_k} p(\mathbf{x}|\tilde{\mathbf{b}})p(\tilde{\mathbf{b}})}{\sum_{\tilde{b}_k:\chi(\tilde{b}_k)=b_k} \sum_{\tilde{\mathbf{b}}:\tilde{b}_k} p(\tilde{\mathbf{b}})}. \tag{4.8}$$

Using the one-to-one component product codebook mapping property and the memoryless channel property, we can write

$$p(\mathbf{x}|\tilde{\mathbf{b}}) = p(\mathbf{x}|\tilde{\mathbf{c}}(\tilde{\mathbf{b}})) = \prod_n p(x_n|\tilde{c}_n(\tilde{\mathbf{b}})) \tag{4.9}$$

and clearly the set of values $p(x_n|\tilde{c}_n)$ for all symbols $\tilde{c}_n$ in the joint alphabet and for all sequence numbers n, i.e. $\{p(x_n|\tilde{c}_n)\}_{\tilde{c}_n,n}$ forms a sufficient statistic. Notice, however, that the nth symbol $\tilde{c}_n$ is a function of *complete* data word $\tilde{\mathbf{b}}$. This relationship represents the codebook. $\qquad\square$

4.3.3 Layered Hierarchical Decoder

The layered H-decoder performs the decoding in a two-step sequence. First, it uses the output of *hierarchical demodulator*, i.e. the *hierarchical* encoded channel symbol metric $\mu_c(x)$, which is a function of the observation x. Then, in a second step, it decodes (potentially with the soft output) the hierarchical *data* symbol measure $\mu_b(\mu_c)$ as a function of μ_c. The choice of the HNC map χ_c related to the measure μ_c must be such that we can exploit the encoded NCM structure in the (soft) decoding of target μ_b. It means that we must be able to find equivalent virtual isomorphic hierarchical encoder codebook mapping b into c based on the knowledge of all used NCM component codebooks. This is achieved, e.g., by using isomorphic layered NCM.

DEFINITION 4.7 (Layered Hierarchical Decoder, Hierarchical SODEM, Hierarchical Codebook) The layered hierarchical decoder is a two-step relay decoding operation. In the first step, a hierarchical soft-output demodulator (H-SODEM), or simply hierarchical demodulator (H-demodulator), observes a single-stage received signal x and evaluates the *hierarchical* information measure (front-end decoding metric) μ_c related to the HNC map of component *encoded* symbols $c = \chi_c(\tilde{c})$ utilizing the knowledge of the encoded symbol H-constellation $\mathcal{U}(c)$. In the second step, the desired hierarchical data measure μ_b is (soft-output) decoded from μ_c. The correspondence between hierarchical data b and hierarchical code symbols c is defined by equivalent H-codebook $\mathcal{C}$.

The layered decoder performs the general relay processing operation $x \mapsto \mu_b$ in two steps $x \mapsto \mu_c \mapsto \mu_b$. The first step performed by the front-end employs the channel symbol H-constellation (H-SODEM) and the second step (relay decoder) exploits

---

[6] More precisely, the channel is memoryless w.r.t. its internal marginalized channel states if $p(\mathbf{x}|\tilde{\mathbf{c}}) = \prod_n p(x_n|\tilde{\mathbf{c}})$. On top of it, it is memoryless w.r.t. channel symbols if $p(\mathbf{x}|\tilde{\mathbf{c}}) = \prod_n p(x_n|\tilde{c}_n)$.

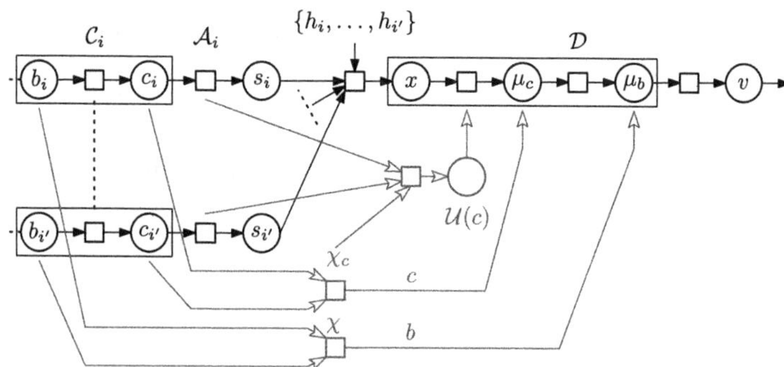

Figure 4.4 Layered NCM with layered H-decoding.

the NCM code structure to get the data information measure (e.g. hard decisions, or likelihoods). Both decoding metric μ_c and processing target μ_b must be *hierarchical*. Also, there must exist the isomorphism for H-codebook and it must be revealed to the decoder (see also Section 5.7.4). Because of the importance of the layered H-decoder using channel symbol H-metric at the input, we dedicate Section 4.4 to various aspects of its construction and performance, including the conditions for the sufficient statistic.

An important observation is the fact that the (soft) decoding $\mu_c \mapsto \mu_b$ turns the decoding process into the *equivalent* one as if we were in a classical *single-user* case. Of course, we must find the equivalent isomorphic hierarchical codebook, which is not necessarily an easy task. Also the choice of χ_c and μ_c is usually not unique, and different variants can have very different performance particularly when we also consider the channel parametrization influence on the H-constellation $\mathcal{U}(c)$.

The layered NCM, however, naturally suggests[7] the choice of the χ_c and μ_c for the layered H-decoder to be the same one as is used on the layered NCM (Figure 4.4). Figure 4.4 also shows the mutual relationships among all involved variables. Component data $\tilde{b}$ with data map χ determine the hierarchical data symbols b. Component encoded symbols $\tilde{c}$ with codesymbol map μ_c determine the hierarchical code symbols c. The codesymbol HNC map χ_c and the component constellation space maps $\mathcal{A}_i$ determine the codesymbol H-constellation $\mathcal{U}(c)$, which in turn is needed by the H-SODEM evaluating the hierarchical code symbol measure from the observation x. All involved variables and maps can be defined at various granularity levels, starting from individual code or channel symbols (scalars), which is the most common, up to being defined over the whole codeword or data vectors. If the layered NCM is isomorphic, then the relay decoder $\mathcal{D}$ is directly given by the equivalent H-codebook $\mathcal{C}$.

4.4 Hierarchical Demodulator

The hierarchical demodulator takes the single H-MAC stage received signal x and produces the H-decoding metric μ_c for encoded hierarchical symbols $c = \chi_c(\tilde{c})$. We may

---

[7] It is not necessarily the only option.

also use the term H-SODEM to stress that the output is a soft metric. As we saw above, the major advantage of this approach is the fact that it *enables* use of *equivalent* H-decoding in the subsequent decoding, which has much lower complexity in comparison with the product codebook decoding of joint-metric decoding and also has the potential for better performance (Section 5.7.4). There are several approaches for designing the H-demodulator and these will be discussed in the following sections.

All H-demodulator forms have a potential effect on the performance by creating the bottleneck μ_c in the processing chain $x \mapsto \mu_c \mapsto \mu_b$ if they are used or designed improperly. We know that the joint-metric codeword-wise likelihoods and, in the case of memoryless channel, also joint-metric symbol-wise likelihoods, form the sufficient statistic. However, the same cannot be *generally* stated about the *hierarchical* metric. Despite this, in the *specific* case of *isomorphic layered NCM*, it leads to performance advantage as is discussed in Section 5.7.4.

4.4.1 H-SODEM with Marginalization

The H-SODEM with *marginalization* is the most straightforward way of obtaining the μ_c metric. It also leads to a simple approximate expression for calculating the metric using the Euclidean distance and thus enabling many simple HW implementations. We use likelihoods as the metric. Specialized, simplified, or approximated forms will be discussed individually.

The joint-metric likelihoods $p(\mathbf{x}|\tilde{\mathbf{c}})$ (or $p(x_n|\tilde{c}_n)$ for a memoryless channel) are uniquely given by the channel model. The marginalization is performed over all component symbols consistent with the given hierarchical symbol. The derivations follow a similar track as used in theorem proofs in Section 4.3.2, so we proceed with fewer details.

Codeword-Wise Marginalization

The codeword-wise H-SODEM marginalized metric is

$$
p(\mathbf{x}|\mathbf{c}) = p\left(\mathbf{x} \,\middle|\, \bigcup_{\tilde{\mathbf{c}}:\chi_c(\tilde{\mathbf{c}})=\mathbf{c}} \tilde{\mathbf{c}}\right)
$$
$$
= \frac{\sum_{\tilde{\mathbf{c}}:\chi_c(\tilde{\mathbf{c}})=\mathbf{c}} p(\mathbf{x}|\tilde{\mathbf{c}})p(\tilde{\mathbf{c}})}{\sum_{\tilde{\mathbf{c}}:\chi_c(\tilde{\mathbf{c}})=\mathbf{c}} p(\tilde{\mathbf{c}})} \tag{4.10}
$$

where we used the disjoint $\tilde{\mathbf{c}}$ event property.

Symbol-Wise Marginalization

The symbol-wise H-SODEM marginalized metric is

$$
p(\mathbf{x}|c_n) = p\left(\mathbf{x} \,\middle|\, \bigcup_{\tilde{\mathbf{c}}:\chi_c(\tilde{c}_n)=c_n} \tilde{\mathbf{c}}\right)
$$
$$
= \frac{\sum_{\tilde{\mathbf{c}}:\chi_c(\tilde{c}_n)=c_n} p(\mathbf{x}|\tilde{\mathbf{c}})p(\tilde{\mathbf{c}})}{\sum_{\tilde{\mathbf{c}}:\chi_c(\tilde{c}_n)=c_n} p(\tilde{\mathbf{c}})}
$$

$$= \frac{\sum_{\tilde{c}_n:\chi_c(\tilde{c}_n)=c_n} \sum_{\tilde{\mathbf{c}}:\tilde{c}_n} p(\mathbf{x}|\tilde{\mathbf{c}})p(\tilde{\mathbf{c}})}{\sum_{\tilde{c}_n:\chi_c(\tilde{c}_n)=c_n} \sum_{\tilde{\mathbf{c}}:\tilde{c}_n} p(\tilde{\mathbf{c}})}$$

$$= \frac{\sum_{\tilde{c}_n:\chi_c(\tilde{c}_n)=c_n} p(\mathbf{x}|\tilde{c}_n)p(\tilde{c}_n)}{\sum_{\tilde{c}_n:\chi_c(\tilde{c}_n)=c_n} p(\tilde{c}_n)}. \tag{4.11}$$

For the memoryless channel $p(\mathbf{x}|\tilde{\mathbf{c}}) = \prod_n p(x_n|\tilde{c}_n)$, the symbol-wise conditioned likelihood of the whole received signal is equivalent to single received component likelihood (up to unimportant scalar scaling) $p(\mathbf{x}|\tilde{c}_n) \equiv p(x_n|\tilde{c}_n)$, and the resulting expression for symbol-wise H-metric becomes

$$p(x_n|c_n) = \frac{\sum_{\tilde{c}_n:\chi_c(\tilde{c}_n)=c_n} p(x_n|\tilde{c}_n)p(\tilde{c}_n)}{\sum_{\tilde{c}_n:\chi_c(\tilde{c}_n)=c_n} p(\tilde{c}_n)}. \tag{4.12}$$

Notice that the resulting H-metric is *not* equivalent to a simple Euclidean distance even if we assume a simple Gaussian channel and make some equivalent manipulations. It depends on a priori PDFs and also on the particular HNC map.

Special Case: Minimal HNC Map and Uniform Common Alphabet Symbols

In the special case of the minimal HNC map and uniform priors $p(c_{i,n}) = 1/M_c$ with all components having the same alphabet size, we have $p(\tilde{c}_n) = 1/M_c^K$ where K is the number of H-MAC stage components. The minimal map implies that

$$p(c_n) = \sum_{\tilde{c}_n:\chi_c(\tilde{c}_n)=c_n} p(\tilde{c}_n) = M_c^{K-1} \frac{1}{M_c^K} = \frac{1}{M_c}. \tag{4.13}$$

The resulting metric is obtained from (4.12)

$$p(x_n|c_n) = \frac{1}{M_c^{K-1}} \sum_{\tilde{c}_n:\chi_c(\tilde{c}_n)=c_n} p(x_n|\tilde{c}_n). \tag{4.14}$$

Special Case: Gaussian Channel

In a special case of AWGN[8] memoryless channel, the observation is

$$x_n = u_n(\tilde{c}_n) + w_n \tag{4.15}$$

where $u_n(\tilde{c}_n)$ is a hierarchical channel-combined symbol, and w_n is a complex valued AWGN with $p_w(w_n)$ PDF with σ_w^2 variance per dimension. The likelihoods for m-dimensional constellation symbols are

$$p(x_n|\tilde{c}_n) = p_w(x_n - u_n(\tilde{c}_n)) = \frac{1}{\pi^m \sigma_w^{2m}} \exp\left(-\frac{1}{\sigma_w^2} \|x_n - u_n(\tilde{c}_n)\|^2\right) \tag{4.16}$$

and (4.14) becomes

---

[8] We assume a complex-valued constellation space system model.

$$p(x_n|c_n) = \frac{1}{M_c^{K-1}} \sum_{\tilde{c}_n:\chi_c(\tilde{c}_n)=c_n} p_w(x_n - u_n(\tilde{c}_n))$$

$$= \frac{1}{\pi^m \sigma_w^{2m} M_c^{K-1}} \sum_{\tilde{c}_n:\chi_c(\tilde{c}_n)=c_n} \exp\left(-\frac{1}{\sigma_w^2}\|x_n - u_n(\tilde{c}_n)\|^2\right). \tag{4.17}$$

We see that the expression cannot be interpreted as a simple function of the Euclidean distance. Notice that we need to know the noise variance in order to evaluate the metric.

Approximation: Hierarchical Minimum Distance

However, the true decoding metric (4.17) can be *approximated* in the medium-to-high SNR regime. In this case the peaks of Gaussian PDFs in the summation are relatively narrow. If the NCM is such that points $\mathcal{U}(c_n)$ are *sufficiently distant and separated* then, for a given x, only *one term* dominates the summation. The dominating term is the one where $u_n(\tilde{c}_n)$ is the point closest to x and it is consistent with c_n. This will be denoted $u_n^{\mathrm{Hmin}}(c_n)$ and named as the minimum hierarchical distance point

$$u_n^{\mathrm{Hmin}}(c_n) = \arg \min_{u_n(\tilde{c}_n):\chi_c(\tilde{c}_n)=c_n} \|x_n - u_n(\tilde{c}_n)\|^2. \tag{4.18}$$

The approximation $p_{\mathrm{Hmin}}(x_n|c_n)$ is then formed by a single exponential

$$p(x_n|c_n) \approx p_{\mathrm{Hmin}}(x_n|c_n) = \frac{1}{\pi^m \sigma_w^{2m} M_c^{K-1}} \exp(-\frac{1}{\sigma_w^2}\|x_n - u_n^{\mathrm{Hmin}}(c_n)\|^2) \tag{4.19}$$

and it is clearly equivalent to the Euclidean distance from the minimal hierarchical distance point $p_{\mathrm{Hmin}}(x_n|c_n) \equiv \rho_{\mathrm{Hmin}}^2(x, c_n)$

$$\rho_{\mathrm{Hmin}}^2(x, c_n) = \min_{u_n(\tilde{c}_n):\chi_c(\tilde{c}_n)=c_n} \|x_n - u_n(\tilde{c}_n)\|^2. \tag{4.20}$$

The (squared) distance $\rho_{\mathrm{Hmin}}^2(x, c_n)$ will be called the *hierarchical distance (H-distance)* *metric*.

Notice that the H-distance $\rho_{\mathrm{Hmin}}^2(x, c_n)$ is still a function of the H-symbol c_n and forms an H-metric $\mu_{c_n}(x)$. The minimization in (4.20) serves purely for selecting the dominant exponential in the likelihood approximation. The H-distance metric, although an approximation, has an advantage in not being dependent on actual SNR.

If the points of $\mathcal{U}(c)$ are not sufficiently spread then, even for high SNR, there is a higher number of significantly non-zero exponentials in (4.17). This also includes a typical case of higher *multiplicity* of the points at given $u(\tilde{c})$, e.g. point zero in the BPSK example. If the multiplicity of the point $u_n^{\mathrm{Hmin}}(c_n)$ is $K(u_n^{\mathrm{Hmin}}(c_n))$ then

$$p(x_n|c_n) \approx p_{\mathrm{Hmin}}(x_n|c_n) = \frac{K(u_n^{\mathrm{Hmin}}(c_n))}{\pi^m \sigma_w^{2m} M_c^{K-1}} \exp(-\frac{1}{\sigma_w^2}\|x_n - u_n^{\mathrm{Hmin}}(c_n)\|^2) \tag{4.21}$$

and taking a properly scaled logarithm gives the H-distance metric correctly respecting the multiplicity

$$\rho_{\mathrm{Hmin}*}^2(x, c_n) = -\sigma_w^2 \ln\left(\pi^m \sigma_w^{2m} M_c^{K-1} p_{\mathrm{Hmin}}(x_n|c_n)\right)$$

$$= \|x_n - u_n^{\mathrm{Hmin}}(c_n)\|^2 - \sigma_w^2 \ln K(u_n^{\mathrm{Hmin}}(c_n)). \tag{4.22}$$

This will be called the *multiplicity-resolving hierarchical distance (H-distance*)* metric. In order to get the equivalent metric, we are allowed to scale or shift it only by such constants that are *not* functions of c_n and x. If the multiplicity $K(u_n^{\text{Hmin}}(c_n))$ depends on c_n we cannot remove it. Another consequence is that the metric *remains* to depend on σ_w^2. Notice that this form of H-distance modification is required even for high SNR. It ignores the tails of Gaussian distribution for distant points but it correctly respects the multiplicity of the close ones. Also notice that

$$\rho_{\text{Hmin}*}^2(x, c_n) = \rho_{\text{Hmin}}^2(x, c_n) - \sigma_w^2 \ln K(u_n^{\text{Hmin}}(c_n)) \qquad (4.23)$$

where the correction term disappears for $K = 1$ and becomes more amplified for low SNR. If the correction is non-zero then it always *degrades* ρ_{Hmin}^2.

Example 4.3 True hierarchical decoding metric vs. hierarchical distance for two BPSK sources and minimal HNC map.

Assume two sources with BPSK alphabets in equal gain ($h_A = h_B = 1$) AWGN (real valued for simplicity) channel $x = h_A s_A(c_A) + h_B s_B(c_B) + w$, $s_i \in \{\pm 1\}$, $c_A, c_B \in \mathbb{F}_2$. The minimal HNC map is XOR (addition in $\mathbb{F}_2$) $c = \chi_c(c_A, c_B) = c_A + c_B$, $c \in \mathbb{F}_2$. The H-constellation mapping is defined in Figure 3.12. Figure 4.5 shows the comparison between the true and the H-distance metric. The true metric is given by the values $p(x|c)$ for $c = 0$ and $c = 1$, whereas the H-distance is given by the distance of H-constellation points from the value x on the horizontal axis. We see that the decision regions for the two forms of the H-metric differ significantly.

4.4.2   H-SODEM Providing Sufficient Statistic

The layered decoding with H-SODEM providing the hierarchical encoded symbol metric μ_c will not suffer any performance loss provided that μ_c is a sufficient statistic for relay processing target data HNC map $\mathbf{b} = \chi(\tilde{\mathbf{b}})$. We now prove an important

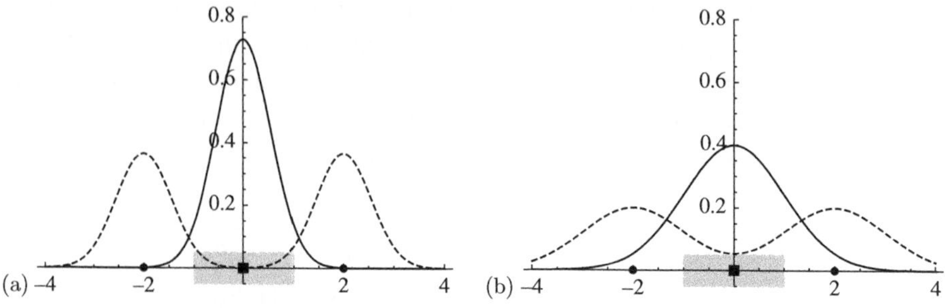

Figure 4.5 True hierarchical decoding metric vs. hierarchical distance approximation for two BPSK sources and two values of noise variance: (a) $\sigma_w^2 = 0.3$, (b) $\sigma_w^2 = 1$. Legend: $p(x|c = 1)$ solid line, $p(x|c = 0)$ dashed line; H-constellation point $u(c = 1)$ rectangle, $u(c = 0)$ circle; gray range on the x-axis denotes H-distance $\rho_{\text{Hmin}}^2(x, c = 1) < \rho_{\text{Hmin}}^2(x, c = 0)$.

theorem stating conditions under which the hierarchical metric μ_c is the sufficient statistic. We first start with the H-SODEM with vector-wise HNC maps and then we explain differences when using symbol-wise maps.

Vector-Wise H-SODEM

THEOREM 4.8 (H-Demodulator Sufficient Statistic) *Assume a layered NCM consisting of outer component codes C_i, $\mathbf{c}_i = C_i(\mathbf{b}_i)$, $\tilde{\mathbf{c}} = \tilde{C}(\tilde{\mathbf{b}})$, and HNC dataword and codeword symbol maps $\mathbf{b} = \chi(\tilde{\mathbf{b}})$, $\mathbf{c} = \chi_c(\tilde{\mathbf{c}})$. H-SODEM metric $\mu_c = p(\mathbf{x}|\mathbf{c})$ associated with HNC map χ_c is a sufficient statistic for hierarchical data $\mathbf{b} = \chi(\tilde{\mathbf{b}})$ if the NCM is* isomorphic *layered NCM.*

Proof We already know that $p(\mathbf{x}|\tilde{\mathbf{c}})$ is a sufficient statistic (Theorem 4.5) for $\mathbf{b} = \chi(\tilde{\mathbf{b}})$. The core of the proof stood on our capability to perform the marginalization of PDFs over the set of all $\tilde{\mathbf{c}}$ consistent with $\tilde{\mathbf{b}}$ and, in turn, all $\tilde{\mathbf{b}}$ consistent with $\mathbf{b}$. The former consistency is given by the one-to-one component codes and therefore it is just a reindexing. The latter is given by the HNC map χ. Since the marginalization is a sum of the PDF values evaluated for all $\tilde{\mathbf{c}}$ consistent with $\mathbf{b}$, the knowledge of the compliant $\tilde{\mathbf{c}}$ uniquely determines the result. We define the set of all $\tilde{\mathbf{c}}$ consistent with $\tilde{\mathbf{b}}$, which is in turn consistent with $\mathbf{b}$,

$$S(\mathbf{b}) = \left\{ \tilde{\mathbf{c}} | \tilde{\mathbf{c}} : \tilde{\mathbf{b}} : \mathbf{b} \right\} = \left\{ \tilde{\mathbf{c}} | \tilde{\mathbf{c}} = \tilde{C}(\tilde{\mathbf{b}}), \mathbf{b} = \chi(\tilde{\mathbf{b}}) \right\}. \tag{4.24}$$

This set defines a partition on the set of all possible $\{\tilde{\mathbf{c}}\}$ and the marginalization is

$$\begin{aligned} p(\mathbf{x}|\mathbf{b}) &= \frac{\sum_{\tilde{\mathbf{b}}:\mathbf{b}} \sum_{\tilde{\mathbf{c}}:\tilde{\mathbf{b}}} p(\mathbf{x}|\tilde{\mathbf{c}})p(\tilde{\mathbf{c}})}{\sum_{\tilde{\mathbf{b}}:\mathbf{b}} \sum_{\tilde{\mathbf{c}}:\tilde{\mathbf{b}}} p(\tilde{\mathbf{c}})} \\ &= \frac{\sum_{\tilde{\mathbf{c}}\in S(\mathbf{b})} p(\mathbf{x}|\tilde{\mathbf{c}})p(\tilde{\mathbf{c}})}{\sum_{\tilde{\mathbf{c}}\in S(\mathbf{b})} p(\tilde{\mathbf{c}})}. \end{aligned} \tag{4.25}$$

The marginalization is a two-step procedure where the first marginalization exploits the component product code structure $\tilde{C}$ and the second one utilizes the HNC map χ.

The marginalized H-SODEM metric $p(\mathbf{x}|\mathbf{c})$ is obtained by the marginalization of PDF values over all $\tilde{\mathbf{c}}$ consistent with $\mathbf{c}$. It is performed in one step and utilizes the code HNC map χ_c

$$p(\mathbf{x}|\mathbf{c}) = \frac{\sum_{\tilde{\mathbf{c}}:\mathbf{c}} p(\mathbf{x}|\tilde{\mathbf{c}})p(\tilde{\mathbf{c}})}{\sum_{\tilde{\mathbf{c}}:\mathbf{c}} p(\tilde{\mathbf{c}})}. \tag{4.26}$$

The marginalization set defines a partition in $\{\tilde{\mathbf{c}}\}$

$$S'(\mathbf{c}) = \left\{ \tilde{\mathbf{c}} | \tilde{\mathbf{c}} : \mathbf{c} \right\} = \left\{ \tilde{\mathbf{c}} | \mathbf{c} = \chi_c(\tilde{\mathbf{c}}) \right\}. \tag{4.27}$$

If the NCM is *isomorphic* then there is a one-to-one mapping $\mathbf{c} = C(\mathbf{b})$. It defines the partition

$$S''(\mathbf{b}) = \left\{ \tilde{\mathbf{c}} | \tilde{\mathbf{c}} : \mathbf{c} : \mathbf{b} \right\} = \left\{ \tilde{\mathbf{c}} | \mathbf{c} = \chi_c(\tilde{\mathbf{c}}), \mathbf{c} = C(\mathbf{b}) \right\} \tag{4.28}$$

and it holds that

$$S'(C(\mathbf{b})) = S''(\mathbf{b}). \tag{4.29}$$

Then we can write

$$p(\mathbf{x}|\mathbf{b}) = \frac{\sum_{c:b} \sum_{\tilde{c}:c} p(\mathbf{x}|\tilde{c})p(\tilde{c})}{\sum_{c:b} \sum_{\tilde{c}:c} p(\tilde{c})} \tag{4.30}$$

where the inner marginalization is done by H-SODEM and the outer one is one-to-one isomorphic H-code. We see, using Neyman–Fisher factorization theorem (Section A.3), that the properly scaled inner marginalization

$$p(\mathbf{x}|\mathbf{c}) = \frac{1}{\sum_{\tilde{c}:c} p(\tilde{c})} \sum_{\tilde{c}:c} p(\mathbf{x}|\tilde{c})p(\tilde{c}) \tag{4.31}$$

is a sufficient statistic.

The isomorphic NCM thus guarantees that the marginalization over all sets (4.24) and (4.28) are equal $\mathcal{S}(\mathbf{b}) = \mathcal{S}''(\mathbf{b})$ and

$$p(\mathbf{x}|\mathbf{b}) = \frac{\sum_{\tilde{c}\in\mathcal{S}''(\mathbf{b})} p(\mathbf{x}|\tilde{c})p(\tilde{c})}{\sum_{\tilde{c}\in\mathcal{S}''(\mathbf{b})} p(\tilde{c})}.$$

The equivalence of the marginalization sets guarantees that it does not matter which option we use: either (1) first exploit the one-to-one product component code and then the HNC data map, or (2) first the HNC code map and then the one-to-one isomorphic H-code (Figure 4.6). □

The core observation revealed by the theorem is the fact that if we guarantee that the partitioning of $\{\tilde{c}\}$ induced by the marginalization consistent with the target data $\mathbf{b}$ is the same then it does not matter how it is performed, in how many steps and in what order. The isomorphic HNC guarantees the equivalence of these sets $\mathcal{S}(\mathbf{b}) = \mathcal{S}''(\mathbf{b})$. A generalization based on the necessary partition set equivalence can slightly relax the requirements on the isomorphic NCM. For example, the H-SODEM map χ'_c might create a deeper nested partitioning consistent with the χ_c map. In fact, it is nothing else than applying the marginalization in three instead of two steps. The multi-step partitioning will work provided that it would lead to the same overall partitioning.

Theorem 4.8 can also be trivially extended into the form providing the sufficient statistic for the data hierarchical *symbol* $b_k = \chi_b(\tilde{b}_k)$. This additional step of marginalization

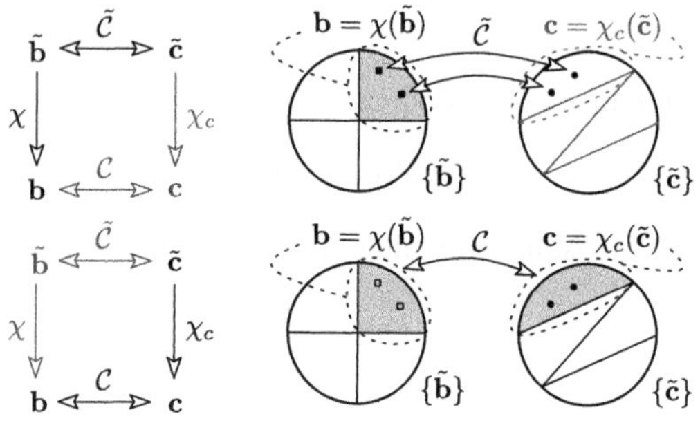

Figure 4.6 Marginalization decoding sets for isomorphic layered HNC.

$$p(\mathbf{x}|b_k) = \frac{\sum_{\mathbf{b}:b_k} p(\mathbf{x}|\mathbf{b})p(\mathbf{b})}{\sum_{\mathbf{b}:b_k} p(\mathbf{b})} \tag{4.32}$$

is common to both lines of obtaining the sufficient statistic and obviously if the measures are sufficient statistics for $\mathbf{b}$ then it holds also for b_k.

Symbol-Wise H-SODEM

In the case of isomorphic NCM, the *vector-wise* H-SODEM HNC map provides the sufficient statistic for both the vector- and symbol-wise *data* HNC map. This statement, however, does not hold for the *symbol-wise* H-SODEM HNC map even under the memoryless channel assumption.

Assuming the memoryless channel, the symbol-wise H-SODEM calculates the marginalization (4.12)

$$p(x_n|c_n) = \frac{\sum_{\tilde{c}_n:\chi_c(\tilde{c}_n)=c_n} p(x_n|\tilde{c}_n)p(\tilde{c}_n)}{\sum_{\tilde{c}_n:\chi_c(\tilde{c}_n)=c_n} p(\tilde{c}_n)}. \tag{4.33}$$

In order for this metric to be a sufficient statistic for b_k, we would have to be able to obtain $p(\mathbf{x}|b_k)$ from $p(x_n|c_n)$. This would require us being capable of performing the marginalization $\sum_{c_n:b_k} p(x_n|c_n)p(c_n)$. Unfortunately, there is no (apart from the trivial uncoded case) relationship between c_n and b_k on which we could identify a subset of c_n values consistent with b_k. Simply, the set is not a function of b_k, $\{c_n|c_n : b_k\} = \{c_n\}$, and the result of marginalization does not depend on b_k. The isomorphic NCM guarantees the one-to-one correspondence for the dataword and the codeword $\mathbf{b} \mapsto \mathbf{c}$ but not for the symbols. The dataword and the codeword do not even need to have the same dimensionality.

However, the marginalization performed on the hierarchical codeword captures the proper isomorphic H-code relationship (Theorem 4.8). The question is whether we can reconstruct (potentially with the help of memoryless channel assumption) the vector-wise H-SODEM metric $p(\mathbf{x}|\mathbf{c})$ *from* the symbol-wise H-SODEM metric $p(x_n|c_n)$. In a standard single-user communication case or in a case of the *joint* metric this would work. However, this is *not* the case for H-symbol metrics. The problem is that, apart from a trivial uncoded case, the conditional channel observations $\{x_n|c_n\}$ are *not independent*, i.e. generally $p(\mathbf{x}|\mathbf{c}) \neq \prod_n p(x_n|c_n)$. The dependency of channel observations conditioned by hierarchical symbols is easily seen from the graph representation (Figure 4.7). Even when we fix values of c_n, the channel observations are still connected through the structure of component codes $\mathcal{C}_A, \mathcal{C}_B$.

The conclusion is that, strictly speaking, symbol-wise H-SODEM does *not* provide the sufficient statistic for the encoded NCM. It only provides a (trivially) sufficient statistic for the uncoded case. As an *approximation*, we may write $p(\mathbf{x}|\mathbf{c}) \approx \prod_n p(x_n|c_n)$. The fidelity of this approximation depends on particular component codes. Some codes that have large-scale spreading of the parity check (e.g. LDPC) might behave favorably in this situation. Then we can practically use the symbol-wise H-SODEM with a subsequent approximate reconstruction of the vector-wise metric, which in turn is used for the decoding of the IH-code. Also, the information-theoretic assessments with IID random

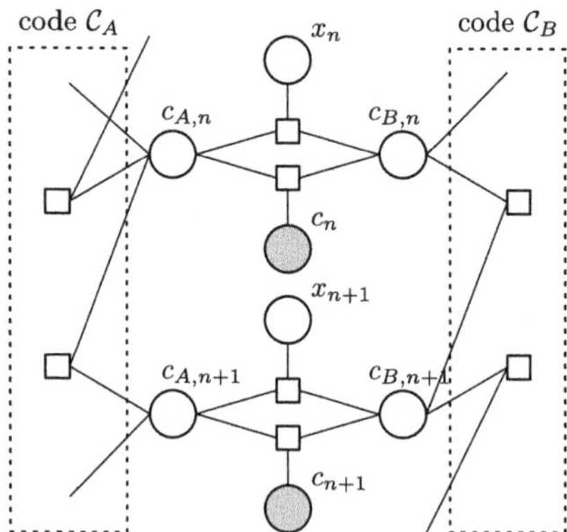

Figure 4.7 The dependency of channel observations conditioned by hierarchical symbols.

codebook constructions (as is used in Chapter 5 and specifically in Section 5.7.4) fall into this category.

4.4.3   Soft-Aided H-SODEM

A classical single-user system demodulator provides the decoding metric of the channel symbol observed at the receiver input. It is formally the likelihood $p(x|c)$ or any other equivalent, e.g. Euclidean distance, form. The metric directly depends on the channel symbol and there is no other degree of freedom.

However, the situation of H-SODEM is very different. The hierarchical likelihood (either symbol-wise (4.12) or vector-wise (4.10)) is already a result of the marginalization over additional degrees of freedom created by superposing multiple component signals. In fact it is a block with some internal structure similar to a simple, and quite specific, code – the HNC map. The marginalization requires the knowledge of the a priori PDF of the hierarchical symbol. H-SODEM connected to some serially or parallel concatenated decoding chain may, and should, utilize the estimates of the a priori PDF – the soft feedback information provided by the H-decoder. The exchange of the information can have an iterative form similar to the iterative soft-information decoding using the Forward–Backward Algorithm.

The codeword-wise form of H-SODEM does not have a large potential in using this, since the messages (unless we properly respect potential correlation of the sources in WPNC network) are typically considered as independent and uniformly distributed. But symbol-wise H-SODEM

$$p_i(x_n|c_n) = \frac{\sum_{\tilde{c}_n : \chi_c(\tilde{c}_n)=c_n} p(x_n|\tilde{c}_n)\hat{p}_i(\tilde{c}_n)}{\sum_{\tilde{c}_n : \chi_c(\tilde{c}_n)=c_n} \hat{p}_i(\tilde{c}_n)} \tag{4.34}$$

can benefit from the updated a priori estimates at ith iteration $\hat{p}_i(\tilde{c}_n)$. Notice that the only expression (similarly as it is for a classical single-user system) that does not depend on a priori soft-information is the *joint* metric $p(x_n|\tilde{c}_n)$. We also nicely see that the H-SODEM can be viewed as a concatenation of joint-metric SODEM $p(x_n|\tilde{c}_n)$, which has a

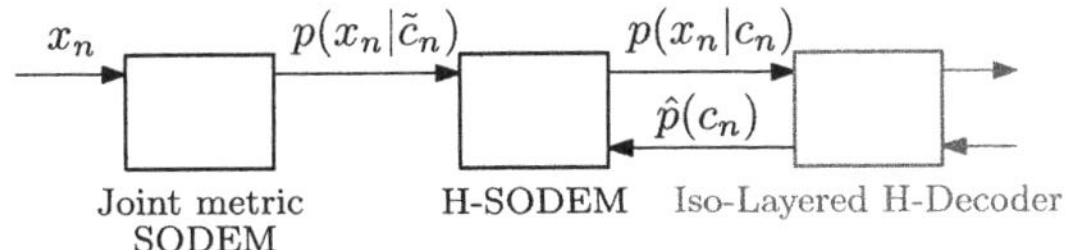

Figure 4.8 Soft-aided H-SODEM.

fixed output for a given observation x, and a specific "decoder" performing the marginalization, which benefits from the a priori soft-information updates from the subsequent H-decoder. The symbol-wise soft-aided H-SODEM thus can utilize the code structure exploited by the connected H-decoder.

A specific problem is how to obtain the a priori PDF $\hat{p}_i(\tilde{c}_n)$ in the case of using an isomorphic layered H-decoder. It can provide only a soft-information iterative estimate on the *hierarchical* symbol c_n, i.e. even in the perfect case it would be $p(c_n)$, see Figure 4.8. The H-SODEM marginalization, however, requires *joint* a priori of all components $p(\tilde{c}_n)$ and it must be reconstructed from $p(c_n)$. There are of course additional dispersion degrees of freedom caused by the fact that the HNC map is a many-to-one map (see also the hierarchical dispersion and the hierarchical equivalent channel in Section 5.7.4). The number of $\tilde{c}_n$ consistent with c_n is larger than one. Generally, this creates a problem.

Special Case of Linear GF HNC Map

In a special case of *linear* HNC map $c_n = \chi_c(\tilde{c}_n)$ defined over some GF and uniformly distributed codesymbols, we can, however, be more specific.

LEMMA 4.9   *Assume a non-singular (all non-zero coefficients and $K \geq 2$) linear HNC map on GF $\mathbb{F}_{M_c}$, $c = \sum_{k=1}^{K} a_k c_k$, and uniformly distributed IID $c_k \in [0 : M_c - 1]$. Then, any subset $\tilde{c}' \subset \tilde{c}$ with J components in $\tilde{c}'$, where $J \leq K - 1$, and the hierarchical symbol c are independent $\tilde{c}' \perp c$, i.e. $p(\tilde{c}'|c) = p(\tilde{c}')$ and this conditional distribution is again IID uniform on $\mathbb{F}_{M_c^J}$.*

Proof   Clearly, the HNC map defines a *hyperplane* with $K - 1$ degrees of freedom. The proof follows lines similar to Lemma 4.14 and Theorem 5.11.

For arbitrary index $i \in [1 : K]$ we can express the component symbol as

$$c_i = \frac{1}{a_i}\left(c - \sum_{k \in [1:K]\backslash i} a_k c_k\right). \tag{4.35}$$

If $K \geq 2$ and coefficients in the sum are non-zero, owing to properties of arithmetics on GF, all elements $a_k c_k$ will be uniformly distributed, as will the sum $\sum_{k \in [1:K]\backslash i} a_k c_k$. For any given fixed c, the expressions $(c - \sum_{k \in [1:K]\backslash i} a_k c_k)$ and $(c - \sum_{k \in [1:K]\backslash i} a_k c_k)/a_i$ will also be uniformly distributed *regardless* of the value c.

For $J > 1$, we define the set of indices $\mathcal{S}'$ of components contained in $\tilde{c}'$. Then for all $i \in \mathcal{S}'$ we get

$$c_i = \frac{1}{a_i}\left(c - \sum_{k \in \mathcal{S}'\backslash i} a_k c_k - \sum_{k \notin \mathcal{S}'} a_k c_k\right). \tag{4.36}$$

If the sum $\sum_{k\notin S'} a_k c_k$ contains *at least one* element (which is guaranteed by the condition $J \leq K - 1$), the distribution of c_i would become uniform for arbitrary fixed value of c or c_k, $k \in S' \setminus i$. Thus the components $\tilde{c}'$ are independent of c and they are IID uniform. $\qquad\qquad\square$

We see that components with indices $k \notin S'$ work as a randomizing scrambler and the size of this set must be at least 1. A practical utilization of this lemma thus allows to set arbitrary $K - 1$ components as IID uniformly distributed and one remaining component then needs to be calculated for the consistency with c.

4.4.4   H-SODEM with Nonlinear Preprocessor

The H-SODEM performing the marginalization, either word- or symbol-wise, and directly on PDFs, is not the only option. We can also obtain the hierarchical soft metric by using a nonlinear preprocessor. In the context of WPNC, the most prominent application is the Compute and Forward technique (Section 5.6) built on lattice codes with modulo lattice nonlinear preprocessor. The resulting soft hierarchical metric does not necessarily need to be equal to the proper PDF marginalization but leads to a simple processing that nicely matches the lattice coding paradigm. The following shows how this preprocessing fits into the concept of H-SODEM.

Modulo Lattice Preprocessor

In the scope of this section, we assume lattice codes in real-valued $\mathbb{R}^N$ space. Component codes $\mathcal{C}_i$ are identical nested lattice codes with a fine lattice Λ_c and a coarse shaping lattice Λ_s carrying uniformly distributed and independent messages with the codebook size M. For simplicity, we also assume an AWGN channel with no channel parametrization (i.e. all channel gains are unity)

$$\mathbf{x} = \sum_{i=1}^{K} \mathbf{c}_i + \mathbf{w}. \tag{4.37}$$

In the first approach, we assume no α scaling and no random dither (see Sections A.5 and 5.6). The HNC map of K components is the *minimal* one $\mathbf{c} = \sum_{i=1}^{K} \mathbf{c}_i \bmod \Lambda_s$, where we assumed that all coefficients are unity for a simplicity. The modulo lattice preprocessor is

$$\mathbf{y}' = \mathbf{x} \bmod \Lambda_s. \tag{4.38}$$

It corresponds (with the above-stated simplifications) to the equivalent modulo lattice channel (see (A.152), (A.153), and (A.149)). Notice, however, that this equivalent model is *enabled only* by the true lattice decoder performing the decoding $\hat{\mathbf{c}} = Q_{\Lambda_c}(\mathbf{x}) \bmod \Lambda_s$ (see (A.148)) containing the *outer modulo lattice* operation. In other words, we cannot apply this model *unless* we use the lattice decoder with the outer $\bmod \Lambda_s$ operation.

The modulo lattice preprocessor output is

$$
\begin{aligned}
\mathbf{y}' &= \left(\sum_{i=1}^{K} \mathbf{c}_i + \mathbf{w} \right) \bmod \Lambda_s \\
&= \left(\left(\sum_{i=1}^{K} \mathbf{c}_i \right) \bmod \Lambda_s + \mathbf{w} \right) \bmod \Lambda_s \\
&= (\mathbf{c} + \mathbf{w}) \bmod \Lambda_s.
\end{aligned}
\tag{4.39}
$$

Clearly, the output is the same for all $\tilde{\mathbf{c}}$ consistent with a given $\mathbf{c}$. The marginalization of the joint metric into the hierarchical one is trivial. All joint-metric cases consistent with $\mathbf{c}$ are identical and with equal probability, giving

$$
\begin{aligned}
p(\mathbf{y}'|\mathbf{c}) &= \frac{1}{p(\mathbf{c})} \sum_{\tilde{\mathbf{c}}:\mathbf{c}} p(\mathbf{y}'|\tilde{\mathbf{c}})p(\tilde{\mathbf{c}}) \\
&= \frac{1}{1/M} \sum_{\tilde{\mathbf{c}}:\mathbf{c}} p(\mathbf{y}'|\tilde{\mathbf{c}}) \frac{1}{M^K} \\
&= p(\mathbf{y}'|\tilde{\mathbf{c}} : \mathbf{c}) \frac{1}{M^{K-1}} \underbrace{\sum_{\tilde{\mathbf{c}}:\mathbf{c}} 1}_{=M^{K-1}} \\
&= p(\mathbf{y}'|\tilde{\mathbf{c}} : \mathbf{c}).
\end{aligned}
\tag{4.40}
$$

Thus the conditional PDF $p(\mathbf{y}'|\mathbf{c})$ is equal to the $p(\mathbf{y}'|\tilde{\mathbf{c}})$ evaluated for *arbitrary* $\tilde{\mathbf{c}}$ consistent with $\mathbf{c}$. We can choose the one belonging to the fundamental Voronoi cell, i.e. such that $\sum_{i=1}^{K} \mathbf{c}_i = \mathbf{c}, \mathbf{c} \in \mathcal{V}_0(\Lambda_s)$.

The modulo lattice operation implies that the output $\mathbf{y}'$ is constrained only into the fundamental Voronoi cell $\mathcal{V}_0(\Lambda_s)$ and the output PDF is a sum of coarse-lattice shifted versions of the density of the argument. Then

$$
p(\mathbf{y}'|\mathbf{c}) = \begin{cases} \sum_{\lambda \in \Lambda_s} p_{\mathbf{w}}(\mathbf{y}' - \mathbf{c} - \lambda), & \mathbf{y}' \in \mathcal{V}_0(\Lambda_s) \\ 0, & \mathbf{y}' \notin \mathcal{V}_0(\Lambda_s) \end{cases}.
\tag{4.41}
$$

The marginalized hierarchical metric of the modulo lattice is obtained by simply behaving as if the channel input were $\mathbf{c}$ and the channel itself were the modulo lattice $(\mathbf{c} + \mathbf{w}) \bmod \Lambda_s$. No other explicit marginalization operations are needed. The modulo lattice preprocessor does that for us automatically. The modulo lattice preprocessor is thus marginalized H-SODEM for the Compute and Forward decoder. However, it holds only under several assumptions: (1) all codes are identical, (2) the HNC map is the minimal one, and (3) the "tail cutting and overlapping" of the noise PDF caused by the modulo lattice operation is ignored (see the next section and the end of Section A.5 for more details). As a consequence, the resulting H-metric does not need to be optimal.

The modulo lattice preprocessor's automatic marginalization can be generalized for a *scaled* preprocessor

$$
\mathbf{y} = (\alpha \mathbf{x} - \mathbf{u}) \bmod \Lambda_s.
\tag{4.42}
$$

With the help of random dithering $\mathbf{u}$, we can find an equivalent modulo channel (see Sections A.5 and 5.6, and also [63])

$$\mathbf{y}_{\mathrm{eq}} = \left(\mathbf{c} + \mathbf{w}_{\mathrm{eq}}\right) \bmod \Lambda_s. \tag{4.43}$$

The equivalent noise (see (5.105) and the discussion there) is additionally affected by the modulo lattice operation.

Lattice Folding

There are two phenomena related to the evaluation of (4.41) that are affected by the modulo lattice operation. The first is the multiplicity of the noise PDFs that significantly overlap each other in (4.41) for some particular $\lambda \in \Lambda_s$. This depends on a particular superposition of the component codewords; the distance from the origin affects the multiplicity. The second phenomenon is the rate of the noise PDF tails decay, which depends on the noise variance. More details and some related comments follow.

A graphical illustration of the lattice shifts of the contributing component PDFs for the H-metric obtained by a proper marginalization (4.10) vs. the modulo lattice operation is shown in Figure 4.9. The modulo lattice operation folds the cells in outer shells onto the fundamental cell. For the expanded superposed lattices, different coarse Voronoi cells, however, correspond to different multiplicities of $\tilde{\mathbf{c}}$ consistent with given $\mathbf{c}$. As we move further from the fundamental cell, the multiplicity becomes smaller. This has an effect on the way the tails of the noise PDF overlap, particularly on their multiplicity, when evaluating either (4.10) or (4.41). Figure 4.10 shows a simple example.

The cut-off and folding of the AWGN noise tails that result from the equivalent noise modulo folding (4.41) present a problem that does not disappear for a high lattice dimension N. In order to show this, we can use simple arguments based on the fine and the coarse shaping cell volume and diameter, assuming that we approximate the cells as spheres. The volumes of the fine and the coarse cells are proportional to the Nth power of their respective radii $V_c = V_{S_c}(N)a_c^N$ and $V_s = V_{S_s}(N)a_s^N$ where $V_{S_c}(N) = \pi^{N/2}/(N/2)!$ is a unit-radius sphere volume in N dimensions. For a given fixed code rate R, the volume ratio must give $V_s/V_c = 2^{NR}$ and thus $a_s/a_c = 2^R$. So the relative size of the fine cell against the shaping cell remains intact. For a fixed noise variance per dimension, the noise tail decay is asymptotically constant relative to the fine cell size. This effect, however, influences only the cells in outer shells where the overlap of the noise tails

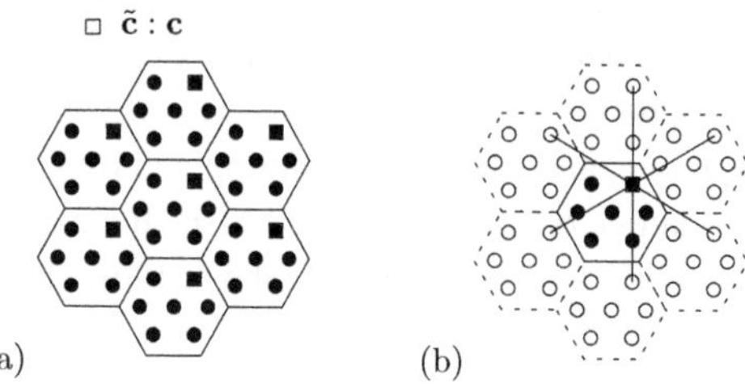

Figure 4.9 Proper marginalization vs. modulo lattice folding: (a) expanded superposed lattices and (b) modulo folded lattice.

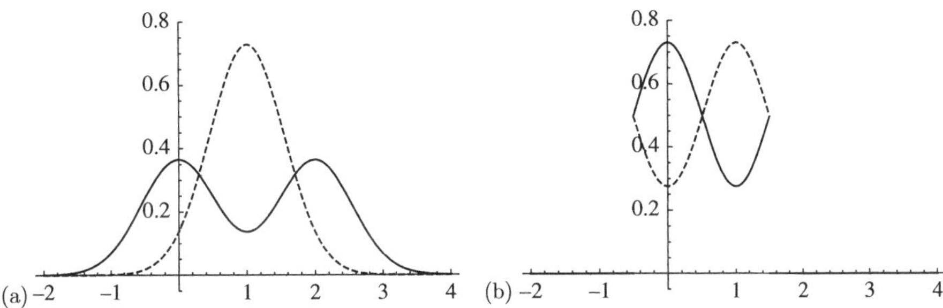

Figure 4.10 The example of one-dimensional binary lattice with (a) proper marginalization and (b) modulo lattice folding. The example shows $p(x|c)$ and $p(y|c)$ for two values $c = 0$ (solid), $c = 1$ (dashed).

from inner and outer shells is asymmetric. The most prone is the border shell, which has no overlapping outer neighbors. The effect becomes negligible for the *high SNR* regime where the tails of the noise decay quickly (see additional details in [63, Section 9.1.2, and Section 9.3.4]).

From a slightly different perspective, the difference between the proper marginalization and the modulo lattice preprocessor is consistent with the theorems identifying the modulo lattice output as a sufficient statistic only for high SNR [63, Section 9.3] or, in the case of the dirty paper coding, for strong interference [63, Theorem 10.6.2].

4.5 Hierarchical Error Probability Performance

4.5.1 Hierarchical Pairwise Error Probability

The error probability is an important performance indicator and also an obvious code and receiver processing optimality goal. The true symbol/frame/bit error probability evaluation based on the true transition probabilities is too complex (apart from trivial uncoded cases) to be practically useful and also gives only a limited insight that could be used for a *synthesis* of the code. A pairwise error probability (see Section A.3) can be used to upper-bound the true error rate. As a very welcome side-effect it also *connects* the performance target with the metric used by the demodulator and decoder. This can be subsequently used for the code synthesis.[9]

4.5.2 Hierarchical Pairwise Error Probability for Isomorphic NCM

When applying the pairwise error probability on the hierarchical demodulator and decoder we, however, need to respect properly all aspects related to the fact that our processing target, the hierarchical symbol/codeword, is generally *a set of multiple* constellation points or codewords $\mathcal{U}(b)$ (i.e. H-constellation) for one given source data

---

[9] For example, a classical single-user AWGN channel using a MAP demodulator metric leads to a classical code design criterion that maximizes the minimum free Euclidean distance of the code.

symbol [56]. In the following, we will define hierarchical pairwise error probability and we will show how this can be used in the NCM design under some specific assumptions. The evaluation of the hierarchical error probability on the relay of course implies that the relay node strategy is HDF.

DEFINITION 4.10 (Hierarchical Pairwise Error Probability)   Assume that the H-data metric is $\mu_b(x)$ and that the decoder *decision* processing *maximizes* its value. Hierarchical pairwise error probability (H-PEP) is

$$P_{2H}(b'|b) = \Pr\left\{\mu_b(x) < \mu_{b'}(x)|b = \chi(\tilde{b}_x)\right\} \tag{4.44}$$

where $\tilde{b}_x$ are transmitted component data and $b \neq b'$ are some given H-messages. We will also use a simplified notation $P_{2H} = \Pr\{\mu_b < \mu_{b'}|b\}$.

The hierarchical pairwise error is thus the probability that the H-metric for the *correct* H-message b is *less* than the one for some *other* message b' provided that the received signal is *consistent* with b, i.e. all component transmitted data are such that $b = \chi(\tilde{b}_x)$. There are several major aspects that complicate the H-PEP evaluation, and that substantially differ from the classical single-user case. First, the conditioning by the *H-message* b still leaves many degrees of *freedom* in the *component* transmitted messages and we must properly address the problem. Similarly it also holds for both the H-metrics staged in the comparison in the probability evaluation. Second, the relation between the codeword and the data error is trivial in the single-user case. However, for the H-codeword and the H-message, this needs to be explicitly assessed by the isomorphism.

We generally apply the $P_{2H}(b'|b)$ on the complete messages **b** (vectors) but it could also be applied on the symbol-wise b_n marginalized metric. We will use a *generic notation b* to cover both. The form of the metric used in H-PEP is arbitrary. It does not even need to be the metric leading to the optimal (e.g. minimum error probability) performance. In such a case it would simply analyze the performance under that suboptimal metric and potentially suggest how to optimize the code for that given (suboptimal) metric. Of course, the most common example is the MAP metric minimizing the error probability.

The evaluation of the H-PEP related directly to the *data* level HNC map metric is a difficult task. It would become much easier when the metric is related to the *code* level HNC map. Also it directly employs the codewords into the calculation which will provide better insight on what the code should optimally look like. Clearly, if the NCM is *isomorphic* then $P_{2H}(b'|b) = P_{2H}(c'|c)$ where $c = \mathcal{C}(b)$ and $c' = \mathcal{C}(b')$. From now on, we will assume isomorphic layered NCM and thus

$$P_{2H}(c'|c) = \Pr\{\mu_c(x) < \mu_{c'}(x)|c\}. \tag{4.45}$$

We will also assume the use of *MAP decoding metric*. The H-metric is a marginalization (4.10)

$$\mu_c(x) = p(x|c) = \frac{1}{p(c)}\sum_{\tilde{c}:c} p(x|\tilde{c})p(\tilde{c}). \tag{4.46}$$

In the evaluation of (4.45), we first focus on the conditioning by the received signal consistent with c. There are multiple component codewords $\tilde{c}_x$ in the received signal consistent with c

$$P_{2H} = \Pr\left\{ \mu_c(x) < \mu_{c'}(x) \,\middle|\, \bigcup_{\tilde{c}_x:c} \tilde{c}_x \right\}. \tag{4.47}$$

Events $\tilde{c}_x$ are *disjoint*. If we assume that they are also *equally probable* $\Pr\{\tilde{c}_x : c\} = \text{const}$ then[10]

$$P_{2H} = \frac{1}{M_{\tilde{c}:c}} \sum_{\tilde{c}_x:c} \Pr\left\{ \mu_c(x) < \mu_{c'}(x) | \tilde{c}_x \right\} \tag{4.48}$$

where $M_{\tilde{c}:c}$ is a size of the subcodebook $\tilde{C}(c)$ which is a subset of $\tilde{C}$ where we take only the entries consistent with c. The H-PEP is the average over all consistent source node component codewords. This is an additional level of averaging over those being present in the traditional pairwise error probability calculation. We denote the H-PEP evaluated for a given H-message c, c' and for some given c-consistent received signal $(\chi_c(\tilde{c}_x) = c)$ as

$$P_{2H_e} = \Pr\left\{ \mu_c(x) < \mu_{c'}(x) | \tilde{c}_x \right\}. \tag{4.49}$$

Similarly as in the classical case, we can upper-bound the H-PEP by *the most probable pairwise H-constellation error event*

$$P_{2H} \leq P_{2H_m} = \max_{\tilde{c}_x:c} \Pr\left\{ \mu_c(x) < \mu_{c'}(x) | \tilde{c}_x \right\}. \tag{4.50}$$

Notice that the hierarchical codewords c, c' are still fixed. We maximized the result only over c-compliant transmitted components $\tilde{c}_x$. The overall error rate behavior can then (similarly as in classical single user code case) be upper-bounded by the *overall most probable pairwise event*

$$P_{2H_{\max}} = \max_{c \neq c', \tilde{c}_x:c} \Pr\left\{ \mu_c(x) < \mu_{c'}(x) | \tilde{c}_x \right\} \tag{4.51}$$

maximized over all $c \neq c'$ and c-compliant transmitted components $\tilde{c}_x$

4.5.3 H-PEP for Gaussian Memoryless Channel

In the next step, we will constrain ourselves to a special case of *Gaussian memoryless channel* (similar to that in (4.15)). The likelihood is then

$$p(x|\tilde{c}) = \frac{1}{\pi^m \sigma_w^{2m}} \exp\left(-\frac{1}{\sigma_w^2} \|x - u(\tilde{c})\|^2 \right) \tag{4.52}$$

---

[10] Assume events A, B_1, B_2, then $\Pr\{A|B_1 \cup B_2\} = \frac{\Pr\{A \cap (B_1 \cup B_2)\}}{\Pr\{B_1 \cup B_2\}} = \frac{\Pr\{(A \cap B_1) \cup (A \cap B_2)\}}{\Pr\{B_1 \cup B_2\}}$. If $B_1 \cap B_2 = \emptyset$ and $\Pr\{B_i\} = \text{const} = \Pr\{B\}$ then $\Pr\{A|B_1 \cup B_2\} = \frac{\Pr\{A \cap B_1\} + \Pr\{A \cap B_2\}}{\Pr\{B_1\} + \Pr\{B_2\}} = \frac{\Pr\{A \cap B_1\} + \Pr\{A \cap B_2\}}{2\Pr\{B\}} = \frac{1}{2}\left(\frac{\Pr\{A \cap B_1\}}{\Pr\{B\}} + \frac{\Pr\{A \cap B_2\}}{\Pr\{B\}} \right) = \frac{1}{2}(\Pr\{A|B_1\} + \Pr\{A|B_2\})$. A generalization for multiple *disjoint equally probable* events clearly gives the average over all conditional probabilities $\Pr\left\{A | \bigcup_{i=1}^{M} B_i \right\} = \frac{1}{M} \sum_{i=1}^{M} \Pr\{A|B_i\}$.

where m is a complete dimensionality of the signals (including symbol dimensionality and the length of the message). On top of assuming uniformly distributed $\tilde{c}$, $\Pr\{\tilde{c}\} = 1/M_{\tilde{c}}$, $M_{\tilde{c}} = |\tilde{\mathcal{C}}|$, we also assume *uniform c*, i.e. $\Pr\{c\} = \text{const} = 1/M_c$, where $M_c = |\mathcal{C}|$ is the size of the H-codebook. Then the metric is

$$\mu_c(x) = \frac{M_c}{M_{\tilde{c}}} \sum_{\tilde{c}:c} p(x|\tilde{c}) = \frac{M_c}{M_{\tilde{c}}} \frac{1}{\pi^m \sigma_w^{2m}} \sum_{\tilde{c}:c} \exp\left(-\frac{1}{\sigma_w^2}\|x - u(\tilde{c})\|^2\right) \tag{4.53}$$

and its normalized form $\dot{\mu}_c(x) = \pi^m \sigma_w^{2m} M_{\tilde{c}}/M_c \mu_c(x)$ is

$$\dot{\mu}_c(x) = \sum_{\tilde{c}:c} \exp\left(-\frac{1}{\sigma_w^2}\|x - u(\tilde{c})\|^2\right). \tag{4.54}$$

The minimum H-distance point for the complete message (similar to that in (4.18) for the symbol) is

$$u^{\text{Hmin}}(c) = \arg \min_{u(\tilde{c}):\chi_c(\tilde{c})=c} \|x - u(\tilde{c})\|^2. \tag{4.55}$$

It is the hierarchical channel-combined symbol, consistent with the given c, which is closest to the received signal. We can factorize (4.54) into

$$\dot{\mu}_c(x) = \exp\left(-\frac{\|x - u^{\text{Hmin}}(c)\|^2}{\sigma_w^2}\right) \sum_{\tilde{c}:c} \exp\left(-\frac{\|x - u(\tilde{c})\|^2 - \|x - u^{\text{Hmin}}(c)\|^2}{\sigma_w^2}\right) \tag{4.56}$$

where all differences in the summation are *non-negative* $\|x-u(\tilde{c})\|^2 - \|x-u^{\text{Hmin}}(c)\|^2 \geq 0$. Finally, taking the negative scaled logarithm $\rho_c^2(x) = -\sigma_w^2 \ln \dot{\mu}_c(x)$, we get the decoder metric

$$\rho_c^2(x) = \|x - u^{\text{Hmin}}(c)\|^2 - \sigma_w^2 \eta_c \tag{4.57}$$

where the correction term is

$$\eta_c = \ln \sum_{\tilde{c}:c} \exp\left(-\frac{1}{\sigma_w^2}\left(\|x - u(\tilde{c})\|^2 - \|x - u^{\text{Hmin}}(c)\|^2\right)\right). \tag{4.58}$$

Clearly, it holds that $0 \leq \eta_c \leq \ln(M_{\tilde{c}}/M_c)$. The correction term is zero $\eta_c = 0$ if exactly one H-constellation point is the minimal H-distance point $u^{(1)}(\tilde{c}) = u^{\text{Hmin}}(c)$ and all others (if any) are at a much larger distance, $\forall i \neq 1$, $\|x-u^{(i)}(\tilde{c})\|^2 \gg \|x-u^{\text{Hmin}}(c)\|^2$. This is, in fact, the condition for the hierarchical minimum distance approximation (4.20) (see also (4.23)). On the other hand, if for all $\tilde{c}:c$ the H-constellation point is the minimal H-distance point, then the correction is non-zero but *constant* and *independent* of x, $\eta_c = \ln(M_{\tilde{c}}/M_c)$. The non-zero value of η_c on its own does not present a problem from the H-PEP evaluation point of view. However, its *dependence* on x is a problem.

The H-PEP for a given H-message c, c' and for some given c-consistent received signal $(\chi_c(\tilde{c}_x) = c)$ (4.49) is then

$$P_{2H_e} = \Pr\left\{\rho_c^2(x) > \rho_{c'}^2(x)|\tilde{c}_x\right\}$$

$$= \Pr\left\{\|x - u^{\text{Hmin}}(c)\|^2 - \sigma_w^2 \eta_c > \|x - u^{\text{Hmin}}(c')\|^2 - \sigma_w^2 \eta_{c'}|\tilde{c}_x\right\}$$

$$= \Pr\left\{\|x - u^{\text{Hmin}}(c)\|^2 - \|x - u^{\text{Hmin}}(c')\|^2 - \sigma_w^2 \eta_{c,c'} > 0|\tilde{c}_x\right\} \tag{4.59}$$

where $\eta_{c,c'} = \eta_c - \eta_{c'}$ and of course the inequality direction changed since we used negative scaled logarithm metric. Let us now denote the c-consistent noiseless part of the received signal for given $\tilde{c}_x$ as $u(\tilde{c}_x)$. It must be a member of the H-constellation set for the c H-symbol, i.e. $u(\tilde{c}_x) \in \mathcal{U}(c)$. The condition of c-consistent received signal is thus reflected in having $x = u(\tilde{c}_x) + w$ and consequently

$$P_{2H_e} = \Pr\left\{ \|u(\tilde{c}_x) + w - u^{\mathrm{Hmin}}(c)\|^2 - \|u(\tilde{c}_x) + w - u^{\mathrm{Hmin}}(c')\|^2 - \sigma_w^2 \eta_{x,c,c'} > 0 \right\}$$
$$(4.60)$$

where the correction terms under this condition are $\eta_{x,c,c'} = \eta_{x,c} - \eta_{x,c'}$

$$\eta_{x,c} = \ln \sum_{\tilde{c}:c} \exp\left(-\frac{1}{\sigma_w^2}\left(\|u(\tilde{c}_x) + w - u(\tilde{c})\|^2 - \|u(\tilde{c}_x) + w - u^{\mathrm{Hmin}}(c)\|^2 \right) \right),$$
$$(4.61)$$

$$\eta_{x,c'} = \ln \sum_{\tilde{c}':c'} \exp\left(-\frac{1}{\sigma_w^2}\left(\|u(\tilde{c}_x) + w - u(\tilde{c}')\|^2 - \|u(\tilde{c}_x) + w - u^{\mathrm{Hmin}}(c')\|^2 \right) \right).$$
$$(4.62)$$

The expression of the distances difference that appears in P_{2H_e} (and with minor modification in $\eta_{x,c}, \eta_{x,c'}$) can be further manipulated

$$\begin{aligned}
&\|u(\tilde{c}_x) + w - u^{\mathrm{Hmin}}(c)\|^2 - \|u(\tilde{c}_x) + w - u^{\mathrm{Hmin}}(c')\|^2 \\
&= \|u(\tilde{c}_x) - u^{\mathrm{Hmin}}(c)\|^2 - \|u(\tilde{c}_x) - u^{\mathrm{Hmin}}(c')\|^2 \\
&\quad + 2\Re\left[\langle u(\tilde{c}_x) - u^{\mathrm{Hmin}}(c); w \rangle \right] - 2\Re\left[\langle u(\tilde{c}_x) - u^{\mathrm{Hmin}}(c'); w \rangle \right] \\
&= \|u(\tilde{c}_x) - u^{\mathrm{Hmin}}(c)\|^2 - \|u(\tilde{c}_x) - u^{\mathrm{Hmin}}(c')\|^2 \\
&\quad - 2\Re\left[\langle u^{\mathrm{Hmin}}(c) - u^{\mathrm{Hmin}}(c'); w \rangle \right].
\end{aligned}$$
$$(4.63)$$

We denote the last term as $\xi = -2\Re\left[\langle u^{\mathrm{Hmin}}(c) - u^{\mathrm{Hmin}}(c'); w \rangle \right]$. It is a Gaussian real-valued scalar zero-mean random variable with variance $\sigma_\xi^2 = 2\sigma_w^2 \|u^{\mathrm{Hmin}}(c) - u^{\mathrm{Hmin}}(c')\|^2$. Then

$$P_{2H_e} = \Pr\left\{ \xi > \|u(\tilde{c}_x) - u^{\mathrm{Hmin}}(c')\|^2 - \|u(\tilde{c}_x) - u^{\mathrm{Hmin}}(c)\|^2 + \sigma_w^2 \eta_{x,c,c'} \right\}.$$
$$(4.64)$$

A similar manipulation can be done for the correction terms

$$\eta_{x,c} = \ln \sum_{\tilde{c}:c} e^{-\frac{1}{\sigma_w^2}\left(\|u(\tilde{c}_x)-u(\tilde{c})\|^2 - \|u(\tilde{c}_x)-u^{\mathrm{Hmin}}(c)\|^2 - 2\Re\left[\langle u(\tilde{c})-u^{\mathrm{Hmin}}(c); w \rangle \right] \right)},$$
$$(4.65)$$

$$\eta_{x,c'} = \ln \sum_{\tilde{c}':c'} e^{-\frac{1}{\sigma_w^2}\left(\|u(\tilde{c}_x)-u(\tilde{c}')\|^2 - \|u(\tilde{c}_x)-u^{\mathrm{Hmin}}(c')\|^2 - 2\Re\left[\langle u(\tilde{c}')-u^{\mathrm{Hmin}}(c'); w \rangle \right] \right)}.$$
$$(4.66)$$

4.5.4 Hierarchical Distance and Self-Distance Spectrum

The properties of the quantities determining the H-PEP clearly depend on two types of the H-constellation/codeword squared distances. The first one is the distance between the points belonging to *different* H-symbols and the second one is the distance between the points belonging to the *same* H-symbol. For this purpose, we define hierarchical distance and self-distance spectrum.

DEFINITION 4.11 (Hierarchical Distance Spectrum)   A hierarchical distance (H-distance) spectrum is a set

$$\mathcal{S}_H(c, c') = \left\{ \|u(\tilde{c}) - u(\tilde{c}')\|^2 : c = \chi_c(\tilde{c}) \neq c' = \chi_c(\tilde{c}') \right\}. \tag{4.67}$$

We also define $\mathcal{S}_H = \bigcup_{c,c'} \mathcal{S}_H(c, c')$.

DEFINITION 4.12 (Hierarchical Self-Distance Spectrum)   A hierarchical self-distance (H-self-distance) spectrum is a set

$$\mathcal{S}_{\bar{H}}(c) = \left\{ \|u(\tilde{c}^{(a)}) - u(\tilde{c}^{(b)})\|^2 : c = \chi_c(\tilde{c}^{(a)}) = \chi_c(\tilde{c}^{(b)}) \wedge \tilde{c}^{(a)} \neq \tilde{c}^{(b)} \right\}. \tag{4.68}$$

We also define $\mathcal{S}_{\bar{H}} = \bigcup_c \mathcal{S}_{\bar{H}}(c)$.

4.5.5   NCM Design Rules Based on H-PEP

We can use (4.64), (4.65), and (4.66) to establish qualitative design rules for NCM that would minimize the H-PEP. The situation is, however, less straightforward than in the classical single-user code. There are several observations we need to keep in our mind before we start. There are multiple mutually correlated random variables in the expression and these cannot be easily factorized into a single one as in the single-user code case. All $\xi, \eta_{x,c}, \eta_{x,c'}$ directly depend on Gaussian noise w and they are continuous valued correlated variables. But the hierarchical minimum distance points $u^{\text{Hmin}}(c), u^{\text{Hmin}}(c')$ depend on the received signal and therefore also on w. These variables are, however, discrete. They are random and dependent on w but *constrained* to be inside the H-constellation and their influence on H-PEP can be thus controlled through the H-distance and H-self-distance spectrum.

In order to minimize H-PEP, we should consider the following.

(1) The distance $\|u(\tilde{c}_x) - u^{\text{Hmin}}(c')\|^2$ in (4.64) should be as large as possible. Notice that $\|u(\tilde{c}_x) - u^{\text{Hmin}}(c')\|^2 \in \mathcal{S}_H(c, c')$ for arbitrary noise w realization.

(2) The self-distance $\|u(\tilde{c}_x) - u^{\text{Hmin}}(c)\|^2$ in (4.64) should be as small as possible. Notice that $\|u(\tilde{c}_x) - u^{\text{Hmin}}(c)\|^2 \in \mathcal{S}_{\bar{H}}(c)$ for arbitrary noise w realization.

(3) The variance of the ξ variable is proportional to $\|u^{\text{Hmin}}(c) - u^{\text{Hmin}}(c')\|^2 \in \mathcal{S}_H(c, c')$, which is constrained by the H-distance spectrum.

(4) The correction term $\eta_{x,c,c'}$ should be as large as possible, which in turn means maximizing $\eta_{x,c}$ and minimizing $\eta_{x,c'}$. Behavior of $\eta_{x,c}$ is dictated by the H-self-distance spectrum while the behavior of $\eta_{x,c'}$ is jointly dictated by both the H-distance and H-self-distance spectra.

(5) The maximum value of $\eta_{x,c}$ is $\ln(M_{\tilde{c}}/M_c)$ as is shown in its original form (4.58), and which must hold also for (4.65). It is reached, for arbitrary x, when arguments of the exponentials are zero, i.e. when all self-distances are zero $\mathcal{S}_{\bar{H}}(c) = \{0\}$. All H-constellation/codeword points for given c are *identical*. We will call this a *self-folded* H-constellation/codebook or *self-folded NCM*. If the H-constellation/codebook is

self-folded then the arguments of the exponentials in $\eta_{x,c'}$ are also all zeros and thus $\eta_{x,c'} = \ln(M_{\tilde{c}}/M_c)$ and the overall correction term is zero regardless of the noise $\eta_{x,c,c'} = 0$. Self-folded NCM also causes the self-distance in (4.64) to be zero and thus

$$P_{2H_e}^{\text{SF}} = \Pr\left\{\xi > \|u(\tilde{c}_x) - u^{\text{Hmin}}(c')\|^2\right\}. \tag{4.69}$$

(6) Now let us have a look on the situation when the NCM is *not* self-folded. Let us assume that the spread in self-distances is symmetric for all c. If it was not a symmetric one then the case that would make an advantage for $P_{2H_e}(c'|c)$ would become a disadvantage for $P_{2H_e}(c|c')$.

(a) Let us also assume that some point pair $\tilde{c}_x$, $\tilde{c}$ in the H-constellation has the self-distance $\rho_{\tilde{H}}^2$. The expression $\eta_{x,c}$ will not be the maximal one (as for the self-folded case) but it will be somewhat smaller. For the given pair of points, the argument $\|u(\tilde{c}_x) - u(\tilde{c})\|^2$ of the exponential in (4.65) increases to the value $\rho_{\tilde{H}}^2$. The second term $\|u(\tilde{c}_x) - u^{\text{Hmin}}(c)\|^2$ will highly likely (at least for high SNR) be zero since the minimum H-distance point is the closest to the received signal. The degradation of the first noiseless term in (4.65) is thus $\rho_{\tilde{H}}^2$ at least for that given point pair.

(b) This degradation can be possibly compensated for by the improvement in the term (4.66). In the most favorable case for the improvement, the points $u(\tilde{c}_x), u(\tilde{c}'), u^{\text{Hmin}}(c')$ lie in the line and the maximal value of $\|u(\tilde{c}_x) - u(\tilde{c}')\|^2 - \|u(\tilde{c}_x) - u^{\text{Hmin}}(c')\|^2$ is $\rho_{\tilde{H}}^2$, where, by the assumption of the symmetry, we have $\|u(\tilde{c}') - u^{\text{Hmin}}(c')\|^2 = \rho_{\tilde{H}}^2$. So the noiseless terms in the exponentials of (4.66) can, at best, just compensate for the degradation of the argument of (4.65) but practically it will be even worse.

(c) The noise terms in both (4.65) and (4.66), i.e. $2\Re\left[\langle u(\tilde{c}) - u^{\text{Hmin}}(c); w\rangle\right]$ and $2\Re\left[\langle u(\tilde{c}') - u^{\text{Hmin}}(c'); w\rangle\right]$, are given by the self-distances only. The LHS in the inner products are different, but under the assumption of symmetric self-distances $\mathcal{S}_{\bar{H}}(c) \approx \mathcal{S}_{\bar{H}}(c')$ they will make the noise term highly correlated and thus both will affect the arguments of the exponentials in (4.65) and (4.66) the same way.

(d) The main expression (4.64) also contains the self-distance. A positive value of $\|u(\tilde{c}_x) - u^{\text{Hmin}}(c)\|^2$ clearly decreases the RHS of the inequality and increases the H-PEP.

(e) As we see, the non-zero spread of the self-distances *cannot* improve the H-PEP and will likely make things worse.

CONJECTURE 4.13 (Self-Folded NCM (H-Constellation/Codebook) Minimizes H-PEP) *Assume isomorphic NCM in Gaussian memoryless channel, decoding MAP H-metric, and uniform component messages and HNC map such that* $\Pr\{\tilde{c}\} = 1/M_{\tilde{c}}$, $\Pr\{c\} = 1/M_c$, $\Pr\{\tilde{c} : c\} = M_c/M_{\tilde{c}}$.

Self-folded NCM, i.e. the one with zero H-self-distance spectrum $\mathcal{S}_{\bar{H}} = \{0\}$, minimizes H-PEP. The resulting H-PEP is then

$$P^{\text{SF}}_{2H_e} = Q\left(\sqrt{\frac{\|u(\tilde{c}_x) - u^{\text{Hmin}}(c')\|^2}{2\sigma_w^2}}\right). \tag{4.70}$$

It is important to note that the self-folding property is expected to be a *natural* one, i.e. naturally performed by the channel combining the component signals into the H-constellation also fully respecting the channel parametrization. Notice that the modulo lattice preprocessing achieves the hierarchical self-folding but it achieves that by the *force*. The price paid for this enforcement is a change of the distortion of the noise which becomes modulo-equivalent Gaussian.

The true *natural* self-folding H-constellation or H-codebook might be difficult to find. However, we can still compare practical H-constellations or H-codebooks in terms of how well they approach or approximate the self-folding property. The design rules above, and particularly the shape and behavior of H-distance and H-self-distance spectrum, allow us to predict the H-BER performance. The following example might be a bit simplistic and artificial, but it clearly demonstrates the role of H-distance and H-self-distance spectra.

Example 4.4   We consider $K = 2$ sources. Each source transmits (for simplicity of the graphical presentation of the H-constellation) an *uncoded* BPSK constellation $c_A, c_B \in \{\pm 1\}$. The channel is assumed to be a linear AWGN channel with the constellation space input–output model

$$x = h_A c_A + h_B c_B + w \tag{4.71}$$

where we set $h_A = 1$ and define relative channel gain $h = h_B/h_A$. The variance of Gaussian noise is σ_w^2. The SNR is set w.r.t. source SA, $\Gamma = |h_A|^2 \mathrm{E}[|c_A|^2]/\sigma_w^2 = \mathcal{E}_b/N_0$.

We will define (ignoring their practical usability at the moment) two variants of the HNC map $c \in \{0, 1\}$ that imply two variants of the H-constellation. The channel parametrization is chosen to be $h = \exp(\mathrm{j}\,\pi/3)$. This specific value is set in such a manner that we can easily compare the H-distance and H-self-distance spectra. Figure 4.11a shows the H-constellation variants.

The H-distance and H-self-distance spectra are in variant #1

$$\mathcal{S}_{H1} = \left\{2^2, 2^2, 2^2, 2^2\right\}, \quad \mathcal{S}_{\bar{H}1} = \left\{2^2, (2\sqrt{3})^2\right\} \tag{4.72}$$

and in variant #2

$$\mathcal{S}_{H2} = \left\{2^2, 2^2, 2^2, (2\sqrt{3})^2\right\}, \quad \mathcal{S}_{\bar{H}2} = \left\{2^2, 2^2\right\}. \tag{4.73}$$

Variant #2 has greater values in H-distance spectrum and lower values in H-self-distance spectrum in comparison with variant #1. The H-constellation design rules identify variant #2 as the better one. The numerical results obtained by a computer simulation confirm that conjecture in Figure 4.11b.

Also notice that the minimum H-distance is the same in both cases. This clearly demonstrates that the minimum H-distance has only a limited descriptive value when evaluating the H-PEP performance.

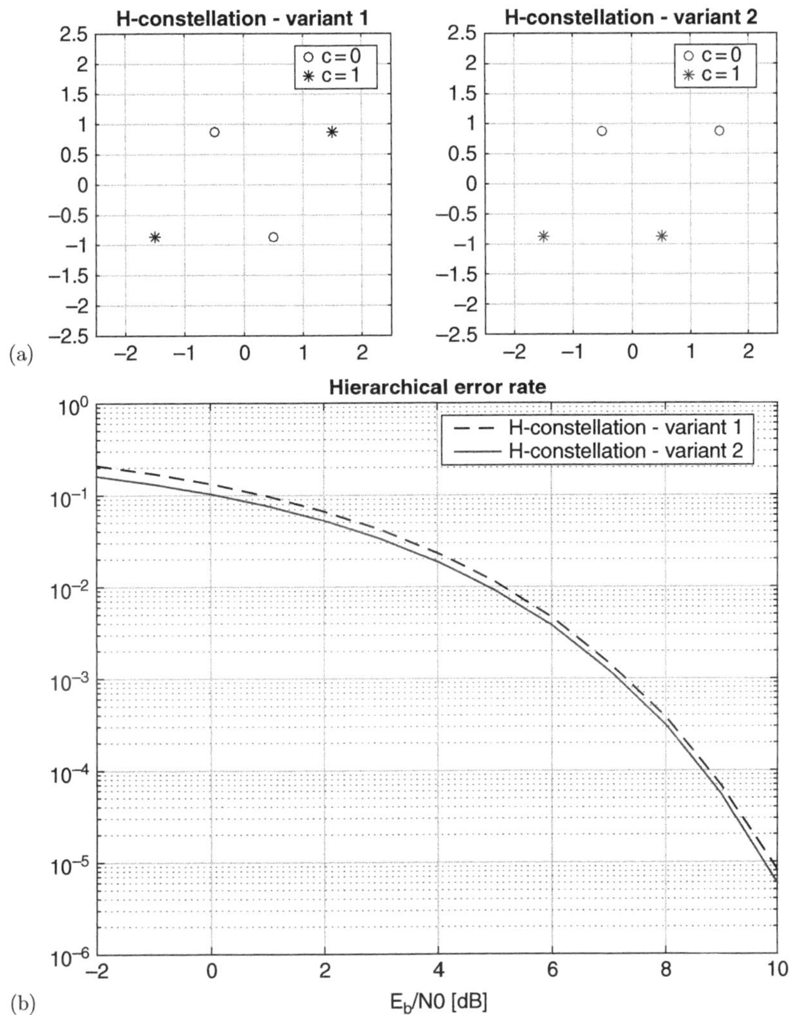

Figure 4.11 H-PEP example: H-distance, H-self-distance, and true simulated hierarchical error rate.

4.6 Hierarchical Side-Information Decoding

This section addresses the design of algorithms for a specific form of multi-stage receiver (typically applicable at the final destination, but not limited to it), which has available several *separate orthogonal* observations of HI and/or HSI observed from multiple stages. We also describe the iterative multi-loop decoder processing that allows the combining of the codebook structure from multiple stages.

4.6.1 Hierarchical Side-Information Decoding – System Model

Hierarchical side-information decoding (HSI-decoding) solves a situation where the node receives multiple independent orthogonal signals, typically coming from multiple stages, each related to some HNC map, and it wants to decode some other HNC map, generally different from the former ones. A fundamental difference between H-decoding and HSI-decoding is that the latter has *multiple orthogonal* observations related to *multiple different HNC message maps* with the goal of obtaining another *target HNC message map*. The H-decoding can form, as we will see later, an auxiliary processing step inside a more complex HSI-decoding. For simplicity, we constrain the treatment to *two* observations in the following text, but all results can be easily generalized. The name "HSI-decoding" reflects the situation where the target is hierarchical message decoding based on multiple observations providing additional information, i.e. side-information, in contrast with single observation H-decoding.

Each observation is associated with a given message HNC map. Assume that the first observation carries H-message $b = \chi(\tilde{b})$ and the second one $\bar{b} = \bar{\chi}(\tilde{b})$. Most typically, b is HI and $\bar{b}$ is HSI but the results hold generally for arbitrary content/form of the map.[11] The components of the maps do not need to be directly the sources $\tilde{b} = \{b_A, b_B, \ldots\}$ but they can be arbitrary H-messages in the *hierarchical encapsulation*, i.e. $\tilde{b}_s = \{b_{s,1}, b_{s,2} \ldots\}$ is a set of arbitrary preceding H-messages. However, we can always recursively substitute all encapsulation levels and express all the maps in terms of the original source set $\tilde{b}$, which provides the finest granularity of the map definition. All observed signals are encoded by some NCMs. The *goal* of HSI decoding is to obtain some other *target* HNC message map $b' = \chi'(\tilde{b})$. In a special case of the final destination node, this target map is one of the source messages $b' = b_i, b_i \in \tilde{b}$.

In this section, we do not try to interpret or guarantee mutual relationships (e.g. the global solvability, see Section 5.8) of maps $\chi, \bar{\chi}, \chi'$. The role of HSI decoding is to get some general H-message b' from the orthogonal received NCM encoded observations of two other different H-messages b and $\bar{b}$.

The simplest application example is the butterfly network (Figure 3.2b). In the first stage, the destination node DA receives the encoded signal carrying a simple (one component only) HNC map $\bar{b} = b_B$, which is the HSI from the perspective of DA. In the second stage, orthogonal with the first one, DA receives the encoded signal carrying HNC map $b = \chi(b_A, b_B)$, e.g. bit-wise modulo-2 sum. The target map for DA is again quite simplistic HNC map $b' = b_A$. In this simple example, both observations are formed by a single component signal, i.e. the NCM has a singular single-user form.

A general system model for two-stage observation is shown in Figure 4.12. We assume that all HNC message maps $\chi, \bar{\chi}, \chi'$ are defined on a *common set* of original source component messages $\tilde{b}$. This is the most generic setup enabling the HSI decoder to decode an arbitrary target map. In special cases, when some components are not involved or some are already hierarchical and common to all maps, we can always formally make a substitution by new formal hierarchical variable.

Single Component Observation

The simplest configuration, where both stages have only a *single component observation*, is in Figure 4.12a. H-message $b = \chi(\tilde{b})$ is encoded into $\mathbf{c} = \mathcal{C}(b)$ and received as

<hr>

[11] Both maps can be arbitrary HI and even neither of them needs to be HSI. But the case of HI complemented by HSI is the most typical.

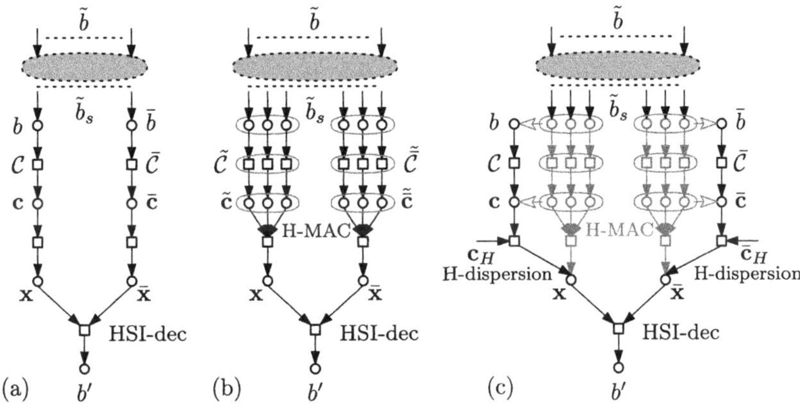

Figure 4.12 System model for two-stage observation HSI-decoding. (a) HSI-decoding with single component observations. (b) HSI-decoding with H-MAC observations. (c) HSI-decoding with H-MAC observations and isomorphic layered NCMs using the *equivalent H-MAC* channel.

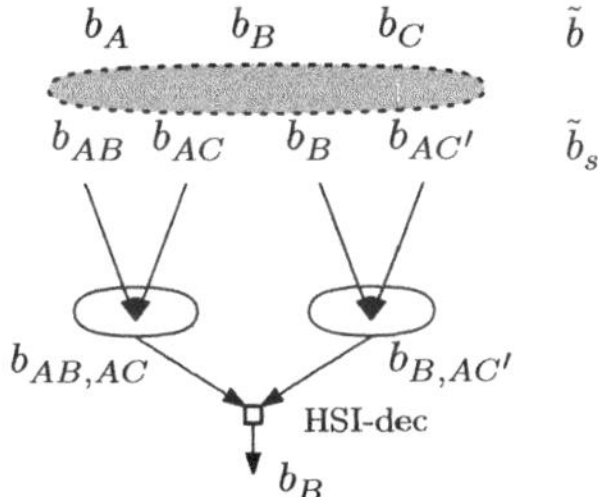

Figure 4.13 Example of HSI-decoding with two H-MAC observations and $b = \chi_s(b_{AB}, b_{AC}) = \chi(b_A, b_B, b_C)$, $\bar{b} = \bar{\chi}_s(b_B, b_{AC'}) = \bar{\chi}(b_A, b_B, b_C)$ and target map $b' = b_B$.

x. A second H-message $\bar{b} = \bar{\chi}(\tilde{b})$ is encoded into $\bar{\mathbf{c}} = \bar{C}(\bar{b})$ and received as $\bar{\mathbf{x}}$, which is conditionally independent with the first signal, $(\mathbf{x} \perp\!\!\!\perp \bar{\mathbf{x}}) | (\mathbf{c}, \bar{\mathbf{c}})$. In the Gaussian channel, it is equivalent to having orthogonal observations $\mathbf{x} \perp \bar{\mathbf{x}}$. The channel observation is jointly described by $p(\mathbf{x}, \bar{\mathbf{x}} | \mathbf{c}, \bar{\mathbf{c}})$. Both stages are standard single-user coding chains and form two Markov chains $\tilde{b} \mapsto \mathbf{c} \mapsto \mathbf{x}$ and $\bar{b} \mapsto \bar{\mathbf{c}} \mapsto \bar{\mathbf{x}}$ and thus

$$p(\mathbf{x}, \bar{\mathbf{x}} | \mathbf{c}, \bar{\mathbf{c}}) = p(\mathbf{x}|\mathbf{c})p(\bar{\mathbf{x}}|\bar{\mathbf{c}}). \tag{4.74}$$

H-MAC Observation

Signals received in both stages, however, can also each form *H-MAC* (Figure 4.12b), i.e. within each stage some signals superimpose and are interpreted only through the H-message. The component-wise codes for both stages are $\tilde{\mathbf{c}} = \tilde{C}(\tilde{b}_{s1})$, $\tilde{\bar{\mathbf{c}}} = \tilde{\bar{C}}(\tilde{b}_{s2})$ where $\tilde{b}_{s1}, \tilde{b}_{s2} \subset \tilde{b}_s$ are some subsets of already hierarchically encapsulated HNC functions. Each component message in $\tilde{b}_{s1}, \tilde{b}_{s2}$ can be itself some HNC map of sources $\tilde{b}$ (see example in Figure 4.13). Joint channel observation is then described by

$$p(\mathbf{x}, \bar{\mathbf{x}} | \tilde{\mathbf{c}}, \tilde{\bar{\mathbf{c}}}) = p(\mathbf{x}|\tilde{\mathbf{c}})p(\bar{\mathbf{x}}|\tilde{\bar{\mathbf{c}}}) \tag{4.75}$$

where the conditioning by complete component sets $\tilde{\mathbf{c}}$ and $\tilde{\bar{\mathbf{c}}}$ is generally required to get the observation independence.

H-MAC with Isomorphic Layered Equivalent Channel

In a special case of both H-MAC stages having *isomorphic layered NCM*, which is the only performance-wise (Section 5.7.4) and practical implementation-wise viable case, we can streamline the model (Figure 4.12c). Both H-MAC stages are modeled by using the equivalent hierarchical channel model for isomorphic layered NCM (see Section 5.7.4 and Figure 5.12). As a consequence, the *hierarchical dispersions* $\mathbf{c}_H$ and $\bar{\mathbf{c}}_H$ are present in the equivalent model. The isomorphic HNC is described by HNC maps $b = \chi(\tilde{b})$ and $\bar{b} = \bar{\chi}(\tilde{b})$. However, we must keep in mind that the component codes are defined against $\tilde{b}_{s1}, \tilde{b}_{s2} \subset \tilde{b}_s$ not against $\tilde{b}$. Code HNC maps are *symbol-wise* $c_n = \chi_c(\tilde{c}_n)$, $\bar{c}_n = \bar{\chi}_c(\tilde{c}_n)$. The isomorphic H-codes are $\mathcal{C}$ and $\bar{\mathcal{C}}$.

The observation model $p(\mathbf{x}, \bar{\mathbf{x}}|\mathbf{c}, \bar{\mathbf{c}})$ properties now depend on the properties of an equivalent isomorphic model (Figure 5.12). The hierarchical dispersions $\mathbf{c}_H$ and $\bar{\mathbf{c}}_H$ are generally not guaranteed to be independent. The only guaranteed observation independence is for the all-component case $p(\mathbf{x}, \bar{\mathbf{x}}|\tilde{\mathbf{c}}, \tilde{\bar{\mathbf{c}}}) = p(\mathbf{x}|\tilde{\mathbf{c}})p(\bar{\mathbf{x}}|\tilde{\bar{\mathbf{c}}})$. The equivalent model marginalization is

$$p(\mathbf{x}, \bar{\mathbf{x}}|\mathbf{c}, \bar{\mathbf{c}}) = \frac{1}{p(\mathbf{c}, \bar{\mathbf{c}})} \sum_{\tilde{\mathbf{c}}:\mathbf{c};\tilde{\bar{\mathbf{c}}}:\bar{\mathbf{c}}} p(\mathbf{x}|\tilde{\mathbf{c}})p(\bar{\mathbf{x}}|\tilde{\bar{\mathbf{c}}})p(\tilde{\mathbf{c}}, \tilde{\bar{\mathbf{c}}}) \tag{4.76}$$

where

$$p(\mathbf{c}, \bar{\mathbf{c}}) = \sum_{\tilde{\mathbf{c}}:\mathbf{c};\tilde{\bar{\mathbf{c}}}:\bar{\mathbf{c}}} p(\tilde{\mathbf{c}}, \tilde{\bar{\mathbf{c}}}). \tag{4.77}$$

The possibility of whether we can factorize this expression into separate $\mathbf{c}, \bar{\mathbf{c}}$ dependent parts, however, depends on mutual stochastic properties of $\tilde{\mathbf{c}}, \tilde{\bar{\mathbf{c}}}$ and $\mathbf{c}, \bar{\mathbf{c}}$.

Equivalent Factorized HSI-Decoding Observation Model

We notice that the single component observation (Figure 4.12a) and the isomorphic layered NCM equivalent channel (Figure 4.12c) have a common system model structure. The only difference is hidden in the observation model $p(\mathbf{x}, \bar{\mathbf{x}}|\mathbf{c}, \bar{\mathbf{c}})$ and the conditions required for its factorization.

The metric factorization is necessary for having two independent H-SODEMs for both stage observations. Having two H-SODEMs allows us in turn to have two decoding chains that can, however, mutually cooperate. Otherwise, the H-SODEM and decoder would have to be *joint* for both stages with a product complexity. There is no special requirement needed for the factorization in the case of single component observation (4.74). Both stages form separate Markov chains for the single variable that appears in the metric. However, in the case of H-MAC with isomorphic equivalent channel, the conditioning variables needed for the factorization $\tilde{\mathbf{c}}, \tilde{\bar{\mathbf{c}}}$, are not the same ones that appear in the required metric, i.e. $\mathbf{c}, \bar{\mathbf{c}}$. The factorization of $p(\mathbf{x}, \bar{\mathbf{x}}|\mathbf{c}, \bar{\mathbf{c}})$ thus depends also on the properties of the involved HNC maps. Generally, the component codewords $\tilde{\mathbf{c}}, \tilde{\bar{\mathbf{c}}}$ are mutually dependent since they can be encoded on mutually dependent sets $\tilde{b}_{s1}, \tilde{b}_{s2}$.

It is clear that in order to guarantee the factorization in both (4.76) and (4.77) we must have independent $\tilde{\mathbf{c}} \perp \bar{\tilde{\mathbf{c}}}$. And since the component codewords are one-to-one codebook functions $\tilde{\mathbf{c}} = \tilde{\mathcal{C}}(\tilde{b}_{s1})$, $\bar{\tilde{\mathbf{c}}} = \bar{\tilde{\mathcal{C}}}(\tilde{b}_{s2})$, of the underlying H-messages $\tilde{b}_{s1}, \tilde{b}_{s2}$, the independence will depend on mutual independence of these H-messages. In a special case of GF-based linear maps $b = \chi(\tilde{b})$ and $\bar{b} = \bar{\chi}(\tilde{b})$, with uniformly distributed sources, we can guarantee this independence by the following lemma.

LEMMA 4.14 (Independence of Linear HNC Maps on GF with Uniform Messages)
Assume linear HNC maps on a common GF $\mathbb{F}_M$, $b = \sum_{k=1}^{K} a_k b_k$, $\bar{b} = \sum_{k=1}^{K} \bar{a}_k b_k$, *with coefficients* $a_k, \bar{a}_k \in [0 : M - 1]$, *and uniformly distributed IID messages* $b_k \in [0 : M - 1]$.
If the coefficient sets $\{a_k\}_k$, $\{\bar{a}_k\}_k$ *differ in at least one coefficient, then* $b \perp \bar{b}$.

Proof Without loss of generality, assume that the coefficient sets differ in the coefficient $a_1 \neq \bar{a}_1$ and the remaining ones are mutually equal. Then

$$b = a_1 b_1 + b_0, \tag{4.78}$$

$$\bar{b} = \bar{a}_1 b_1 + b_0, \tag{4.79}$$

where

$$b_0 = \sum_{k=2}^{K} a_k b_k = \sum_{k=2}^{K} \bar{a}_k b_k. \tag{4.80}$$

An elimination of b_0 from the equations above gives

$$b = \bar{b} + b_1(a_1 - \bar{a}_1). \tag{4.81}$$

If $a_1 \neq \bar{a}_1$ then $b_1(a_1 - \bar{a}_1)$ is uniformly distributed on $[0 : M - 1]$. Also $\bar{b} + b_1(a_1 - \bar{a}_1)$ will be uniformly distributed on $[0 : M - 1]$ regardless of the value $\bar{b}$ and thus $p(b|\bar{b}) = p(b)$.

If the linear functions differ in more than one coefficient, we can use mathematical induction and the fact that adding whatever number to uniform independent variables does not change the independence. $\quad\square$

Clearly, if all HNC maps involved in $\tilde{b}_{s1}, \tilde{b}_{s2}$ are, e.g. by using linear GF maps, pairwise independent, i.e. $b_{s1} \perp b_{s2}$ for all $b_{s1} \in \tilde{b}_{s1}$, $b_{s2} \in \tilde{b}_{s2}$, then $\tilde{b}_{s1} \perp \tilde{b}_{s2}$, $\tilde{\mathbf{c}} \perp \bar{\tilde{\mathbf{c}}}$ and also $\mathbf{c} \perp \bar{\mathbf{c}}$, $\mathbf{c}_H \perp \bar{\mathbf{c}}_H$. As a consequence, the HSI-decoding metric (4.76) and (4.77) can be factorized

$$p(\mathbf{x}, \bar{\mathbf{x}}|\mathbf{c}, \bar{\mathbf{c}}) = p(\mathbf{x}|\mathbf{c})p(\bar{\mathbf{x}}|\bar{\mathbf{c}}) \tag{4.82}$$

where

$$p(\mathbf{x}|\mathbf{c}) = \frac{1}{p(\mathbf{c})} \sum_{\tilde{\mathbf{c}}:\mathbf{c}} p(\mathbf{x}|\tilde{\mathbf{c}})p(\tilde{\mathbf{c}}), \tag{4.83}$$

$$p(\mathbf{c}) = \sum_{\tilde{\mathbf{c}}:\mathbf{c}} p(\tilde{\mathbf{c}}), \tag{4.84}$$

and similarly for $p(\bar{\mathbf{x}}|\bar{\mathbf{c}})$. We can have two independent H-SODEMs for both stages producing soft decoding metrics $p(\mathbf{x}|\mathbf{c})$ and $p(\bar{\mathbf{x}}|\bar{\mathbf{c}})$.

On the other hand, the independent factorized metric was obtained under the price of complete independence of both stages. Therefore both decoding branches will be completely independent and do not provide each other with any extrinsic information. The mutual *cooperation* thus does *not* have any sense. If the two observations are *not* independent, we lose the possibility of having independent H-SODEMs but both decoding branches have mutually extrinsic information and can help each other by the cooperation.

4.6.2   HSI-Decoding Processing Structure

As we discussed above, there are several major aspects affecting the overall processing structure of HSI-decoding. In the following treatment, we assume that the target map $b' = \chi'(b, \bar{b})$ can be *directly* obtained from $b, \bar{b}$ maps.

Separate H-SODEMs The first one reflects the possibility to factorize the observation metric into *per-stage* separate metrics, which in turn means two separate H-SODEMs. This is enabled by having *independent* component codewords in stages $\tilde{\mathbf{c}} \perp \tilde{\bar{\mathbf{c}}}$, which is in turn enabled by *independent* messages $b_{s1} \perp b_{s2}$. This can be achieved by having GF linear HNC maps. On the other side, the independent messages and codewords mean that the two decoding branches cannot provide each other with the extrinsic information and the cooperation of decoders does not help.

H-SODEM marginalization The second important aspect is the *marginalization* of the joint metric to produce the H-metric in the case of using isomorphic layered NCM. Since the channel symbol HNC map is many-to-one mapping, the set of $\tilde{\mathbf{c}}$ consistent with $\mathbf{c}$ still has some degrees of freedom. However, the H-SODEM needs to reconstruct the joint a priori metric (4.77). The usage of GF linear HNC maps, now at the code symbol level (see Section 4.4.3), can help.

Layered H-decoding If the NCMs used in both observed stages are *isomorphic layered* NCMs, then we have the relationship between H-messages and H-codewords fully defined, $b \mapsto \mathbf{c}$, $\bar{b} \mapsto \bar{\mathbf{c}}$, and we can use H-SODEMs.

Independent HSI-Decoding

Independent HSI decoding is performed by two independent non-cooperating decoders for both stages (Figure 4.14). It is the optimal solution in any situation when $\tilde{\mathbf{c}} \perp \tilde{\bar{\mathbf{c}}}$, e.g. by having linear independent message HNC component maps $\tilde{b}_{s1} \perp \tilde{b}_{s2}$. The final hard decisions on H-messages b and $\bar{b}$ are then entered to the χ' target HNC map. Independent decoders use separate observation models $p(\mathbf{x}|\tilde{\mathbf{c}})$, $p(\bar{\mathbf{x}}|\tilde{\bar{\mathbf{c}}})$, and neither utilizes any potential advantage, nor struggles with potential problems of their coupling. The decoder itself can, but does not have to, be the layered one.

Joint and Cooperative HSI-Decoding

If the observation model $p(\mathbf{x}|\tilde{\mathbf{c}})p(\bar{\mathbf{x}}|\tilde{\bar{\mathbf{c}}})p(\tilde{\mathbf{c}}, \tilde{\bar{\mathbf{c}}})$ cannot be factorized, we must use joint or cooperative HSI-decoding structure (Figure 4.15). The joint decoding is performed by

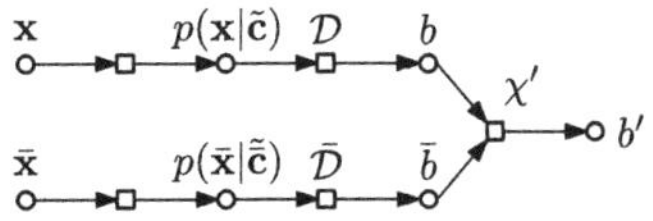

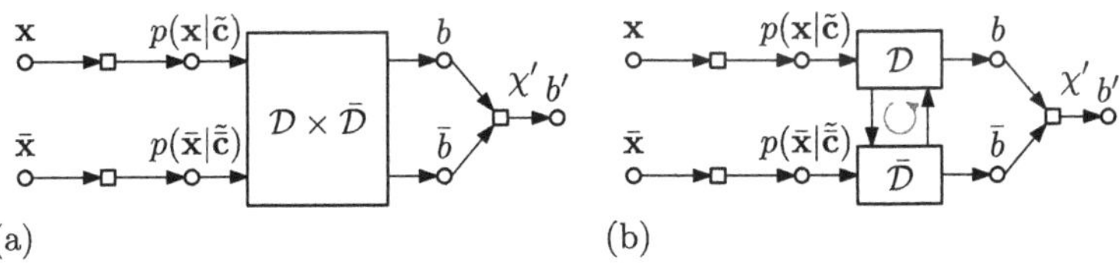

Figure 4.14 Independent HSI-decoding.

(a)   (b)

Figure 4.15 Joint (a) and cooperative (b) HSI-decoding.

a product decoder $\mathcal{D} \times \bar{\mathcal{D}}$ for all component codes in both stages. The cooperative structure achieves the same processing goal but with two mutually iteratively cooperating decoders. Because the individual decoders themselves typically also use internal iterative decoding, this structure leads to a multi-loop iterative receiver. The internal structure and finer structuring of the components of the iterative receiver, e.g. utilizing the soft-aided H-SODEMs, can have many forms.

Doubly Layered HSI-Decoding

A very specific, and practically useful, form of the HSI-decoder uses the isomorphic layered principle in *two* layers. Assume that both stages use isomorphic layered NCMs forming IH-codebooks $\mathbf{c} = \mathcal{C}(b)$, $\bar{\mathbf{c}} = \bar{\mathcal{C}}(\bar{b})$. If, on top of it, the target HSI-decoding message map can be directly obtained from $b, \bar{b}$ maps $b' = \chi'(b, \bar{b})$ and similarly *also* the codeword map $\mathbf{c}' = \chi'_c(\mathbf{c}, \bar{\mathbf{c}})$, and if they are such that there exists *isomorphic second layer codebook C'*, $\mathbf{c}' = \mathcal{C}'(b')$, we call involved HNCs *doubly isomorphic layered NCMs*, and the corresponding decoding is *doubly layered* HSI-decoding (Figure 4.16).

The equivalent doubly layered isomorphic channel is a processing chain $b' \mapsto \mathbf{c}' \mapsto (\mathbf{x}, \bar{\mathbf{x}})$. The joint channel observation can be formally denoted as $\mathbf{x}' = (\mathbf{x}, \bar{\mathbf{x}})$. We simply concatenate two observations into one longer one. The hierarchical dispersion is $\mathbf{c}'_H$. Clearly, the model now has exactly the same form as single layer H-decoding isomorphic equivalent channel (see Section 5.7.4 and Figure 5.12). This formal equivalence of the doubly isomorphic layered model with the single layer one allows us to use all coding and processing theorems in both situations.

Example 4.5   Assume all component codes being identical $\mathcal{C}_k = \mathcal{C}_0$ linear (N_b, N_c) codes $\mathbf{c}_k = \mathcal{C}_0(\mathbf{b}_k)$ over GF $\mathbb{F}_M$, i.e. $\mathbf{b}_k \in \mathbb{F}_{M^{N_b}}$, $\mathbf{c}_k \in \mathbb{F}_{M^{N_c}}$. Assume HNC linear maps over GF with all coefficients from $\mathbb{F}_M$, $\mathbf{b} = \sum_k a_k \mathbf{b}_k$, $\bar{\mathbf{b}} = \sum_k \bar{a}_k \mathbf{b}_k$, $\mathbf{c} = \sum_k a_k \mathbf{c}_k$, $\bar{\mathbf{c}} = \sum_k \bar{a}_k \mathbf{c}_k$. Clearly, the system forms isomorphic layered NCMs, $\mathbf{c} = \mathcal{C}_0(\mathbf{b})$, $\bar{\mathbf{c}} = \mathcal{C}_0(\bar{\mathbf{b}})$.

If the second layer of the maps is again linear over the same GF, $\mathbf{b}' = a'\mathbf{b} + \bar{a}'\bar{\mathbf{b}}$, $\mathbf{c}' = a'\mathbf{c} + \bar{a}'\bar{\mathbf{c}}$, the system is doubly isomorphic layered NCM, where $\mathbf{c}' = \mathcal{C}_0(\mathbf{b}')$.

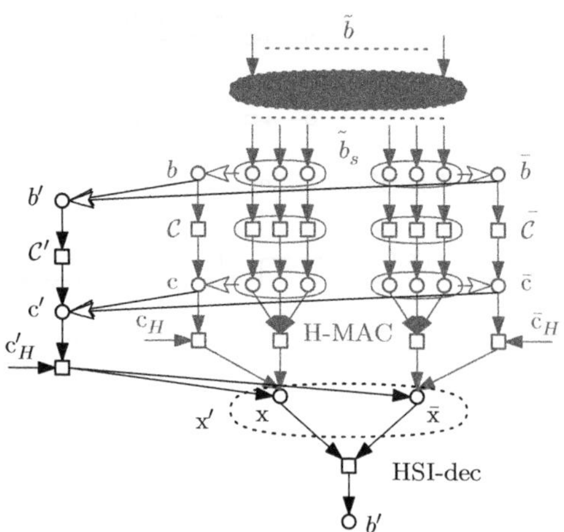

Figure 4.16 Doubly isomorphic layered NCMs and doubly layered HSI-decoding.

4.7　Hierarchical Network Code Map

The HNC map defines what "content" is processed by the relay nodes and how it fits into a global solvability condition. The H-processing operation defines the representation of the "contents", e.g. hard decisions (data, codesymbols), compression, soft information, etc. (see Section 5.8.4). This section sets a common background for the map design and shows only canonical (mostly linear) solutions. Particular design options, further details, and more complicated designs will come in Sections 6.3.4 and 5.8.

4.7.1　Linear HNC Map Designs

General Form and Notation for Linear Maps

Linear HNC maps can be defined in two contexts. In the first one, they are applied at the level of *messages*, and in the second one, they are applied at the level of *codewords*. The former is essential from the perspective of information flow through the network. The latter is important for isomorphic layered NCM design. Both individually, however, share common principles and we choose to demonstrate these principles only on the message level. Apart from the individual map design on the message and codeword level, the isomorphic layered NCM also requires proper attention to be paid to their mutual relation. It is solved in Section 4.7.2.

We will use the following notation. Linear maps associated with messages $\mathbf{b}$ will use coefficients q, vectors $\mathbf{q}$, and matrices $\mathbf{Q}$, whereas the maps associated with codewords $\mathbf{c}$ will use a, $\mathbf{a}$, and $\mathbf{A}$. All will be demonstrated for message maps. The codeword maps can be obtained by a simple change in the notation.

Symbol-Wise Maps

First, we start with a single relay evaluating the map $\mathbf{b}$. We assume that the component messages are represented by vectors $\mathbf{b}_k = [b_{k,1}, \ldots, b_{k,N_b}]^{\mathrm{T}}$, $k \in [1 : K]$, $b_{k,n} \in \mathbb{F}_M$.

All operations are assumed to be on GF $\mathbb{F}_M$. Symbol-wise maps share common scaling coefficients for each symbol in the sequence, i.e. the nth H-message symbol is

$$b_n = \sum_{k=1}^{K} q_k b_{k,n}. \tag{4.85}$$

When applied to whole message vectors, we have

$$\mathbf{b} = \sum_{k=1}^{K} q_k \mathbf{b}_k = (\mathbf{q}^{\mathrm{T}} \boxtimes \mathbf{I}_{N_b})\tilde{\mathbf{b}} \tag{4.86}$$

where $\mathbf{q} = [q_1, \ldots, q_K]^{\mathrm{T}}$ and message vectors are concatenated into one "tall" column $\tilde{\mathbf{b}} = [\mathbf{b}_1^{\mathrm{T}}, \ldots, \mathbf{b}_K^{\mathrm{T}}]^{\mathrm{T}}$.

We can do a simple extension of linear maps with scalar coefficients over $\mathbb{F}_M$ by grouping both component and H-message M-ary symbols into groups of m symbols. The resulting elements of the vectors and also the scaling coefficients then become *extended GF* $\mathbb{F}_{M^m}$; however, the overall linear formalism and notation remains. Notice that preserving *scalar* scaling coefficients, whether in $\mathbb{F}_M$ or $\mathbb{F}_{M^m}$, allows a simple implementation of linear isomorphic NCM (Section 4.7.2).

Vector-Wise Maps

On the other hand, the $\tilde{\mathbf{b}} \mapsto \mathbf{b}$ linear mapping does not have to have a diagonalized symbol-wise structure. In this case we talk about a *vector-wise* linear map with general matrix structure

$$\mathbf{b} = \sum_{k=1}^{K} \mathbf{Q}_k \mathbf{b}_k = [\mathbf{Q}_1, \ldots, \mathbf{Q}_K]\tilde{\mathbf{b}} = \tilde{\mathbf{Q}}\tilde{\mathbf{b}} \tag{4.87}$$

where $\mathbf{Q}_k \in \mathbb{F}_M^{N_b \times N_b}$ and $\tilde{\mathbf{Q}} \in \mathbb{F}_M^{N_b \times (KN_b)}$. In a special case of symbol-wise maps, it is $\mathbf{Q}_k = q_k \mathbf{I}_{N_b}$.

Here, the component matrices $\mathbf{Q}_k$ are assumed to have a *square* shape. It means that they capture only a *combination* of the messages and do *not* perform any compression, error protection redundancy encoding, or other extended cardinality related features. The non-square $\mathbf{Q}_k$ matrices would open the way to incorporate a *discrete* GF coding/compression in a classical discrete NC sense.

Maps for Multiple Relays in One Stage

The situation can be now extended to multiple relays receiving a *common* set of component messages $\tilde{\mathbf{b}} = [\mathbf{b}_1^{\mathrm{T}}, \ldots, \mathbf{b}_K^{\mathrm{T}}]^{\mathrm{T}}$ in one stage. HNC messages $\mathbf{b}'_j, j \in [1 : K']$, at K' relays will be concatenated into $\tilde{\mathbf{b}}' = [\mathbf{b}_1'^{\mathrm{T}}, \ldots, \mathbf{b}_{K'}'^{\mathrm{T}}]^{\mathrm{T}}$. This one-stage multiple relay map captures the hierarchical relationship in one processing step from $\tilde{\mathbf{b}}$ to $\tilde{\mathbf{b}}'$. Of course, the messages $\tilde{\mathbf{b}}$ can themselves be the result of previous hierarchical stage steps.

The whole idea stands on a block-structured matrix $\tilde{\mathbf{Q}}$ where the block rows correspond to individual receiving nodes (relays) and the block columns correspond to transmitted component messages

$$
\underbrace{\begin{bmatrix} \mathbf{b}'_1 \\ \vdots \\ \mathbf{b}'_{K'} \end{bmatrix}}_{\tilde{\mathbf{b}}'} = \underbrace{\begin{bmatrix} \mathbf{Q}_{11} & \cdots & \mathbf{Q}_{1K} \\ \vdots & \ddots & \vdots \\ \mathbf{Q}_{K'1} & \cdots & \mathbf{Q}_{K'K} \end{bmatrix}}_{\tilde{\mathbf{Q}}} \underbrace{\begin{bmatrix} \mathbf{b}_1 \\ \vdots \\ \mathbf{b}_K \end{bmatrix}}_{\tilde{\mathbf{b}}}
\tag{4.88}
$$

where $\mathbf{Q}_{jk} \in \mathbb{F}_M^{N_b \times N_b}$ and $\tilde{\mathbf{Q}} \in \mathbb{F}_M^{(K'N_b) \times (KN_b)}$. In a special case of symbol-wise maps, it is $\mathbf{Q}_{jk} = q_{jk}\mathbf{I}_{N_b}$ where q_{jk} is a combination coefficient between kth source and jth receive node.

4.7.2 HNC Maps for Linear Isomorphic Layered NCM

The HNC map design has, apart from its impact on the H-constellation itself, another equally important aspect. It is the mutual relation between the H-message map and the H-codeword map required for the isomorphic layered NCM. The isomorphic layered NCM requires a uniquely defined one-to-one mapping between H-message and H-codeword (or H-message symbol and H-code symbol). However, its particular *form* can vary.

Linear mappings are an important special class of isomorphic layered NCM and they are formed by component codes and corresponding message and code HNC maps. Particularly, in the case when all involved mapping (component codes, message HNC map, and code HNC map) is linear, one can expect a relatively straightforward solution. In a basic concept, i.e. the linear system preserves the linear combination of excitations as a linear combination of responses, which is of course the core and fundamental principle, the situation might seem easy. However, there are some aspects that make the situation more complicated. Particularly, the involved alphabets or message/code spaces do *not* need to be identical and even might *not* share a common alphabet.

DEFINITION 4.15 (Linear Isomorphic Layered NCM)   Assume an isomorphic layered NCM with component codes $c_k = C_k(b_k)$ and message and code HNC maps $b = \chi(\tilde{b})$ and $c = \chi_c(\tilde{c})$ respectively. If *all* involved component codes C_k and maps χ, χ_c are linear, we call this NCM *Linear Isomorphic Layered NCM*.

Symbol-Wise Maps Over Common GF

An almost trivial case is the one with *identical* component *linear* (N_b, N) codes and *identical* $q_k = a_k$ linear *symbol-wise* HNC maps all defined on a *common GF*, i.e.

$$
\mathbf{c}_k = C(\mathbf{b}_k) = \mathbf{G}\mathbf{b}_k, \quad \mathbf{b}_k \in \mathbb{F}_M^{N_b}, \quad \mathbf{c}_k \in \mathbb{F}_M^N, \quad \forall k \in [1:K],
\tag{4.89}
$$

$$
\mathbf{b} = \sum_{k=1}^{K} a_k \mathbf{b}_k,
\tag{4.90}
$$

$$
\mathbf{c} = \sum_{k=1}^{K} a_k \mathbf{c}_k,
\tag{4.91}
$$

where the *scalar* (which corresponds to symbol-wise maps) coefficients are $a_k \in \mathbb{F}_M$ and all operations are on GF. Using straightforward properties of GF arithmetic, we get

$$
\begin{aligned}
\mathbf{c} &= \sum_{k=1}^{K} a_k \mathbf{G} \mathbf{b}_k \\
&= \mathbf{G} \sum_{k=1}^{K} a_k \mathbf{b}_k
\end{aligned}
\tag{4.92}
$$

and the isomorphic relation clearly holds $\mathbf{c} = \mathbf{G}\mathbf{b}$.

Notice that this result required a *commutativity* of map coefficient *multiplication*. This is trivially guaranteed for a *scalar* coefficient on GF but *cannot* be extended to general vector-wise maps using matrix operations.

Maps for Nested Lattice NCM

Linear HNC maps over nested lattice codes used in CF NCM strategy (Section 5.6) are an important case of maps defined over different spaces. CF is a particular form of isomorphic layered NCM where all component codes are identical (see also Footnote 7 in Section 5.6) nested lattice codes $\mathbf{c}_k = \mathcal{C}_0(\mathbf{b}_k)$ and we define the code HNC map as a linear combination over nested lattice codewords

$$
\mathbf{c} = \left(\sum_{k=1}^{K} a_k \mathbf{c}_k \right) \bmod \Lambda_s
\tag{4.93}
$$

where $a_k \in \mathbb{Z}_j$ are *complex integers*.

In order to have an *isomorphic* CF, we need to find whether, and in what form, there exists a corresponding HNC map for the messages $\mathbf{b} = \chi(\tilde{\mathbf{b}})$. A quite generic algebraic approach based on finding a proper form of lattice generating matrices allowing us to build isomorphic relation is derived in [17] (particularly the core results in Section V.5 and Theorem 6 therein). Another possibility (which originally appeared in [45]) relies heavily on a very specific form of nested lattice code, *Construction A* (see Section A.5), but in turn it is quite simple, as will be shown in the next.

The HNC map design relies on the linearity property of Construction A nested lattice code. The lattice codebook is constructed according to (A.143) with a modification for complex lattices

$$
\mathcal{C}_0 = \left\{ \mathbf{c}_k : \mathbf{c}_k = \left(\mathbf{G}_s \left(\frac{1}{M} \mathbf{G}_0 \mathbf{b}_k + \mathbf{z}_k \right) \right) \bmod \Lambda_s, \ \mathbf{b}_k \in \mathbb{F}_M^{N_b}, \ \mathbf{z}_k \in \mathbb{Z}_j^N \right\}
\tag{4.94}
$$

where $\mathbf{G}_0$ is an underlying linear block code on GF $\mathbb{F}_M$ and $\mathbf{G}_s$ is a generating matrix of the coarse lattice Λ_s. The generating matrix is invertible and we express codewords in the derotated space as $\mathbf{c}_k' = \mathbf{G}_s^{-1} \mathbf{c}_k$. Clearly, the shaping lattice in this derotated space is $\Lambda_s' = \mathbb{Z}_j^N$ and

$$
\mathbf{c}_k' = \left(\frac{1}{M} \mathbf{G}_0 \mathbf{b}_k + \mathbf{z}_k \right) \bmod \mathbb{Z}_j^N.
\tag{4.95}
$$

The HNC map is

$$
\mathbf{c} = \left(\sum_{k=1}^{K} a_k \mathbf{G}_s \mathbf{c}'_k \right) \bmod \Lambda_s
$$

$$
= \left(\mathbf{G}_s \sum_{k=1}^{K} a_k \mathbf{c}'_k \right) \bmod \Lambda_s
$$

$$
= \mathbf{G}_s \left(\left(\sum_{k=1}^{K} a_k \mathbf{c}'_k \right) \bmod \mathbb{Z}_j^N \right) \tag{4.96}
$$

where the derotated HNC map is

$$
\mathbf{c}' = \left(\sum_{k=1}^{K} a_k \mathbf{c}'_k \right) \bmod \mathbb{Z}_j^N . \tag{4.97}
$$

The analysis can thus be done equivalently for the derotated maps. Then we use the properties of mod operation (see Appendix A) and

$$
\mathbf{c}' = \left(\sum_{k=1}^{K} a_k \left(\left(\frac{1}{M} \mathbf{G}_0 \mathbf{b}_k + \mathbf{z}_k \right) \bmod \mathbb{Z}_j^N \right) \right) \bmod \mathbb{Z}_j^N
$$

$$
= \left(\sum_{k=1}^{K} \left(a_k \frac{1}{M} \mathbf{G}_0 \mathbf{b}_k \right) \bmod \mathbb{Z}_j^N \right) \bmod \mathbb{Z}_j^N
$$

$$
= \frac{1}{M} \left(\sum_{k=1}^{K} (a_k \mathbf{G}_0 \mathbf{b}_k) \bmod M \mathbb{Z}_j^N \right) \bmod M \mathbb{Z}_j^N
$$

$$
= \frac{1}{M} \left(\sum_{k=1}^{K} a_k \mathbf{G}_0 \mathbf{b}_k \right) \bmod M \mathbb{Z}_j^N
$$

$$
= \frac{1}{M} \left(\mathbf{G}_0 \sum_{k=1}^{K} a_k \mathbf{b}_k \right) \bmod M \mathbb{Z}_j^N
$$

$$
= \frac{1}{M} \left(\mathbf{G}_0 \sum_{k=1}^{K} (a_k \bmod M) \, \mathbf{b}_k \right) \bmod M \mathbb{Z}_j^N . \tag{4.98}
$$

Clearly we can define the H-message HNC map

$$
\mathbf{b} = \sum_{k=1}^{K} q_k \mathbf{b}_k, \quad \text{where } q_k = a_k \bmod M \tag{4.99}
$$

and the derotated H-codeword corresponds to the H-message

$$\mathbf{c}' = \frac{1}{M} \left(\mathbf{G}_0 \mathbf{b} \right) \bmod M\mathbb{Z}_j^N$$

$$= \left(\frac{1}{M} \mathbf{G}_0 \mathbf{b} \right) \bmod \mathbb{Z}_j^N. \tag{4.100}$$

We see that the resulting scheme is an isomorphic layered NCM with the H-message map being linear with modulo mapping to the original CF coefficients.

5 WPNC in Cloud Communications

5.1 Introduction

This chapter focusses on a *global whole-network perspective* of WPNC networks. It is in contrast with Chapter 4, which rather focussed on individual building blocks and a local neighborhood of the node. The WPNC network is a cloud network that serves its outer terminals/nodes (sources and destinations) with a communication service that does not necessarily need to reveal all its internals to the terminal nodes, hence the name *WPNC cloud*.

The quality of the WPNC cloud service has two basic aspects: (1) delivering the desired information to the given destination node, and (2) quantitative performance aspects of this delivery. Both aspects have to respect various network constraints and design decisions, e.g. half-duplex constraint, relay node strategies, channel parametrization, etc. From a practical perspective, we usually prefer to build a complex WPNC cloud from the smaller building blocks sharing some common relay strategy. Following the *hierarchical* principle, the smallest building blocks are the ones given by a node interacting with its direct local radio neighborhood as described in Chapter 4.

We first show how to build the cloud hierarchically from the local components while making sure that the desired information from the sources makes its path to its destination. Second, we focus on the overall end-to-end performance metric and information-theoretic limits. This will include information-theoretic general bounds and also a performance assessment of some particular encoding strategies, namely Noisy Network Coding, Compute and Forward, and Hierarchical Decode and Forward of Layered NCM. Finally, we analyze conditions of end-to-end solvability of HNC maps including the H-processing operation aspects.

5.2 Hierarchical Structure and Stages of Wireless Cloud

5.2.1 Hierarchical Network Transfer Function

Let us assume that the building blocks of the WPNC cloud are *half-duplex constrained* relays performing their front-end strategy (multiple-stage H-MAC), back-end strategy (H-BC), and the relay processing strategy described by the HNC map χ and the associated information measure $\mu(x)$ (see Chapter 3). We assume that we are also given a connectivity map of the network. The goal is to develop a technique that will allow us to find the whole encapsulation hierarchy of the information flow between the source and its target destination respecting all involved HNC maps, Tx activity stages, received signals participating in a given HNC map, mixed-stage flows, and potential buffering at nodes.

We will develop a *hierarchical network transfer function* using a polynomial formalism. It will be used to identify the end-to-end solvability for the information flow including all hierarchical encapsulations, in scheduling of the stages, and it will help to identify the critical bottlenecks. Since complex networks usually have a high diversity potential and the processing and scheduling provide many possible options, it will also establish the model for *optimization* of node operations.

We first develop the polynomial formalism and the hierarchical network transfer function for given and known network stage scheduling. Then we show how to use this technique for a half-duplex scheduling design [55].

Polynomial Formalism and Hierarchical Network Transfer Function

We assume a WPNC cloud network with nodes numbered by integers $\mathcal{S} = \{1, \ldots, K\}$, where K is a total number of the nodes. Sources and destinations are included in this set. The set of source nodes indices is $\mathcal{S}_S = \{i_1, \ldots, i_{K_S}\} \subset \mathcal{S}$ and the corresponding (ordered in a correspondence with the associated sources) destination nodes are $\mathcal{S}_D = \{\hat{i}_1, \ldots, \hat{i}_{K_S}\} \subset \mathcal{S}$. Indices in $\mathcal{S}_D$ can repeat if a given node is a destination for multiple sources. In a correspondence with Chapter 3 notation, we also (on top of numerical indices) denote the sources and destinations with letter indices, $S_{i_1} \equiv S_A, S_{i_2} \equiv S_B, \ldots$ and $D_{\hat{i}_1} = D_A, D_{\hat{i}_2} = D_B, \ldots$ We assume a given stage definition. Nodes participating in the ℓth stage have indices from the set $\mathcal{S}_\ell \subset \mathcal{S}, \ell \in \{1, \ldots, L\}$, where L is the number of stages.

A directed connectivity $K \times K$ matrix for the ℓth stage is denoted as $\mathbf{H}_\ell$ where columns correspond to Tx activity and rows to Rx activity. Its ith row and jth column entry $H_{\ell,ij}$ is equal to 1 if the ith node receives the signal from the jth node in the ℓth stage, otherwise it is 0. Notice that the connectivity matrix under the half-duplex constraint has zeros on the main diagonal. We also define a global directed connectivity matrix $\mathbf{H}_0$, which describes the connectivity *regardless of the stage* and its entries are

$$H_{0,ij} = \mathrm{U}\left(\sum_{\ell=1}^{L} H_{\ell,ij}\right), \tag{5.1}$$

i.e. it is equal to 1 if any of the per-stage directed connectivity elements is 1.

For each stage, we define a per-stage $K \times K$ network transfer matrix $\mathbf{G}_\ell$,

$$\mathbf{G}_\ell = W_\ell \left(\mathbf{X}_\ell \mathbf{H}_\ell \mathbf{V}_\ell + \mathbf{B}_\ell\right). \tag{5.2}$$

The polynomial formalism represents the passing of the network flow through the ℓth stage by W_ℓ variable. The $K \times K$ diagonal matrix $\mathbf{X}_\ell$ represents the event of the network flow passing through the ith receiver's HNC map in the ℓth stage by the polynomial variable $X_{\ell,i}$, i.e.

$$\mathbf{X}_\ell = \begin{bmatrix} X_{\ell,1} & 0 & \cdots & 0 \\ 0 & X_{\ell,2} & \cdots & 0 \\ \vdots & \vdots & \ddots & \vdots \\ 0 & 0 & \cdots & X_{\ell,K} \end{bmatrix}. \tag{5.3}$$

Similarly, the $K \times K$ diagonal matrix represents the transmit activity of the jth node in the ℓth stage by the polynomial variable $V_{\ell,j}$, i.e.

$$\mathbf{V}_\ell = \begin{bmatrix} V_{\ell,1} & 0 & \cdots & 0 \\ 0 & V_{\ell,2} & \cdots & 0 \\ \vdots & \vdots & \ddots & \vdots \\ 0 & 0 & \cdots & V_{\ell,K} \end{bmatrix}. \tag{5.4}$$

Both $\mathbf{X}_\ell, \mathbf{V}_\ell$ matrices have a fixed form. The left-hand matrix multiplication of $\mathbf{X}_\ell$ and $\mathbf{H}_\ell$ leaves $X_{\ell,i}$ on the ith row only if the entry of $\mathbf{H}_\ell$ matrix on the ith row indicates that node is receiving. Similarly, $V_{\ell,j}$ appears on the jth column only if the jth node is transmitting. The diagonal $K \times K$ matrix $\mathbf{B}_\ell$ represents the buffering at the jth node at stage ℓ. It adds diagonal entries on $\mathbf{G}_\ell$ matrix, i.e. the node virtually receives its own transmitted signal

$$\mathbf{B}_\ell = \begin{bmatrix} B_{\ell,1} & 0 & \cdots & 0 \\ 0 & B_{\ell,2} & \cdots & 0 \\ \vdots & \vdots & \ddots & \vdots \\ 0 & 0 & \cdots & B_{\ell,K} \end{bmatrix}. \tag{5.5}$$

In summary, the matrix $\mathbf{G}_\ell$ has non-zero entries (1) on the main diagonal $W_\ell B_{\ell,j}$ that represent buffering, and (2) on the ith row and jth column $W_\ell X_{\ell,i} V_{\ell,j}$ if node i receives the signal from node j in the ℓth stage.

Hierarchical Network Transfer Matrix and Function

DEFINITION 5.1 (Hierarchical Network Transfer Matrix and Function)   The hierarchical network transfer matrix (H-NTM) is defined as a compound network transfer matrix over all stages[1,2]

$$\mathbf{F} = \sum_{\ell=1}^{L} \prod_{m=1}^{\ell} \mathbf{G}_m = \mathbf{G}_1 + \mathbf{G}_2\mathbf{G}_1 + \cdots + \mathbf{G}_L\mathbf{G}_{L-1}\ldots\mathbf{G}_2\mathbf{G}_1. \tag{5.6}$$

The hierarchical network transfer function (H-NTF) is a multi-stage network response $\mathbf{z}$ on the excitation from the sources $\mathbf{s}$ evaluated at the selected destination indices $\mathbf{z} = [\tilde{z}_{\hat{i}_1}, \ldots, \tilde{z}_{\hat{i}_{K_S}}]^\mathsf{T}$ where destination nodes have indices $\mathcal{S}_D = \{\hat{i}_1, \ldots, \hat{i}_{K_S}\} \subset \mathcal{S}$, the full response for all nodes is

$$\tilde{\mathbf{z}} = \mathbf{F}\mathbf{s}, \tag{5.7}$$

---

[1] In the per-stage network transfer matrix product, the subsequent stages (the later stages) are left-hand matrix multiplications. However, we do not use an explicit notation for this in the product operator.

[2] We also assume a finite number L of stages and no pipelining over several repetitions of the L-stage sequence.

and $\mathbf{s}$ is the source excitation vector with entries S_i on positions $i \in \mathcal{S}_S = \{i_1, \ldots, i_{K_S}\} \subset \mathcal{S}$, otherwise the entries are zeros.

The hierarchical network transfer matrix describes the network response combining the results from all stages, e.g. $\mathbf{G}_1\mathbf{s}$ is the response after the first stage with the source excitation $\mathbf{s}$, $\mathbf{G}_2\mathbf{G}_1\mathbf{s}$ is the response after the first and the second stage, etc. The H-NTF contains complex information about the network flows. It can be simplified and interpreted in various ways as we explain later, but first, let us show a basic example of constructing H-NTF.

Example 5.1   We consider the butterfly network (Figure 3.2b) with a half-duplex constrained Tx activity exactly[3] according to Figure 3.2b. We index nodes SA, SB, R, DA, DB by $\{1, 2, 3, 5, 4\}$. Stage 1 and 2 directed connectivity matrices are

$$\mathbf{H}_1 = \begin{bmatrix} 0 & 0 & 0 & 0 & 0 \\ 0 & 0 & 0 & 0 & 0 \\ 1 & 1 & 0 & 0 & 0 \\ 1 & 0 & 0 & 0 & 0 \\ 0 & 1 & 0 & 0 & 0 \end{bmatrix}, \quad \mathbf{H}_2 = \begin{bmatrix} 0 & 0 & 0 & 0 & 0 \\ 0 & 0 & 0 & 0 & 0 \\ 0 & 0 & 0 & 0 & 0 \\ 0 & 0 & 1 & 0 & 0 \\ 0 & 0 & 1 & 0 & 0 \end{bmatrix}. \tag{5.8}$$

Per-stage transfer matrices are

$$\mathbf{G}_1 = W_1 \begin{bmatrix} 0 & 0 & 0 & 0 & 0 \\ 0 & 0 & 0 & 0 & 0 \\ V_{1,1}X_{1,3} & V_{1,2}X_{1,3} & 0 & 0 & 0 \\ V_{1,1}X_{1,4} & 0 & 0 & 0 & 0 \\ 0 & V_{1,2}X_{1,5} & 0 & 0 & 0 \end{bmatrix}, \tag{5.9}$$

$$\mathbf{G}_2 = W_2 \begin{bmatrix} 0 & 0 & 0 & 0 & 0 \\ 0 & 0 & 0 & 0 & 0 \\ 0 & 0 & 0 & 0 & 0 \\ 0 & 0 & V_{2,3}X_{2,4} & 0 & 0 \\ 0 & 0 & V_{2,3}X_{2,5} & 0 & 0 \end{bmatrix}. \tag{5.10}$$

The resulting H-NTM is defined by its columns $\mathbf{F} = [\mathbf{f}_1, \ldots, \mathbf{f}_K]$

$$\mathbf{f}_1 = \begin{bmatrix} W_1 B_{1,1} \left(1 + W_2 B_{2,1}\right) \\ 0 \\ W_1 \left(1 + W_2 B_{2,3}\right) V_{1,1}X_{1,3} \\ W_1 V_{1,1} \left(\left(1 + W_2 B_{2,4}\right) X_{1,4} + W_2 V_{2,3}X_{1,3}X_{2,4}\right) \\ W_1 W_2 V_{1,1} V_{2,3}X_{1,3}X_{2,5} \end{bmatrix}, \tag{5.11}$$

---

[3] Other scheduling schemes are possible.

$$
\mathbf{f}_2 = \begin{bmatrix} 0 \\ W_1 B_{1,2} \left(1 + W_2 B_{2,2}\right) \\ W_1 \left(1 + W_2 B_{2,3}\right) V_{1,2} X_{1,3} \\ W_1 W_2 V_{1,2} V_{2,3} X_{1,3} X_{2,4} \\ W_1 V_{1,2} \left(\left(1 + W_2 B_{2,5}\right) X_{1,5} + W_2 V_{2,3} X_{1,3} X_{2,5}\right) \end{bmatrix}, \qquad (5.12)
$$

$$
\mathbf{f}_3 = \begin{bmatrix} 0 \\ 0 \\ W_1 B_{1,3} \left(1 + W_2 B_{2,3}\right) \\ W_1 W_2 B_{1,3} V_{2,3} X_{2,4} \\ W_1 W_2 B_{1,3} V_{2,3} X_{2,5} \end{bmatrix}, \qquad (5.13)
$$

$$
\mathbf{f}_4 = \begin{bmatrix} 0 \\ 0 \\ 0 \\ W_1 B_{1,4} \left(1 + W_2 B_{2,4}\right) \\ 0 \end{bmatrix}, \qquad (5.14)
$$

$$
\mathbf{f}_5 = \begin{bmatrix} 0 \\ 0 \\ 0 \\ 0 \\ W_1 B_{1,5} \left(1 + W_2 B_{2,5}\right) \end{bmatrix}. \qquad (5.15)
$$

The source excitation vector is $\mathbf{s} = [S_1, S_2, 0, 0, 0]^{\mathrm{T}}$, $S_1 \equiv S_A, S_2 \equiv S_B$, the destination indices are $\mathcal{S}_D = \{5, 4\}$, $D_5 \equiv D_A, D_4 \equiv D_B$, and the resulting H-NTF is

$$
\mathbf{z} = \begin{bmatrix} W_1 \left(S_A W_2 V_{1,1} V_{2,3} X_{1,3} X_{2,5} + S_B V_{1,2} \left(\left(1 + W_2 B_{2,5}\right) X_{1,5} + W_2 V_{2,3} X_{1,3} X_{2,5}\right)\right) \\ W_1 \left(S_B W_2 V_{1,2} V_{2,3} X_{1,3} X_{2,4} + S_A V_{1,1} \left(\left(1 + W_2 B_{2,4}\right) X_{1,4} + W_2 V_{2,3} X_{1,3} X_{2,4}\right)\right) \end{bmatrix}
$$

$$(5.16)$$

where $\mathbf{z} = [z_A, z_B]^{\mathrm{T}}$.

The result above can be now analyzed. We have chosen to factorize and sort the polynomial according to the source variables. This allows us to see at a first glance whether and how the source data reach the destination. For example in the z_A entry, we see that source S_A reached the destination in two stages (W_1, W_2) where the stage 1 used transmitter in node 1 ($V_{1,1}$) and the receiver in node 3 ($X_{1,3}$), and the second stage used transmitter in node 3 ($V_{2,3}$) and receiver in node 5 ($X_{2,5}$). This was the only way data S_A got to the destination.

The data S_B have, however, more complicated paths. There are three ways S_B reached the destination: (1) $W_1 S_B V_{1,2} X_{1,5}$, (2) $W_1 S_B V_{1,2} W_2 B_{2,5} X_{1,5}$, and (3) $W_1 S_B V_{1,2} W_2 V_{2,3} X_{1,3} X_{2,5}$. Path (1) reached the destination within one stage and used transmit node 2 and receive node 5. Path (2) reached the destination in two stages where the first stage is common with path (1) and in stage 2 the data were buffered at node 5 ($B_{2,5}$). This particular buffering is only internal at node 5 and not transmitted anywhere. Path (3) used two stages: the first stage transmitter was node 2 and receiver was node

3, and the second stage transmitter was node 3 and receiver was node 5. Notice that the buffering of path (2) was essential for obtaining "synchronized" data epochs (i.e. both delayed by the same number of stages) from sources S_A, S_B which is in turn needed for their capability to support each other as HI or HSI. In this example it is a common term $(S_A + S_B)X_{1,3}V_{2,3}X_{2,5}W_1 W_2$ that reaches the destination in the second stage and we need S_B in that second stage to solve the HNC map for the desired S_A.

Using H-NTF to Analyze and Optimize WPNC

The complete H-NTF $\mathbf{z}$ contains complex information about the data flow in the WPNC network. For specific purposes, we can simplify it in order to reveal some specific particular properties. Also we can optimize the cloud operation by selectively switching on/off various functionalities and subsequently analyzing the result.

(1) The most important information contained in H-NTF is whether the desired source data found their way to the intended destination. This is fulfilled if the source variable appears in the corresponding destination of H-NTF. However, on its own, this does not guarantee the solvability (generalized exclusive law), it is only a necessary condition.

(2) Buffering can be switched off by simply setting all $\tilde{B} = 0$. We can also do that selectively for individual nodes and/or stages. We define no-buffering H-NTF

$$\mathbf{z}_{\tilde{B}} = \mathbf{z}|_{\tilde{B}=0}. \qquad (5.17)$$

(3) The hierarchical encapsulation of the WPNC network is revealed by identifying the HNC maps at particular nodes and particular stages where the source participates. For this purpose we evaluate

$$\mathbf{z}_{\tilde{B},X} = \mathbf{z}_{\tilde{B}}|_{\tilde{W}=1,\tilde{V}=1}, \qquad (5.18)$$

which removes the superfluous information about the transmission activity and the identification of the stages, and leaves only $X_{\ell,i}$ variables, which identify the H-MAC components processed by the ith receiver in the ℓth stage. For that purpose, we usually factorize the H-NTF w.r.t. individual $X_{\ell,i}$ variables. This clearly collects the component signals belonging to the given receiving operation.

(4) If the network has high hierarchical information path diversity, we can selectively switch off some nodes' transmission by setting $V_{\ell,j} = 0$ for a given ℓ,j

$$\mathbf{z}_{\tilde{B},X,\tilde{V}_{\ell,j}} = \mathbf{z}_{\tilde{B}}|_{\tilde{W}=1,\tilde{V}=1,V_{\ell,j}=0}. \qquad (5.19)$$

(5) In a network with high diversity of the end-to-end flows having a non-uniform number of stage activity over the paths, we need a tool allowing us to recognize the role of signals in terms of HI and HSI. Components (or sub-components) of H-NTF can be HI or HSI only if they have the same number of W_ℓ variables. This indicates that the source data come from the same epoch (defined by the stages) and therefore have a chance to support themselves (HI) or to help resolve a friendly interference (HSI). Otherwise the data come from different epochs and they are independent (if the source data are IID per stage).

Example 5.2   We continue using the butterfly network from Example 5.1. A no-buffering H-NTF is

$$\mathbf{z}_{\bar{B}} = \left[\begin{array}{c} S_B W_1 V_{1,2} X_{1,5} + W_1 W_2 \left(S_A V_{1,1} + S_B V_{1,2} \right) V_{2,3} X_{1,3} X_{2,5} \\ S_A W_1 V_{1,1} X_{1,4} + W_1 W_2 \left(S_A V_{1,1} + S_B V_{1,2} \right) V_{2,3} X_{1,3} X_{2,4} \end{array} \right], \tag{5.20}$$

and the hierarchical encapsulation is revealed by evaluating

$$\mathbf{z}_{\bar{B},X} = \left[\begin{array}{c} S_B X_{1,5} + \left(S_A + S_B \right) X_{1,3} X_{2,5} \\ S_A X_{1,4} + \left(S_A + S_B \right) X_{1,3} X_{2,4} \end{array} \right]. \tag{5.21}$$

We can see that the D_A has two-stage observations. In the first stage, only S_B is received by node 5, as we see from the term $S_B X_{1,5}$. In the second stage, the term $(S_A + S_B) X_{1,3} X_{2,5}$ shows that we received the function of S_A and S_B first processed in the first stage by HNC map $X_{1,3}$ at node 3 and then in stage 2 it was received with no other additional components by node 5.

5.2.2   Half-Duplex Constrained Stage Scheduling

The polynomial formalism of the network transfer matrices can be used for the *design* of the half-duplex constrained stage scheduling with enforced latency-critical causal sequence. There are many possible half-duplex scheduling possibilities. Apart from the half-duplex constraint, we impose additional requirements to reduce the number of possible solutions. A natural additional requirement is to minimize the *latency* while keeping the multi-stage data flow *causal*.

The following algorithm solves the half-duplexing systematically while the latency and causality is solved by enforcing an ad-hoc solution which, however, in many cases, gives the minimum latency solution. Essentially, we will identify the per-stage network transfer matrices $\mathbf{G}_\ell$ fulfilling the half-duplex constraint while enforcing the critical transmission sequence. It guarantees that the data flow on the critical path (typically the longest path) will causally find its way to the desired destination with minimal latency. The algorithm is doubly greedy in the sense that (1) all receivers that can hear transmitters on the critical path are set to the reception mode on a given stage, and (2) all transmitters that do not violate the half-duplex constraint (dictated by the previous point) are allowed to transmit. We can later switch them off selectively after analyzing and optimizing the H-NTF. Putting the greedy reception before the greedy transmission attempts to minimize the number of interacting signals in the WPNC cloud.

Half-Duplex Rx–Tx Greedy Stage Scheduling Algorithm

The synthesis of the *half-duplex Rx–Tx greedy stage scheduling procedure with enforced latency-critical causal path* is given by the following steps.

(1) **Global directed connectivity** The node global radio visibility is defined by global directed connectivity matrix $\mathbf{H}_0$ and we assume that it is known.

(2) **Minimum latency causal path** We identify the minimum latency causal path. It is the longest directed and sequentially numbered path in the network graph between any of the sources and their corresponding destinations. Thus it is the minimum number of hops if we respect only the directed connectivity regardless of the half-duplex constraint. This can be obtained by observing the source flow propagation through the network with increasing number of the hops. We observe the response

$$(\mathbf{H}_0 \mathbf{V}_m) \times \cdots \times (\mathbf{H}_0 \mathbf{V}_2)(\mathbf{H}_0 \mathbf{V}_1)\mathbf{s} \qquad (5.22)$$

with sequentially increasing $m = 1, 2, \ldots$ The smallest m (denoted by M_{min}) such that all sources find their way, at least for some $m \leq M_{min}$, to their corresponding destinations becomes the longest path ensuring causal delivery of source flow to the destination. The sequential multi-hop and causality principle also guarantees that the nodes on one individual path of given source flow (ignoring other sources and paths) are consistent with half-duplex constraint. The corresponding ordered set $\mathcal{S}_{min}$ of transmitting nodes can be easily identified from the set of variables $\{V_{1,i_1}, \ldots, V_{m,i_m}\}$ associated with the given source variable S_i. It will be the *minimum latency causal path* and it defines *mandatory* transmit activity of the nodes. If there are several of them, we randomly choose one and cross-check the end-to-end flow for all sources in the following steps. If it fails, we choose a different one until all options are exploited.

(3) **Critical sequence** The previous step, however, does *not* generally guarantee, when we later impose the half-duplex constraint, that the all other sources find their way to their destinations in the number of half-duplex hops limited by M_{min}. If this happens, we must ad-hoc choose another enforced and possibly longer sequence of the transmitting nodes (not violating the half-duplex) and cross-check that the subsequent half-duplex schedule guarantees the end-to-end flow. This sequence $\mathcal{S}_c$ of transmitting nodes will be called *enforced latency-critical causal sequence*, or simply the *critical sequence*. The critical sequence guarantees the minimum latency causal network if the step #2 succeeded, i.e. $\mathcal{S}_c = \mathcal{S}_{min}$. Otherwise the minimum latency network does not exist and the critical path becomes an ad-hoc solution. However, if the minimum latency path from the step #2 remains a subset of the critical path we get a solution which is close to the minimum latency one.

(4) **Mapping the critical sequence on stages** The critical sequence of transmitting nodes $\mathcal{S}_c = \{m_1, m_2, \ldots, m_L\}$ defines the stages. The node m_ℓ belongs to the stage $\ell \in \{1, \ldots, L\}$. It means that there is mandatory (by the critical sequence) transmission by the node m_ℓ in stage ℓ. This mandatory transmission is represented by multiplying the m_ℓth column (corresponding to the m_ℓth Tx activity) of the matrix $\mathbf{H}_0$ by the stage variable W_ℓ. The critical sequence transfer matrix is

$$\mathbf{G}_{cr} = \mathbf{H}_0 \begin{bmatrix} w_1 & 0 & \cdots & 0 \\ 0 & w_2 & \cdots & 0 \\ \vdots & \vdots & \ddots & \vdots \\ 0 & 0 & \cdots & w_K \end{bmatrix} \qquad (5.23)$$

where $w_i = W_\ell$ if $m_\ell = i$ otherwise $w_i = 0$, $i \in \{1, \ldots, K\}$, $\ell \in \{1, \ldots, L\}$. Columns of $\mathbf{G}_{cr}$ that belong to mandatory transmissions are labeled by the corresponding W_ℓ. Columns that do not participate in mandatory critical sequence transmission are set to zero. Since the critical sequence was set as a causal Tx activity sequentially mapped on the stages, each stage appears only once in the matrix $\mathbf{G}_{cr}$.

(5) **Critical sequence Rx nodes (greedy Rx)** All nodes that can receive the signals from the critical sequence are set to the receive mode in the corresponding stage. These nodes, regardless of whether they are on the critical path (i.e. greedy Rx), can be found by evaluating

$$\mathbf{r}_{cr} = \mathbf{G}_{cr}\mathbf{i} \tag{5.24}$$

where $\mathbf{i} = [1, 1, \ldots, 1]^\mathrm{T}$. The ith component of $\mathbf{r}_{cr}$ contains the sum of variables W_ℓ representing the stages received by the ith receiver from the transmitters on the critical path.

(6) **Half-duplex constrained Tx (greedy Tx)** Nodes that do not receive in the given stage are allowed to transmit in that stage (greedy Tx). The set of allowed transmission half-duplex stages is simply obtained by subtracting (in the polynomial representation) the reception vector $\mathbf{r}_{cr}$ from the vector containing all stages

$$\mathbf{v}_{hd} = \left(\sum_{\ell=1}^{L} W_\ell \right) \mathbf{i} - \mathbf{r}_{cr}. \tag{5.25}$$

The allowed Tx stages are then mapped onto the *half-duplex transfer generating matrix*

$$\tilde{\mathbf{G}} = \mathbf{H}_0 \, \mathrm{diag}(\mathbf{v}_{hd}). \tag{5.26}$$

The generating matrix has non-zero entries on the positions inherited from the directed global connectivity matrix $\mathbf{H}_0$ and each non-zero entry is a sum of W_ℓ variables representing the half-duplex consistent allowed Tx stages.

(7) **Per-stage connectivity matrices** The per-stage directed connectivity matrix $\mathbf{H}_\ell$ is simply obtained by taking the generator matrix and setting $W_{\ell'} = 0$ for all stages $\ell' \neq \ell$

$$\mathbf{H}_\ell = \tilde{\mathbf{G}}|_{W_\ell=1,\, W_{\ell'}=0 \text{ for } \ell' \neq \ell}. \tag{5.27}$$

The half-duplex Rx–Tx greedy scheduling procedure with enforced latency-critical causal path will be demonstrated on example network topologies (Figure 5.1).

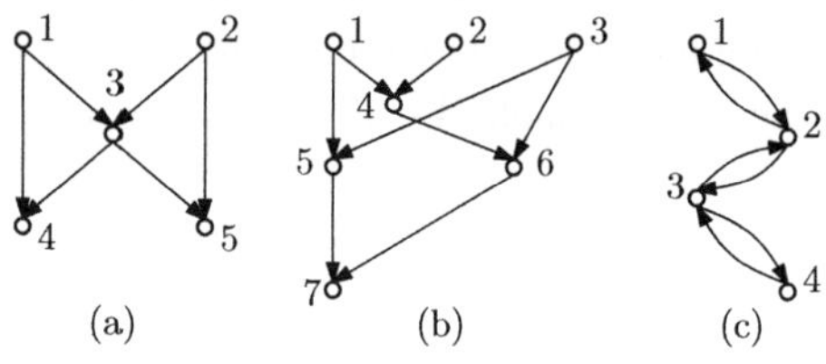

Figure 5.1 Half-duplex scheduling examples.

Example 5.3   Butterfly network (Figure 5.1a) Sources are $\mathcal{S}_S = \{1,2\}$, $\mathbf{s} = [S_1, S_2, 0, 0, 0]^T$, destinations are $\mathcal{S}_D = \{5,4\}$, $S_1 \equiv S_A, S_2 \equiv S_B, D_5 \equiv D_A, D_4 \equiv D_B$. The global directed connectivity is

$$\mathbf{H}_0 = \begin{bmatrix} 0 & 0 & 0 & 0 & 0 \\ 0 & 0 & 0 & 0 & 0 \\ 1 & 1 & 0 & 0 & 0 \\ 1 & 0 & 1 & 0 & 0 \\ 0 & 1 & 1 & 0 & 0 \end{bmatrix}. \tag{5.28}$$

The minimum latency causal path is obtained from

$$(\mathbf{H}_0\mathbf{V}_1)\mathbf{s} = \begin{bmatrix} 0 \\ 0 \\ S_1 V_{1,1} + S_2 V_{1,2} \\ S_1 V_{1,1} \\ S_2 V_{1,2} \end{bmatrix}, \tag{5.29}$$

$$(\mathbf{H}_0\mathbf{V}_2)(\mathbf{H}_0\mathbf{V}_1)\mathbf{s} = \begin{bmatrix} 0 \\ 0 \\ 0 \\ S_1 V_{1,1} V_{2,3} + S_2 V_{1,2} V_{2,3} \\ S_1 V_{1,1} V_{2,3} + S_2 V_{1,2} V_{2,3} \end{bmatrix}, \tag{5.30}$$

and clearly the minimum latency causal path (and also the critical sequence) is $\mathcal{S}_c = \{1,3\}$ (other option would be $\{2,3\}$) and the number of stages is $L = 2$.

The critical sequence transfer matrix is

$$\mathbf{G}_{\text{cr}} = \mathbf{H}_0 \begin{bmatrix} W_1 & 0 & 0 & 0 & 0 \\ 0 & 0 & 0 & 0 & 0 \\ 0 & 0 & W_2 & 0 & 0 \\ 0 & 0 & 0 & 0 & 0 \\ 0 & 0 & 0 & 0 & 0 \end{bmatrix} = \begin{bmatrix} 0 & 0 & 0 & 0 & 0 \\ 0 & 0 & 0 & 0 & 0 \\ W_1 & 0 & 0 & 0 & 0 \\ W_1 & 0 & W_2 & 0 & 0 \\ 0 & 0 & W_2 & 0 & 0 \end{bmatrix} \tag{5.31}$$

and

$$\mathbf{r}_{\text{cr}} = \mathbf{G}_{\text{cr}}\mathbf{i} = \begin{bmatrix} 0 \\ 0 \\ W_1 \\ W_1 + W_2 \\ W_2 \end{bmatrix}, \tag{5.32}$$

$$\mathbf{v}_{\text{hd}} = \left(\sum_{\ell=1}^{L} W_\ell \right)\mathbf{i} - \mathbf{r}_{\text{cr}} = \begin{bmatrix} W_1 + W_2 \\ W_1 + W_2 \\ W_2 \\ 0 \\ W_1 \end{bmatrix}. \tag{5.33}$$

The half-duplex transfer generating matrix is

$$\tilde{\mathbf{G}} = \mathbf{H}_0 \, \mathrm{diag}(\mathbf{v}_{\mathrm{hd}}) = \begin{bmatrix} 0 & 0 & 0 & 0 & 0 \\ 0 & 0 & 0 & 0 & 0 \\ W_1 + W_2 & W_1 + W_2 & 0 & 0 & 0 \\ W_1 + W_2 & 0 & W_2 & 0 & 0 \\ 0 & W_1 + W_2 & W_2 & 0 & 0 \end{bmatrix}. \tag{5.34}$$

Notice that the Rx–Tx greedy generating matrix allows additional Tx activity in comparison with Example 5.1.

Now we cross-check the end-to-end flow for all sources. The critical sequence is, in this case, also a minimum latency causal path for the source S_A; however, we need to check the source S_B end-to-end flow by evaluating H-NTF. The H-NTF is

$$z_A = S_A W_1 W_2 V_{1,1} V_{2,3} X_{1,3} X_{2,5} \tag{5.35}$$
$$+ S_B W_1 \left(W_2 B_{1,2} V_{2,2} X_{2,5} + V_{1,2} \left(\left(1 + W_2 B_{2,5}\right) X_{1,5} + W_2 V_{2,3} X_{1,3} X_{2,5} \right) \right),$$

$$z_B = S_B W_1 W_2 V_{1,2} V_{2,3} X_{1,3} X_{2,4} \tag{5.36}$$
$$+ S_A W_1 \left(W_2 B_{1,1} V_{2,1} X_{2,4} + V_{1,1} \left(\left(1 + W_2 B_{2,4}\right) X_{1,4} + W_2 V_{2,3} X_{1,3} X_{2,4} \right) \right)$$

and the no-buffering form is

$$z_{\bar{B},A} = S_A W_1 W_2 V_{1,1} V_{2,3} X_{1,3} X_{2,5} + S_B W_1 V_{1,2} \left(X_{1,5} + W_2 V_{2,3} X_{1,3} X_{2,5} \right), \tag{5.37}$$

$$z_{\bar{B},B} = S_B W_1 W_2 V_{1,2} V_{2,3} X_{1,3} X_{2,4} + S_A W_1 V_{1,1} \left(X_{1,4} + W_2 V_{2,3} X_{1,3} X_{2,4} \right). \tag{5.38}$$

In both cases, both sources reach their destinations in the second stage.

In contrast with Example 5.1, nodes 1 and 2 are also allowed to transmit in stage 2. However, a proper interpretation of z_A (and similarly for z_B) reveals that the S_B transmission at stage 2, defined by the terms containing $V_{2,2}$, in this case $S_B W_1 W_2 B_{1,2} V_{2,2} X_{2,5}$, becomes HSI w.r.t. $V_{2,3}$ only when buffering $B_{1,2}$ takes place. Only on this condition will the data S_B in $V_{2,3}$ and $V_{2,2}$ be identical. It is recognized by the presence of common term $W_1 W_2$ indicating a common "stage delay." We can selectively switch off this transmission by setting $V_{2,2} = 0$. Similarly, the term $S_B W_1 V_{1,2} W_2 B_{2,5} X_{1,5}$ becomes HSI w.r.t. $V_{2,3}$ only due to buffering $B_{2,5}$. No-buffering term $S_B W_1 V_{1,2} X_{1,5}$ on its own, w.r.t. any stage 1 activity of node 3 (e.g. when pipelining the stages), is only the interference. Also notice that Example 5.1 requires implicit buffering (at least at destinations) in order to obtain HSI.

Example 5.4   Three-source three-relay one-destination network (Figure 5.1b)
Sources are $S_S = \{1, 2, 3\}$, $\mathbf{s} = [S_1, S_2, S_3, 0, 0, 0, 0]^{\mathrm{T}}$, destinations are $S_D = \{7, 7, 7\}$, $S_1 \equiv S_A, S_2 \equiv S_B, S_3 \equiv S_C, D_7 \equiv D_A, D_7 \equiv D_B, D_7 \equiv D_C$. The global directed connectivity matrix is

$$\mathbf{H}_0 = \begin{bmatrix} 0 & 0 & 0 & 0 & 0 & 0 & 0 \\ 0 & 0 & 0 & 0 & 0 & 0 & 0 \\ 0 & 0 & 0 & 0 & 0 & 0 & 0 \\ 1 & 1 & 0 & 0 & 0 & 0 & 0 \\ 1 & 0 & 1 & 0 & 0 & 0 & 0 \\ 0 & 0 & 1 & 1 & 0 & 0 & 0 \\ 0 & 0 & 0 & 0 & 1 & 1 & 0 \end{bmatrix}. \tag{5.39}$$

The minimum latency causal path is obtained from

$$(\mathbf{H}_0\mathbf{V}_1)\mathbf{s} = \begin{bmatrix} 0 \\ 0 \\ 0 \\ S_1 V_{1,1} + S_2 V_{1,2} \\ S_1 V_{1,1} + S_3 V_{1,3} \\ S_3 V_{1,3} \\ 0 \end{bmatrix}, \tag{5.40}$$

$$(\mathbf{H}_0\mathbf{V}_2)(\mathbf{H}_0\mathbf{V}_1)\mathbf{s} = \begin{bmatrix} 0 \\ 0 \\ 0 \\ 0 \\ 0 \\ S_1 V_{1,1} V_{2,4} + S_2 V_{1,2} V_{2,4} \\ S_1 V_{1,1} V_{2,5} + S_3 \left(V_{1,3} V_{2,5} + V_{1,3} V_{2,6} \right) \end{bmatrix}, \tag{5.41}$$

$$(\mathbf{H}_0\mathbf{V}_3)(\mathbf{H}_0\mathbf{V}_2)(\mathbf{H}_0\mathbf{V}_1)\mathbf{s} = \begin{bmatrix} 0 \\ 0 \\ 0 \\ 0 \\ 0 \\ 0 \\ S_1 V_{1,1} V_{2,4} V_{3,6} + S_2 V_{1,2} V_{2,4} V_{3,6} \end{bmatrix}, \tag{5.42}$$

and we see that S_1, S_3 reach the destination in two steps while S_2 needs three steps. Then the minimum latency causal path (and also the critical sequence) is dictated by S_2 and it is $\mathcal{S}_c = \{2, 4, 6\}$ (there is no other option) and the number of stages is $L = 3$.

The critical sequence transfer matrix is

$$\mathbf{G}_{\mathrm{cr}} = \begin{bmatrix} 0 & 0 & 0 & 0 & 0 & 0 & 0 \\ 0 & 0 & 0 & 0 & 0 & 0 & 0 \\ 0 & 0 & 0 & 0 & 0 & 0 & 0 \\ 0 & W_1 & 0 & 0 & 0 & 0 & 0 \\ 0 & 0 & 0 & 0 & 0 & 0 & 0 \\ 0 & 0 & 0 & W_2 & 0 & 0 & 0 \\ 0 & 0 & 0 & 0 & 0 & W_3 & 0 \end{bmatrix} \tag{5.43}$$

and

$$\mathbf{r}_{cr} = \mathbf{G}_{cr}\mathbf{i} = \begin{bmatrix} 0 \\ 0 \\ 0 \\ W_1 \\ 0 \\ W_2 \\ W_3 \end{bmatrix}, \tag{5.44}$$

$$\mathbf{v}_{hd} = \left(\sum_{\ell=1}^{L} W_\ell\right)\mathbf{i} - \mathbf{r}_{cr} = \begin{bmatrix} W_1 + W_2 + W_3 \\ W_1 + W_2 + W_3 \\ W_1 + W_2 + W_3 \\ W_2 + W_3 \\ W_1 + W_2 + W_3 \\ W_1 + W_3 \\ W_1 + W_2 \end{bmatrix}. \tag{5.45}$$

The half-duplex transfer generating matrix is

$$\tilde{\mathbf{G}} = \begin{bmatrix} 0 & 0 & 0 & 0 & 0 & 0 & 0 \\ 0 & 0 & 0 & 0 & 0 & 0 & 0 \\ 0 & 0 & 0 & 0 & 0 & 0 & 0 \\ W_1+W_2+W_3 & W_1+W_2+W_3 & 0 & 0 & 0 & 0 & 0 \\ W_1+W_2+W_3 & 0 & W_1+W_2+W_3 & 0 & 0 & 0 & 0 \\ 0 & 0 & W_1+W_2+W_3 & W_2+W_3 & 0 & 0 & 0 \\ 0 & 0 & 0 & 0 & W_1+W_2+W_3 & W_1+W_3 & 0 \end{bmatrix}. \tag{5.46}$$

The cross-check of the end-to-end flow is simplified by the fact that all sources have a common destination and therefore only one component of H-NTF needs to be evaluated

$$\begin{aligned}
z_{ABC} = &\, S_A W_1 W_2 \Big(W_3 B_{1,1} V_{2,1} V_{3,5} X_{2,5} X_{3,7} \\
&+ V_{1,1} \left(\left(1 + W_3 B_{3,7}\right) V_{2,5} X_{1,5} X_{2,7} \right. \\
&+ W_3 \left(B_{2,5} V_{3,5} X_{1,5} + V_{2,4} V_{3,6} X_{1,4} X_{2,6} \right) X_{3,7} \Big) \Big) \\
&+ S_B W_1 W_2 W_3 V_{1,2} V_{2,4} V_{3,6} X_{1,4} X_{2,6} X_{3,7} \\
&+ S_C W_1 W_2 \Big(W_3 B_{1,3} V_{2,3} \left(V_{3,5} X_{2,5} + V_{3,6} X_{2,6} \right) X_{3,7} \\
&+ V_{1,3} \left(\left(1 + W_3 B_{3,7}\right) V_{2,5} X_{1,5} X_{2,7} + W_3 \left(B_{2,5} V_{3,5} X_{1,5} + B_{2,6} V_{3,6} X_{1,6} \right) X_{3,7} \right) \Big).
\end{aligned} \tag{5.47}$$

Clearly all three sources reach the destination and we can use the H-NTF to analyze and optimize stage activity, buffering, and HNC maps.

Example 5.5   Two-way two-relay line network (Figure 5.1c) This rather singular example will serve as a demonstration of the case where the minimum latency causal scheduling does not exist. Sources are $\mathcal{S}_S = \{1,4\}$, $\mathbf{s} = [S_1,0,0,S_4]^{\mathrm{T}}$, destinations are $\mathcal{S}_D = \{4,1\}$, $S_1 \equiv S_A$, $S_e \equiv S_B$, $D_4 \equiv D_A$, $D_1 \equiv D_B$. The global directed connectivity matrix is

$$\mathbf{H}_0 = \begin{bmatrix} 0 & 1 & 0 & 0 \\ 1 & 0 & 1 & 0 \\ 0 & 1 & 0 & 1 \\ 0 & 0 & 1 & 0 \end{bmatrix}. \tag{5.48}$$

(a) We start with the minimum latency causal path. We immediately see, even without formally evaluating the network response $(\mathbf{H}_0\mathbf{V}_m) \times \cdots \times (\mathbf{H}_0\mathbf{V}_2)(\mathbf{H}_0\mathbf{V}_1)\mathbf{s}$, that it is $\mathcal{S}_c = \{1,2,3\}$ and the other one is $\{4,3,2\}$. Owing to the full symmetry it does not matter which one is chosen, and we pick the first one. The critical sequence transfer matrix is then

$$\mathbf{G}_{\mathrm{cr}} = \begin{bmatrix} 0 & W_2 & 0 & 0 \\ W_1 & 0 & W_3 & 0 \\ 0 & W_2 & 0 & 0 \\ 0 & 0 & W_3 & 0 \end{bmatrix} \tag{5.49}$$

and

$$\mathbf{r}_{\mathrm{cr}} = \mathbf{G}_{\mathrm{cr}}\mathbf{i} = \begin{bmatrix} W_2 \\ W_1 + W_3 \\ W_2 \\ W_3 \end{bmatrix}, \tag{5.50}$$

$$\mathbf{v}_{\mathrm{hd}} = \left(\sum_{\ell=1}^{L} W_\ell\right)\mathbf{i} - \mathbf{r}_{\mathrm{cr}} = \begin{bmatrix} W_1 + W_3 \\ W_2 \\ W_1 + W_3 \\ W_1 + W_2 \end{bmatrix}. \tag{5.51}$$

The half-duplex transfer generating matrix is

$$\tilde{\mathbf{G}} = \begin{bmatrix} 0 & W_2 & 0 & 0 \\ W_1 + W_3 & 0 & W_1 + W_3 & 0 \\ 0 & W_2 & 0 & W_1 + W_2 \\ 0 & 0 & W_1 + W_3 & 0 \end{bmatrix}. \tag{5.52}$$

The end-to-end flow cross-check reveals that even with arbitrary buffering the source S_B does *not* find its way to its destination

$$z_A = S_A W_1 W_2 W_3 V_{1,1} V_{2,2} V_{3,3} X_{1,2} X_{2,3} X_{3,4} + S_B W_1 \Big(W_2 W_3 B_{2,3} V_{1,4} V_{3,3} X_{1,3} X_{3,4}$$

$$+ B_{1,4} \left(1 + W_2 \left(B_{2,4} \left(1 + W_3 B_{3,4}\right) + W_3 V_{2,4} V_{3,3} X_{2,3} X_{3,4}\right)\right)\Big), \tag{5.53}$$

$$z_B = S_A W_1 \left(B_{1,1} \left(1 + W_2 B_{2,1} \left(1 + W_3 B_{3,1}\right)\right)\right.$$

$$\left. + W_2 \left(1 + W_3 B_{3,1}\right) V_{1,1} V_{2,2} X_{1,2} X_{2,1}\right). \tag{5.54}$$

(b) The critical path needs to be adjusted ad hoc. We extend it to be $\mathcal{S}_c = \{1,2,3,4\}$. Obviously, the system is not minimum latency any more, since there is one additional stage that adds one epoch delay to the path for the source S_1. However, the system now gains the end-to-end flow also for the other source S_4 as can be seen in the following. The critical sequence transfer matrix is then

$$
\mathbf{G}_{\mathrm{cr}} = \begin{bmatrix} 0 & W_2 & 0 & 0 \\ W_1 & 0 & W_3 & 0 \\ 0 & W_2 & 0 & W_4 \\ 0 & 0 & W_3 & 0 \end{bmatrix} \tag{5.55}
$$

and

$$
\mathbf{r}_{\mathrm{cr}} = \mathbf{G}_{\mathrm{cr}}\mathbf{i} = \begin{bmatrix} W_2 \\ W_1 + W_3 \\ W_2 + W_4 \\ W_3 \end{bmatrix}, \tag{5.56}
$$

$$
\mathbf{v}_{\mathrm{hd}} = \left(\sum_{\ell=1}^{L} W_\ell \right) \mathbf{i} - \mathbf{r}_{\mathrm{cr}} = \begin{bmatrix} W_1 + W_3 + W_4 \\ W_2 + W_4 \\ W_1 + W_3 \\ W_1 + W_2 + W_4 \end{bmatrix}. \tag{5.57}
$$

The half-duplex transfer generating matrix is

$$
\tilde{\mathbf{G}} = \begin{bmatrix} 0 & W_2 + W_4 & 0 & 0 \\ W_1 + W_3 + W_4 & 0 & W_1 + W_3 & 0 \\ 0 & W_2 + W_4 & 0 & W_1 + W_2 + W_4 \\ 0 & 0 & W_1 + W_3 & 0 \end{bmatrix}. \tag{5.58}
$$

The H-NTF is

$$
\begin{aligned}
z_A = {} & S_A W_1 W_2 W_3 \left(1 + W_4 B_{4,4} \right) V_{1,1} V_{2,2} V_{3,3} X_{1,2} X_{2,3} X_{3,4} \\
& + S_B W_1 \Big(W_2 W_3 B_{2,3} \left(1 + W_4 B_{4,4} \right) V_{1,4} V_{3,3} X_{1,3} X_{3,4} \\
& + B_{1,4} \Big(1 + W_2 \Big(B_{2,4} \left(1 + W_3 B_{3,4} \left(1 + W_4 B_{4,4} \right) \right) \\
& + W_3 \left(1 + W_4 B_{4,4} \right) V_{2,4} V_{3,3} X_{2,3} X_{3,4} \Big) \Big) \Big),
\end{aligned} \tag{5.59}
$$

$$
\begin{aligned}
z_B = {} & S_B W_1 W_2 W_3 W_4 V_{3,3} V_{4,2} \left(B_{2,3} V_{1,4} X_{1,3} + B_{1,4} V_{2,4} X_{2,3} \right) X_{3,2} X_{4,1} \\
& + S_A W_1 \Big(B_{1,1} \left(1 + W_2 B_{2,1} \left(1 + W_3 \left(B_{3,1} \left(1 + W_4 B_{4,1} \right) + W_4 V_{3,1} V_{4,2} X_{3,2} X_{4,1} \right) \right) \right) \\
& + W_2 V_{1,1} X_{1,2} \Big(W_3 W_4 B_{2,2} B_{3,2} V_{4,2} X_{4,1} \\
& + V_{2,2} \Big(W_3 W_4 V_{3,3} V_{4,2} X_{2,3} X_{3,2} X_{4,1} \\
& + X_{2,1} \left(1 + W_3 \left(B_{3,1} \left(1 + W_4 B_{4,1} \right) + W_4 V_{3,1} V_{4,2} X_{3,2} X_{4,1} \right) \right) \Big) \Big) \Big)
\end{aligned} \tag{5.60}
$$

and clearly both sources get to their destinations.

5.3   Information-Theoretic Limits

5.3.1   Information-Theoretic Assessment of WPNC

An information-theoretic assessment of WPNC gives us guidelines of the idealized (in the information-theory sense) performance and also gives us valuable hints and directions for how to design practical coding schemes that mimic the idealized behavior. The performance assessment can be generally divided into the outer bounds on the information rates and inner achievable rate bounds for some given coding and decoding strategies (see Figure A.2 and also the background material in Section A.4). An example of the outer bound is the cut-set bound. Unfortunately we do not know generally under what conditions, and whether or not, it is achievable. In the context of WPNC, we will analyze some examples of the coding and decoding strategies that form the inner bounds. In particular, we focus on (a) Noisy Network Coding as a very generic concept, (b) the Compute and Forward technique which heavily utilizes lattice coding concepts, and finally (c) an HDF strategy with layered NCM, which leads to quite easily applicable practical schemes. We need to stress in particular that the first one is a pure information theoretic that does not lead to particular practical code design. However, it sets the lines on the playground and also gives us a number of hints that could be utilized in the practical code design.

5.3.2   Information-Theoretic System Model

Memoryless Network

The information-theoretic assessment requires a streamlined mathematically rigorous notation. For this purpose we use a traditional (see [13], [18], and Section A.4) notation used in information theory in this section. A random variable will be denoted by a capital X, its realization by a lower-case letter x, and its PDF/PMF by $p_X(x)$ or, if no ambiguity is possible, by $p(x)$. A sequence of variables will be denoted as $X^K = \{X_1, X_2, \ldots, X_K\}$. A notation $x(\mathcal{S})$ means a tuple containing all x_k such that $k \in \mathcal{S}$.

We also use a streamlined system model with all nodes simply numbered by $k \in [1 : K]$ and not particularly distinguishing their role (source, destination, relay). The total number of the nodes is K. We assume that each node has its own uniformly distributed message $b_k \in [1 : 2^{NR_k}]$ with rate R_k. The node transmits the signal s_k using $(2^{NR_k}, N)$ code (unless specified differently) and the received signal of the node is x_k. Generally, the nodes are *full duplex*. The network is assumed to be a memoryless one and with a common coded symbol alignment. It means that the nth received symbol in the sequence depends only on the nth transmitted symbol in the sequence and that all received symbols are perfectly aligned. An important consequence is the fact that we can describe the network by single symbol (single letter) properties. Most typically the sequence number corresponds to the temporal-sequence position but from the information-theoretic perspective it does not matter, it can also correspond for example to individual subcarriers in the frequency multiplex. The input–output relationship of the network is then given by symbol-wise input and output K-tuples $p(x^K | s^K)$ transmitted and received

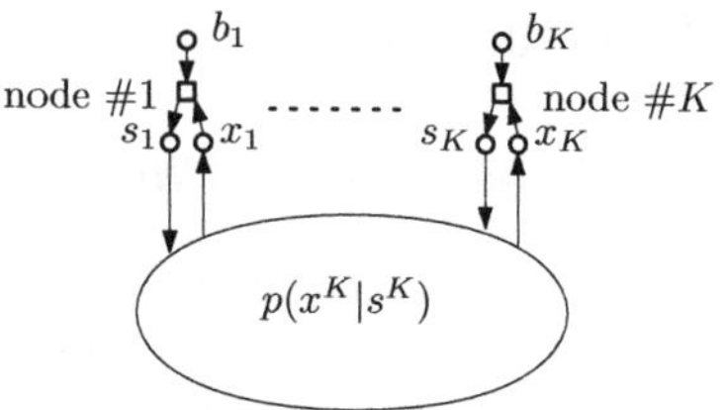

Figure 5.2 System model for information-theoretic assessment.

by the K-tuple of nodes (Figure 5.2). This conditional PDF/PMF captures both – the connectivity structure, and stochastic properties of the observations.

The networks that evolve in time or the networks that must have multiple states (e.g. half-duplex stages) are typically modeled by conditioning the model and all results by a "time sharing" random variable Q denoting individual states. The proportions of the states are defined by the PMF $p_Q(q)$. It can also model more subtle adjustments of the node strategies. Particularly, the half-duplex property in a given network state $Q = q$ and some given node k can be modeled by properly adjusting the definition of $p(x^K|s^K)$ for that given state where we make its kth output independent of the kth input. It means that there is no direct connectivity between the node's input and output.

Complex-Valued Codebooks

Many statements in information-theoretic treatment of the coding or codebooks are typically stated for real-valued codewords/codebooks. However, the majority of practical systems use complex-valued codebooks applied in the complex envelope constellation space. We show how, and under what conditions, these results are mutually related.

The complex-valued codebook means a pair of IID component codebooks for real and imaginary parts. Both component codebooks have the same size 2^{NR}. The complex-valued codeword symbol $\tilde{s}_n = (s_{R,n}, s_{I,n})$ is a two-dimensional symbol with components representing the real and imaginary parts. The codewords are drawn according to the distribution $(s_{R,n}, s_{I,n}) \sim \prod_{n=1}^{N} p(s_{R,n})p(s_{I,n})$, where $p(.)$ is a common shared PMF. The resulting complex codebook thus has a double rate $\tilde{R} = 2R$, where $2^{2NR} = 2^{N\tilde{R}}$. On the other hand, each component codebook can use only half of the resources. In the particular case of Gaussian codebooks, it means using half of the total power in each component $\sigma_{s_R}^2 = \sigma_{s_I}^2 = \sigma_{\tilde{s}}^2/2$.

Having a codebook with *IID real and imaginary components* has an important consequence – the codebook is rotationally invariant. We assume a complex valued unity-magnitude rotation (e.g. the channel rotation) $h = e^{j\varphi}$ applied on the codesymbols $\tilde{s}' = h\tilde{s}$. In the component-wise notation, we get

$$\begin{bmatrix} s'_R \\ s'_I \end{bmatrix} = \begin{bmatrix} \cos\varphi & -\sin\varphi \\ \sin\varphi & \cos\varphi \end{bmatrix} \begin{bmatrix} s_R \\ s_I \end{bmatrix}. \tag{5.61}$$

It is clearly a unitary transform. As a consequence, s'_R, s'_I are again uncorrelated and with the same second-order moment as the original s_R, s_I. If the original codeword $\tilde{s}$ is Gaussian then the rotated one, $\tilde{s}$, has exactly the same properties as the original one and can be considered as an *equivalent* one. The codebook rotated by the channel thus has the same properties as the original one and we can treat the system as if there was no rotation at all.

Notice, however, that the above statement holds only in the information-theoretic sense when the codebook is defined in terms of the codeword distribution. The codewords are mutually independent and symbol-wise IID (see also the "random codebook" principle in Section A.4). Also, the distribution must be such that it is fully defined by its second-order moments. The *Gaussian distribution* fulfills that. But practically used codes, having e.g. finite discrete-valued alphabet, or with a non-Gaussian alphabet, are still generally *dependent* on the rotation.

5.3.3　Cut-Set Bound for Multicast Network

The cut-set bound is an outer bound on the code rates. Any achievable rate must be inside the cut-set bound region; however, the bound does not need to be tight, i.e. there might be a "gap" between the capacity region and the outer cut-set bound (see more details in Section A.4). The cut-set bound is relatively easy to evaluate and sets the outer performance limits. Next, we state the cut-set bound in a form reflecting a typical WPNC network setup where each source node has *one* data message that is potentially multicasted to *multiple* destinations (see also the multi-message unicast form in Theorem A.3).[4]

THEOREM 5.2 (Cut-Set Bound for Multicast Network)　*Assume a multi-message mul-ticast memoryless network defined by $p(x^K|s^K)$ where each node has the message b_k encoded with rate R_k and the set of destination nodes for this message is $S_D(k)$. Any achievable rate K-tuple $(R_1, \ldots, R_K)$ is upper bounded by*

$$\sum_{k \in S, \bar{S} \cap S_D(k) \neq \emptyset} R_k \leq I\left(s(S); x(\bar{S})|s(\bar{S})\right) \tag{5.62}$$

for all cut-sets $S \subset [1:K]$, $\bar{S} = [1:K] \setminus S$ and for some joint input distribution $p(s^K)$.

Proof　The exact proof using Fano's inequality can be found for a common set of destinations $S_D(k) = S_D$ in [18, Section 18.1] and an extension for general multicast in [18, Section 18.4]. □

Here we provide only a high-level interpretation of the cut-set bound (see also the discussion below Theorem A.3). On the left-hand side, we sum only the rates that have the source node in S and the corresponding destination on the other side of the cut $\bar{S}$. The sum of the rates then represents the total number of messages $\prod_k 2^{NR_k} = 2^{N\sum_k R_k}$ that need to be distinguished over the cut. The mutual information on the right-hand side represents the maximum communication rate between the virtual cooperative super-transmitter $s(S)$ and cooperative super-receiver $x(\bar{S})$ with a perfect interference

---

[4]　Notice that having a specific multicast form of the cut-set bound allows us to easily find a tighter bound than the unicast form would allow. For example, one common rate R_1 to destinations 2 and 3 (both in $\bar{S}$ set) would have to be modeled as $R_{12} + R_{13}$ in the unicast form. This is clearly too strict, particularly when we adopt the super-Rx interpretation principle as explained in Section A.4.

neutralization of $s(\bar{\mathcal{S}})$. The allowed cooperation is indicated by assuming a *joint* input distribution $p(s^K)$. Intuitively, no real *non-cooperative* set of transmit and receive nodes under interference $s(\bar{\mathcal{S}})$ can exceed this rate.

5.4   Noisy Network Coding

A noisy network coding (NNC) ([36], [18]) is a particular encoding strategy defined in terms that are an information-theoretic concept rather than a practical encoding scheme. However, it allows us to find *achievable* rates for the WPNC communication. These rates are not tight (i.e. not reaching the cut-set bound) except for a very special type of the networks (e.g. a deterministic network). The NNC strategy builds on a theoretical framework using randomly generated codebooks drawn according to some given PMF and not paying any attention to the codebook internal structure. Clearly, it is not intended for a practical use – encoding/decoding would have an exponential complexity. Also the decoding relies on joint typicality decoding, which is again an information-theoretic concept for proving the coding theorems rather than a practical algorithm for constructing a decoder. Nevertheless, the NNC concept is very useful for setting the achievable rates, i.e. the inner bound of the capacity region (see Figure A.2).

5.4.1   Core Principle

The core principle of the NNC can be summarized in the following points. We start by providing a global picture that will later be described in more detail.

(1) Assume a general memoryless network with arbitrary connectivity, full-duplex nodes, and complex-valued codebooks according to the system model described in Section 5.3.2.

(2) The operation of the NNC is split into the L blocks indexed by $\ell \in [1 : L]$. Inside each block, the codewords of a common length N are exchanged among nodes.

(3) Each node has its own *data* message b_k to be sent.

(4) In each block with index $\ell \in [1 : L]$, the kth node represents (encodes, approximates) the superposition of all received signals from all other nodes (except the node's own transmitted signal) by a compression codeword $\hat{x}_k^N$. The individual codewords are indexed by the *compression* message $a_k(\ell)$. The compression codebook is designed to match the node's own message b_k and the previous compression index $a_k(\ell-1)$ used in the previous block. The compression $\hat{x}_k^N$ simply represents (approximates) the received signal x_k^N and does *not* attempt to relate it in any way to the individual data contents messages of the received signals, either individually or to any function of them.

(5) At the ℓth block, each node transmits jointly its own message b_k and a compression message obtained in the previous block $a_k(\ell - 1)$ using independent codebooks per block. The node's own message is sent *repeatedly* in L consecutive blocks. The compression message $a_k(\ell - 1)$ *varies* over the blocks depending on the signals received in the *previous* block. At the beginning, it is bootstrapped by $a_k(0) = 1$.

(6) The final destination, based on all collected received signals from all blocks, reconstructs all messages b^K and all compression indices $a^K(1), \ldots, a^K(L)$ of all the other nodes in all slots. This reconstruction can be viewed, in the perspective of this book, as finding a consistent solution matching all mutual relationships among all received signals and codewords in the network. The solution is formally performed in terms of joint typicality decoding. It can be interpreted as an information-theoretic concept of "solving the set of equations" represented by own message and compression codebooks.

In the following, we introduce individual components and processing in more rigorous detail. Although we provide deeper details here, at some points, in order to keep the explanation accessible, we relax the details, particularly those referring to the asymptotic behavior for large N used in the proofs. The reader can find the full rigor in [36], [18].

5.4.2 Block Structure

The NNC operates in a per-block sequentially activated structure. The total number of the blocks L is assumed to be large. At each block, the node transmits a codeword depending on node's own message and the compression index representing what was received by the node in the *previous* block. At the first block $\ell = 1$, the previous compression index is set to the default value $a_k(0) = 1$. Each block carries the codewords of the common length N assumed to be large to support asymptotic behavior of the code.

(1) The block structure solves the problem of propagating the information flow between an arbitrary pair of nodes when the connectivity requires more than one hop. The causal processing in the network is supported by the processing using the compressed received signal from the previous block.

(2) The node's *own* message b_k is repeatedly sent in all blocks using independent codebooks. Its information contents "dissolves" with the received signal compression messages. After a large number of block transmissions, each node will collect a large number of independently encoded mixtures of (a) all other nodes' own data messages, and (b) all other nodes' received signal compression messages. The compression messages are many-to-one compression functions of the superposed received signals coming from the individual repeatedly transmitted messages b_k.

(3) Provided that the network has full connectivity and L is large, all nodes will be able to reconstruct compression messages of all other nodes at the end of all blocks. This is because the compression messages become a part of the information flow propagating through the network. The compression messages together with the node's own repeatedly used data b_k are encoded using independent codebooks in each block. The transmission step (see below) will have to dimension the transmission codebook rate to support reliable transmission of both the own message and the compression message. At the end, all nodes thus have a compression "image" of received signals of all other nodes in some form and potentially after many subsequent "compression wrappings" performed by nodes when combining it with own data b_k.

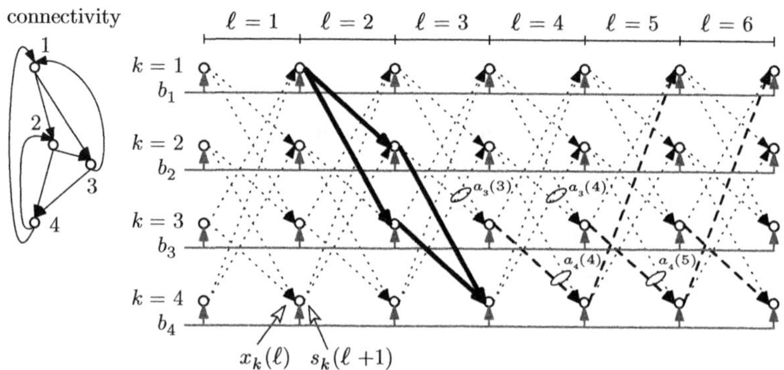

Figure 5.3 Block structure of NNC, in time-expanded form: example for a four-node network.

The block structure of NNC in time-expanded form is shown in Figure 5.3 for an example of a four-node network. Thick solid lines show the flow from node #1 to node #4 that carries the data message b_1 and the compression message $a_1(1)$. Thick dashed lines show the flow of compression message $a_3(\ell)$ wrapped inside $a_4(\ell)$ from node #3 to node #1. For example (thick solid line in Figure 5.3), we can see that the compression message $a_1(1)$ describing the received signal at node $k = 1$ in block $\ell = 1$ gets to node $k = 4$ by two paths each with two hops. In the first hop of the first path, from node #1 to node #3, it is a regular part of the combined node #1 message (containing also b_1). In the second hop of the same path, from node #3 to node #4, the received signal carrying the message $a_1(1)$ will be together with $(b_2, a_2(1))$ represented by the compression message $a_3(2)$ and sent to node #4. Also we see that the own node messages b_k are dissolved by repeated usage over large frame $L \to \infty$. They are encoded with the rates related to the overall length LN. The compression messages flow, which is the "contents" of the forwarding, is superposed on repeated own message pattern. A proper adjustment of all involved codebook rates matched to the "quality" of the observation is, of course, required. This will be discussed later.

It is also important to state that *all* nodes will have some form of compressed information about *received signals* at other nodes. But only those nodes that are in the *set of destinations* are guaranteed to have this auxiliary information of such *quality* that it suffices for reliable decoding of *data messages* themselves. The codebook rates are matched w.r.t. destination set nodes only. Also, the statement deals with the availability of the *compression* information not the *data* information.

5.4.3   Transmission Step Codebooks and Encoding

The purpose of this step is jointly to transmit data message and compression messages of the node. Each node $k \in [1 : K]$ at each block $\ell \in [1 : L]$ transmits the node's own data message $b_k \in [1 : 2^{LNR_k}]$, which does not change over the blocks, and the compression messages $a_k(\ell) \in [1 : 2^{N\hat{R}_k}]$ per each block. The compression message transmitted at the ℓth block is a result of the compression step of the previous block which describes the received signal in the previous block. The transmitted codeword in the ℓth block is $s_k^N(b_k, a_k(\ell - 1))$ and it is taken from the codebook $\mathcal{C}_{s_k}(\ell)$.

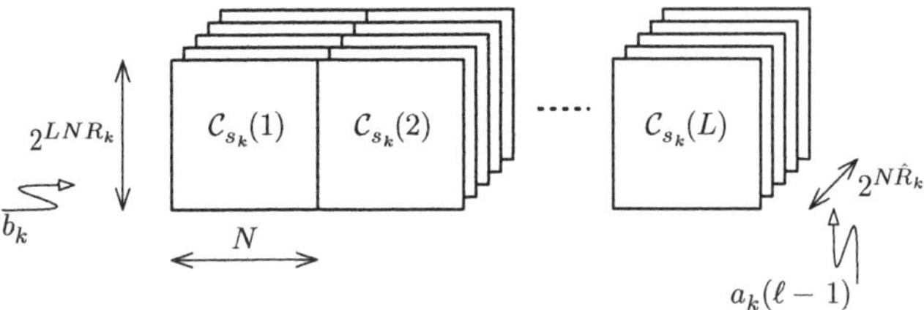

Figure 5.4 NNC – transmission step codebooks. The diagram shows codebooks each having the length N per block. The codebook has two-dimensional input $b_k \in [1 : 2^{LNR_k}]$ and $a_k(\ell - 1) \in [1 : 2^{N\hat{R}_k}]$. The former is shown as a vertical codebook size, the latter as the depth size.

The codebook is generated randomly and independently for each block according to the PMF $\prod_{n=1}^{N} p(s_{k,n})$ by generating sequences s_k^N for each $b_k \in [1 : 2^{LNR_k}]$ and $a_k(\ell - 1) \in [1 : 2^{N\hat{R}_k}]$; see Figure 5.4. The compression index initial value is set as $a_k(0) = 1$. Notice that the data message b_k is dissolved into L blocks and mixed with L compression messages $a_k(\ell)$.

5.4.4 Compression Step Codebooks and Encoding

The purpose of this step is to approximate (compress) the received signal of a node in the ℓth block given the knowledge of a node's own transmitted signal at that block. This allows a *full-duplex* operation and NNC, as an information-theoretic concept, generally assumes it. For the purpose of the received signal compression, a so-called test channel is used to model the approximation. The compression variable $\hat{x}_k$ models the actual observation x_k by a stochastic model $p(\hat{x}_k|x_k, s_k)$. The fidelity of the compression model can be adjusted by a proper choice of $p(\hat{x}_k|x_k, s_k)$ and the corresponding compression codebook. The higher the fidelity, the higher the description rate of the compression code needs to be.

The compression codebook is used to represent the received signal. The index of the codeword, i.e. the compression message, is later used together with the own message b_k to form the transmitted signal (see Figure 5.4). The compression code description rate must satisfy

$$\hat{R}_k > I(\hat{X}_k; X_k|S_k) \tag{5.63}$$

where the mutual information is evaluated for a *given* test channel $p(\hat{x}_k|x_k, s_k)$. This is dictated by a covering lemma [18]. If the condition is fulfilled then the probability of making an error in joint typicality encoding (see below) is asymptotically zero. Essentially it sets the minimum size of the compression codebook needed successfully to find the compression index a_k of codeword $\hat{x}_k^N$ approximating x_k^N given the knowledge of s_k^N. The condition guarantees that the compression codebook will be able to represent the actual observation in the fidelity modeled according to the test channel.

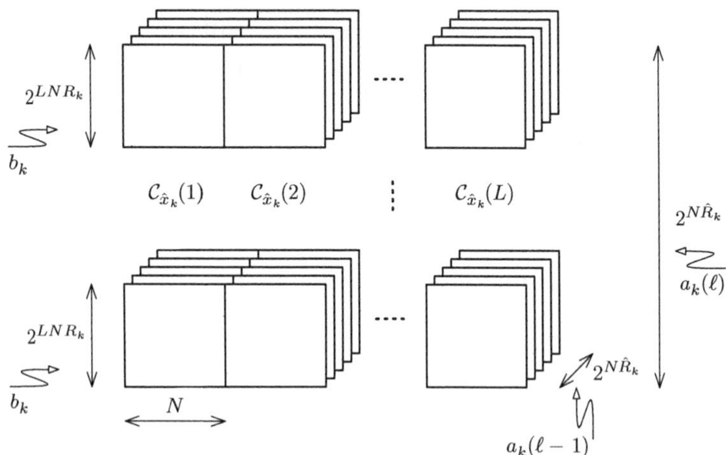

Figure 5.5 NNC – compression step codebooks. The diagram shows compression codebooks each of the length N per block. The codebook has three-dimensional input $b_k \in [1 : 2^{LNR_k}]$, $a_k(\ell-1) \in [1 : 2^{N\hat{R}_k}]$, $a_k(\ell) \in [1 : 2^{N\hat{R}_k}]$. The first is shown as a vertical codebook size in each row, the second as the depth size, and the third as a row of codebooks.

The compression codebook $\mathcal{C}_{\hat{x}}(\ell)$ (see Figure 5.5) is generated randomly and independently for each block $\ell \in [1 : L]$ according to PMF

$$\prod_{n=1}^{N} p(\hat{x}_{k,n}|s_{k,n}(b_k, a_k(\ell-1))) \tag{5.64}$$

by generating the sequences $\hat{x}_k^N$ for each compression message $a_k(\ell) \in [1 : 2^{N\hat{R}_k}]$, data message $b_k \in [1 : 2^{LNR_k}]$, and the previous compression message $a_k(\ell-1) \in [1 : 2^{N\hat{R}_k}]$. The latter two determine the own transmitted signal s_k^N. The resulting codeword is $\hat{x}_k^N(a_k(\ell)|b_k, a_k(\ell-1))$.

The compression encoding is done by finding a compression message index $a_k(\ell)$ such that the received signal, the own transmitted signal, and the compression codeword form a typical set

$$\left(x_k^N(\ell), \hat{x}_k^N(a_k(\ell)|b_k, a_k(\ell-1)), s_k^N(b_k, a_k(\ell-1))\right) \in \mathcal{T}. \tag{5.65}$$

The own transmitted signal s_k^N with the actual received signal x_k^N are "matched" by the compression signal to make them all jointly typical. The data message b_k and the compression messages $a_k(\ell-1)$ and $a_k(\ell)$ then fully represent the actual received signal x_k^N in the fidelity given by the test channel choice. Notice that, although the mutual information condition looks similar, we do not use Wyner–Ziv theorem. Particularly, we do not explicitly bin the indices. The joint typicality encoding directly finds the compression indices in the reduced rate codebook.

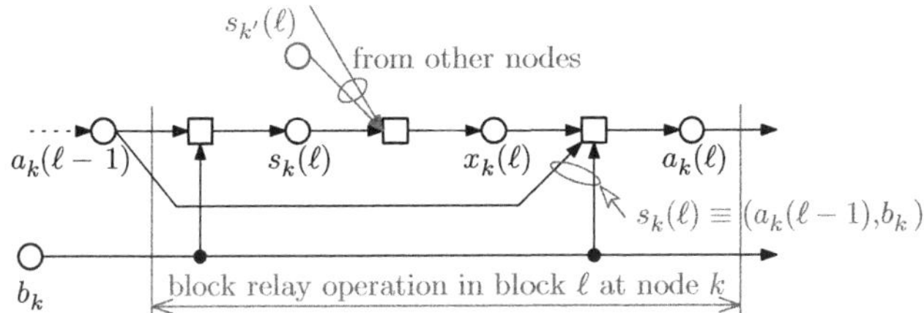

Figure 5.6 NNC – node block relay processing. The diagram shows the processing in the ℓth block. Before we start processing, we need (1) the compression message $a_k(\ell-1)$ from the previous block which represents the received signal in the previous block, and (2) the own message b_k repeatedly used in all blocks. The own message and the previous compression message uniquely determine the signal $s_k(\ell)$ transmitted in the current block. The signals from other nodes together with the own transmitted signal determine the current received signal $x_k(\ell)$. The compression message $a_k(\ell)$ is matched, in the sense of joint typicality encoding, with the received signal, given the knowledge of the own transmitted signal, i.e. only extrinsic information is used. The compression index $a_k(\ell)$ is then used together with b_k to form the transmitted signal in the next block.

5.4.5 Node Block Relay Processing

Each node combines the compression step and the transmission encoding step. The compression message $a_k(\ell)$ depends on (it is matched to) the current transmitted signal s_k, which in turn depends on a previous compression message $a_k(\ell-1)$ and the own data message b_k. The compression message $a_k(\ell)$ together with data message b_k are used to form a codeword for the next block. The relay processing thus forms a Markov chain (Figure 5.6).

5.4.6 Final Destination Decoding

The final destination decoding takes place at the end of all L blocks' relay transmissions. The jth destination node collects all observations $x_j(\ell)$, $\ell \in [1:L]$. Then it uses the joint typicality decoding (see Section A.4). The node finds estimates of all data messages $\hat{\mathbf{b}}_j = [\hat{b}_{j,1}, \ldots, \hat{b}_{j,K}]$ (the estimates might differ for different j) such that the set

$$\begin{aligned}
\Big(& s_1^N(\hat{b}_{j,1}, \hat{a}_1(\ell)), \ldots, s_K^N(\hat{b}_{j,K}, \hat{a}_K(\ell)), \\
& \hat{x}_1^N(\hat{a}_1(\ell)|\hat{b}_{j,1}, \hat{a}_1(\ell-1)), \ldots, \hat{x}_K^N(\hat{a}_K(\ell)|\hat{b}_{j,K}, \hat{a}_K(\ell-1)), \\
& x_j^N(\ell) \Big) \in \mathcal{T}
\end{aligned}$$

(5.66)

is jointly typical for all $\ell \in [1:L]$ and some estimates of the compression messages $\hat{\mathbf{a}}(\ell) = [\hat{a}_1(\ell), \ldots, \hat{a}_K(\ell)]$, $\ell \in [1:L]$.

The joint typicality decoding procedure can be viewed as a "soft" solution of the "equations" given by the *compression* and *transmission* codebooks. The side effect is that, apart from the desired data messages $\hat{\mathbf{b}}_j$, we also get a "consistent" set of compression messages. The set is consistent in the sense that it complies with all relationships imposed by the compression and transmission codebooks, but this solution does not necessarily need to be unique.

5.4.7   Achievable Rates

The NNC technique sets the inner achievable bound of the rate region that is given by the following theorem. The theorem assumes that all destinations want to decode all messages – multi-message common-multicast. The generalization for specific destination subsets $\mathcal{S}_D(k) \neq \mathcal{S}_D$ can be found in [18, Section 18.4.2].

THEOREM 5.3 (Noisy Network Coding Achievable Rates)   *In the multi-message common-multicast network described by $p(x^K|s^K)$ and having a common destination set $\mathcal{S}_D$, the data message rates $(R_1, \ldots, R_K)$ are achievable by the noisy network coding strategy if, for all $\mathcal{S} \subset [1 : K]$ such that the destination is in the complement set $\bar{\mathcal{S}} = [1 : K] \setminus \mathcal{S}$, it holds that*

$$\sum_{i \in \mathcal{S}, \bar{\mathcal{S}} \cap \mathcal{S}_D \neq \emptyset} R_i < \min_{j \in \bar{\mathcal{S}} \cap \mathcal{S}_D} I\left(S(\mathcal{S}); \hat{X}(\bar{\mathcal{S}}), X_j | S(\bar{\mathcal{S}})\right) - I\left(X(\mathcal{S}); \hat{X}(\mathcal{S}) | S^K, \hat{X}(\bar{\mathcal{S}}), X_j\right) \quad (5.67)$$

for some PMF $\prod_{k=1}^{K} p(s_k) p(\hat{x}_k | x_k, s_k)$.

Proof   The proof is in [18, Section 18.4.1].   □

The theorem is stated in a slightly simplified form, reflecting a single state of the network, i.e. the one that uses only one given strategy all the time. The more elaborate form uses time-sharing, allowing the network to have multiple states used in a proportion of the total time. Formally, it is done by conditioning all PMFs and mutual information expressions by time-sharing random variable Q with some PMF $p_Q(q)$. The time-sharing random variable allows *convexization* of the achievable rate region. For the clarity of the treatment, we drop it from the expressions and assume that the time-sharing, if needed, is done separately at the end.

The achievable rate bound is generally *not* tight to the cut-set outer bound. However, in some special cases, notably the deterministic network without the interference and deterministic GF network, the inner bound is tight [18, 36].

5.4.8   Equivalent Model

While skipping an exact proof of the NNC achievable rates, we rather focus on the interpretation of the theorem. We explain that using an equivalent model (Figure 5.7).

(1) The equivalent model is built on the core idea that the NNC network is, from the perspective of the target node j, in fact a MAC channel with dual data and compression messages. In Figure 5.7, we denote the data messages by double-head arrows and compression messages by single-head arrows. A proper interpretation of this fact allows us to reuse some facts from the classical MAC channel rates' achievability. Essentially, we explain the achievability in NNC realizing that the classical

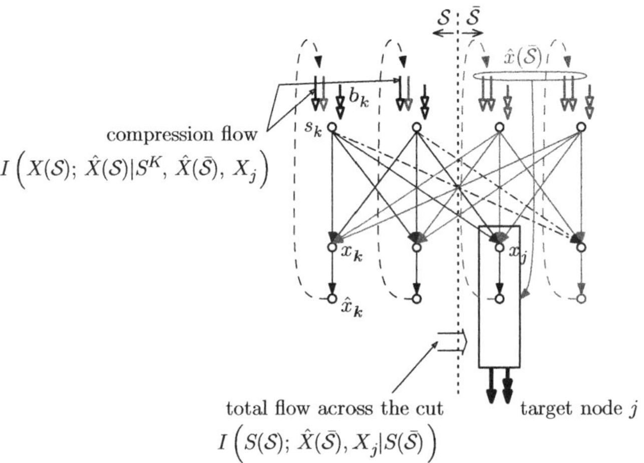

Figure 5.7 NNC – equivalent model.

MAC attains the cut-set bound and thus the achievable rates can be found by properly evaluating the information flow across the $\mathcal{S}, \bar{\mathcal{S}}$ cut. This is shown as total flow across the dotted line cut in Figure 5.7.

(2) The equivalent model is a MAC channel from the perspective of each destination node. Since all messages are supposed to get to all destination nodes, the node with the least favorable situation creates the bottleneck for the rates. This explains the $\min_j$ operation in the theorem. We will next focus on a given individual target node j. This is illustrated by solid lines to the jth node in the destination set while the others are dash-dotted.

(3) The overall rate of the combined codebook for all messages that are supposed to be decodable across the cut is given by the mutual information between sources $S(\mathcal{S})$ and the observation (details will be described in points (4) and (5)) at the target node j conditioned by all transmitted signals in the destination half of the cut, i.e. $S(\bar{\mathcal{S}})$. The signals from the destination half of the cut are shown in gray in Figure 5.7. The conditioning is equivalent to perfect interference cancellation. In Figure 5.7, all gray-color transmitted signals are assumed to be known.

(4) The observation at node j has two parts. The first part is obviously the received signal X_j on the node itself.

(5) The second part is less obvious. Compression messages represent the received signals of other nodes. As the compression messages are part of the overall message transmitted by nodes, the conditioning by $S(\bar{\mathcal{S}})$ is equivalent to the perfect knowledge of all messages in $\bar{\mathcal{S}}$, particularly the compression messages $A(\bar{\mathcal{S}})$ and consequently also $\hat{X}(\bar{\mathcal{S}})$. The compressed observations of other nodes in $\bar{\mathcal{S}}$ are thus available to node j as additional mediated observations. These compressed messages are in fact the HSI. In Figure 5.7, this is represented by gray color $\hat{x}(\bar{\mathcal{S}})$, which is available to the target node – the oval with the arrow towards the node j.

(6) The previous three points together determine the total combined (compression and data) message rate across the cut. The maximum achievable rate is

$$I\left(S(\mathcal{S}); \hat{X}(\bar{\mathcal{S}}), X_j | S(\bar{\mathcal{S}})\right). \tag{5.68}$$

(7) The total information flow comprises both data and compression messages. In order to find the maximum achievable rate for data-only messages, we need to subtract the compression rate. Mutual information $I(X(\mathcal{S}); \hat{X}(\mathcal{S}) | S^K)$ is the required compression rate describing all received signals in $\mathcal{S}$ according to the test channel model $p(\hat{x}_k | x_k, s_k)$ at each node. The conditioning by all transmitted signals S^K reflects that we are interested only in the received signal compression model while zeroing all random influences from transmitted signals. However, part of this received information is correlated with $\hat{X}(\bar{\mathcal{S}}), X_j$, which are available at the destination j as the observation (see points (4) and (5)). Therefore the description rate can be further reduced to $I(X(\mathcal{S}); \hat{X}(\mathcal{S}) | S^K, \hat{X}(\bar{\mathcal{S}}), X_j)$. This term needs to be subtracted from the total combined rate over the cut in order to obtain the rate available for the data-only messages. The compression messages, due to the above-described correlation, need to have only a reduced rate (solid black single-head arrow). The discarded rate (due to the correlation) is shown as a gray single-head arrow. Data and compression codebook rates are properly defined to represent information per one symbol and the above-stated mutual information thus correctly relates to the involved rates. Notice that the data codeword has effective length LN while the compression codeword has length N and it is duly reflected in the definition of rates.

5.4.9   Noisy Network Coding in the Perspective of WPNC

NNC is a specific form of WPNC. In the following list we comment on the major features and aspects.

(1) NNC is an *information-theoretic coding strategy* mainly used to prove achievable rate theorems. It is not a practical coding scheme.

(2) The data messages are repeatedly coded over all blocks and the number of blocks L is assumed to be large.[5] As a consequence of the large block frame, the compression messages are decodable in all destinations at the end. This can be understood as a *"flooding" schedule* with huge temporal (block) diversity. Practical WPNC schemes are likely to be much more specific and constrained in this aspect.

(3) The compression variable $\hat{x}$ is the hierarchical information measure that is related to the *observation* x and *not* to the individual messages b_i nor to the transmitted symbols $x_{i,n}$ of the component source nodes. The front-end metric is purely a function of observation $\mu = \mu(x)$ (Section 3.3.5). The node processing can be classified under *Quantize and Forward* (Section 3.3.3) where the quantization is understood in a rather generalized manner. It is performed over the whole observation x_i^N using a compression codebook. But still, it is a source coding of the observation not

---

[5] Notice, however, that there are some recent works on Short Message NNC (e.g. [23], [6]) where, while keeping the same rate, many independent short messages are sent over the blocks.

exploiting any internal channel coding structure of the signal. The back-end strategy (Section 3.3.4) is direct H-BC, where the quantized observation together with the own data message is encoded for further transmission.

5.5 Gaussian Networks

The multicast cut-set bound (Section 5.3.3) and the NNC (Section 5.4) are quite generally applicable to the network with an arbitrary input–output stochastic model $p(x^K|s^K)$. However, for the Gaussian models, we can obtain much simpler results that also allow easier interpretations including relatively straightforward numerical evaluations. This section revisits the multicast cut-set bound and the NNC in the Gaussian networks.

5.5.1 Gaussian Networks

The Gaussian network is the network where each link between the ith transmitter and the jth receiver is the memoryless linear AWGN channel. The overall input–output relationship for each symbol can thus be easily described using matrix notation

$$\mathbf{x} = \mathbf{H}\mathbf{s} + \mathbf{w} \tag{5.69}$$

where $\mathbf{s} = [s_1, \ldots, s_K]^{\mathrm{T}}$ is the vector of symbols transmitted by all nodes at time sequence index n, and similarly the received signal $\mathbf{x} = [x_1, \ldots, x_K]^{\mathrm{T}}$ and the Gaussian noise $\mathbf{w} = [w_1, \ldots, w_K]^{\mathrm{T}}$. To simplify the notation, we drop the temporal index n. The channel transfer matrix $\mathbf{H} \in \mathbb{C}^{K \times K}$ contains complex-valued link coefficients h_{ji} which are assumed to be constant. The Gaussian noise is assumed to be IID over nodes and complex-valued zero mean rotationally invariant with variance σ_w^2 per node. Transmitted signals s_i are assumed to be zero mean with mean power $E[|s_i|^2] = P$.

5.5.2 Cut-Set Bound for Multicast Gaussian Network

The input–output model (5.69) closely resembles the MIMO channel. We can directly reuse some of the MIMO channel capacity results, through interpreting and adjusting their parameters with some caution. It will allow us to get a cut-set bound for Gaussian Network by a simple adaptation of (5.62).

THEOREM 5.4 (Cut-Set Bound for Multicast Gaussian Network) *Assume a memoryless multicast Gaussian network* $\mathbf{x} = \mathbf{H}\mathbf{s} + \mathbf{w}$ *with sources s_i each having power P where each node has the message b_k encoded with rate R_k and the set of the destination nodes for this message is $\mathcal{S}_D(k)$. Any achievable rates are upper bounded by*

$$\sum_{k \in \mathcal{S}, \bar{\mathcal{S}} \cap \mathcal{S}_D(k) \neq \emptyset} R_k \leq \lg \det \left(\mathbf{I} + \frac{P}{\sigma_w^2} \mathbf{H}(\mathcal{S}) \mathbf{H}^{\mathrm{H}}(\mathcal{S}) \right) \tag{5.70}$$

for all cut-sets $\mathcal{S} \subset [1:K]$, $\bar{\mathcal{S}} = [1:K] \setminus \mathcal{S}$. The matrix $\mathbf{H}(\mathcal{S})$ denotes the punctured matrix $\mathbf{H}$ with deleted columns for all $i \notin \mathcal{S}$ and rows for $j \notin \bar{\mathcal{S}}$.

Proof We need to maximize the general form (5.62) of the mutual information $I\left(s(\mathcal{S}); x(\bar{\mathcal{S}})|s(\bar{\mathcal{S}})\right)$ for the Gaussian case. The result is a straightforward adaptation of the classical MIMO capacity result (e.g. [19]). With no channel state information on the transmitter side, the mutual information is maximized by independent Gaussian inputs with equal power. The punctured columns in matrix $\mathbf{H}(\mathcal{S})$ remove the component links corresponding to perfect interference neutralization implied by the conditioning in $I\left(s(\mathcal{S}); x(\bar{\mathcal{S}})|s(\bar{\mathcal{S}})\right)$. The punctured rows correspond to non-active receivers $x(\mathcal{S})$. $\qquad\square$

5.5.3 NNC Achievable Rates for Gaussian Network

Similarly as for the cut-set bound, the Gaussian network assumption simplifies the evaluation of the mutual information in (5.67). However, the optimizing distribution $\prod_{k=1}^{K} p(s_k)p(\hat{x}_k|x_k, s_k)$ in *not* known. Therefore we evaluate the bound for a chosen fixed distribution. In order to get an accessible result, we also lower and upper bound the mutual information expressions in (5.67). This will lead to a *more strict bound* on achievable rates but it results in a much simpler form.

We start with the compression rate term $I(X(\mathcal{S}); \hat{X}(\mathcal{S})|S^K, \hat{X}(\bar{\mathcal{S}}), X_j)$. This term will be upper bounded using the chain rule for the mutual information $I(X; Y, Z) = I(X; Y|Z) + I(X; Z)$, and the Markov chain property of the compression processing. The compression codeword is uniquely given by the observation and the transmitted signal. The mutual information between any variable and $\hat{X}(\mathcal{S})$ conditioned by $(X(\mathcal{S}), S^N)$ will be zero $I(\hat{X}(\mathcal{S}); (.)|X(\mathcal{S}), S^N) = 0$ since $\mathcal{H}\left[\hat{X}(\mathcal{S})|X(\mathcal{S}), S^N\right] = 0$ and also additional conditioning $\mathcal{H}\left[\hat{X}(\mathcal{S})|(.), X(\mathcal{S}), S^N\right]$ only reduces entropy. Using this and the chain property, we get

$$I\left(\hat{X}(\mathcal{S}); X(\mathcal{S})|S^K, \hat{X}(\bar{\mathcal{S}}), X_j\right)$$

$$= I\left(\hat{X}(\mathcal{S}); X(\mathcal{S}), S^K, \hat{X}(\bar{\mathcal{S}}), X_j\right) - I\left(\hat{X}(\mathcal{S}); S^K, \hat{X}(\bar{\mathcal{S}}), X_j\right)$$

$$= I\left(\hat{X}(\mathcal{S}); X(\mathcal{S}), S^K\right) + \underbrace{I\left(\hat{X}(\mathcal{S}); \hat{X}(\bar{\mathcal{S}}), X_j|X(\mathcal{S}), S^K\right)}_{=0} - I\left(\hat{X}(\mathcal{S}); S^K, \hat{X}(\bar{\mathcal{S}}), X_j\right)$$

$$= I\left(\hat{X}(\mathcal{S}); S^K\right) + I\left(\hat{X}(\mathcal{S}); X(\mathcal{S})|S^K\right) - I\left(\hat{X}(\mathcal{S}); S^K, \hat{X}(\bar{\mathcal{S}}), X_j\right)$$

$$= I\left(\hat{X}(\mathcal{S}); X(\mathcal{S})|S^K\right) - \underbrace{\left(I\left(\hat{X}(\mathcal{S}); S^K, \hat{X}(\bar{\mathcal{S}}), X_j\right) - I\left(\hat{X}(\mathcal{S}); S^K\right)\right)}_{=I\left(\hat{X}(\mathcal{S}); \hat{X}(\bar{\mathcal{S}}), X_j|S^K\right)\geq 0}$$

$$\leq I\left(\hat{X}(\mathcal{S}); X(\mathcal{S})|S^K\right). \tag{5.71}$$

The overall cut flow rate term $I\left(S(\mathcal{S}); \hat{X}(\bar{\mathcal{S}}), X_j|S(\bar{\mathcal{S}})\right)$ can be easily lower bounded by

$$I\left(S(\mathcal{S}); \hat{X}(\bar{\mathcal{S}}), X_j|S(\bar{\mathcal{S}})\right)$$

$$= I\left(S(\mathcal{S}); \hat{X}(\bar{\mathcal{S}})|S(\bar{\mathcal{S}})\right) + I\left(S(\mathcal{S}); X_j|\hat{X}(\bar{\mathcal{S}}), S(\bar{\mathcal{S}})\right)$$

$$\geq I\left(S(\mathcal{S}); \hat{X}(\bar{\mathcal{S}})|S(\bar{\mathcal{S}})\right). \tag{5.72}$$

The achievable rates (using the same notation as in (5.67)) are thus

$$\sum_{i\in\mathcal{S},\bar{\mathcal{S}}\cap\mathcal{S}_D\neq\emptyset} R_i < I\left(S(\mathcal{S});\hat{X}(\bar{\mathcal{S}})|S(\bar{\mathcal{S}})\right) - I\left(\hat{X}(\mathcal{S});X(\mathcal{S})|S^K\right) \tag{5.73}$$

where the right-hand side is no longer a function of the destination node j and the minimization $\min_{j\in\bar{\mathcal{S}}\cap\mathcal{S}_D}$ can be *dropped*. Notice that this expression still holds for a general input–output network stochastic model but provides a slightly simpler form. However, it is at the price of being a stricter bound than (5.67). The mutual information expressions now need to be evaluated for the Gaussian network.

First, we need to set the compression test channel model. The optimal distribution is not known. We set the test channel to be Gaussian

$$\hat{X}_k = X_k + Z_k \tag{5.74}$$

where Z_k are independent complex-valued zero-mean rotationally invariant Gaussian random variables with variance σ_z^2. The choice of Gaussian test channel is motivated by the assumed Gaussian distribution of X_k. For simplicity, we also assume a constant variance across the nodes. Some additional rate gains can be obtained by optimizing the individual variance values.

The compression mutual information term can be easily obtained by realizing that the fixation of S^K leaves the only ambiguity generated by the channel and/or the compression noise

$$\begin{aligned}
I\left(\hat{X}(\mathcal{S});X(\mathcal{S})|S^K\right) &= \mathcal{H}\left[\hat{X}(\mathcal{S})|S^K\right] - \mathcal{H}\left[\hat{X}(\mathcal{S})|X(\mathcal{S}),S^K\right] \\
&= |\mathcal{S}|\lg\left(\pi\,\mathrm{e}(\sigma_w^2+\sigma_z^2)\right) - |\mathcal{S}|\lg\left(\pi\,\mathrm{e}\,\sigma_z^2\right) \\
&= |\mathcal{S}|\lg\left(1+\frac{\sigma_w^2}{\sigma_z^2}\right).
\end{aligned} \tag{5.75}$$

The total flow mutual information term is

$$I\left(S(\mathcal{S});\hat{X}(\bar{\mathcal{S}})|S(\bar{\mathcal{S}})\right) = \mathcal{H}\left[\hat{X}(\bar{\mathcal{S}})|S(\bar{\mathcal{S}})\right] - \mathcal{H}\left[\hat{X}(\bar{\mathcal{S}})|\underbrace{S(\mathcal{S}),S(\bar{\mathcal{S}})}_{=S^K}\right]. \tag{5.76}$$

The first term $\mathcal{H}\left[\hat{X}(\bar{\mathcal{S}})|S(\bar{\mathcal{S}})\right]$ is the entropy of the compression words on the $\bar{\mathcal{S}}$ side of the cut conditioned by transmitted signals $S(\bar{\mathcal{S}})$. The randomness will be given by sources $S(\mathcal{S})$ transformed across the cut by $\mathbf{H}(\mathcal{S})$ and the observation noise $W(\bar{\mathcal{S}})$ and the compression model noise $Z(\bar{\mathcal{S}})$ on the $\bar{\mathcal{S}}$ side of the cut. We will assume Gaussian independent sources S^K. The entropy of a Gaussian complex n-dimensional zero-mean vector is $\mathcal{H}[\mathbf{U}] = \lg((\pi\,\mathrm{e})^n\det\mathrm{E}[\mathbf{U}\mathbf{U}^H])$. This gives

$$\begin{aligned}
&\mathcal{H}\left[\hat{X}(\bar{\mathcal{S}})|S(\bar{\mathcal{S}})\right] \\
&= \lg\left((\pi\,\mathrm{e})^{|\bar{\mathcal{S}}|}\det\mathrm{E}\left[(\mathbf{H}(\mathcal{S})S(\mathcal{S})+\mathbf{W}(\bar{\mathcal{S}})+\mathbf{Z}(\bar{\mathcal{S}}))(\mathbf{H}(\mathcal{S})S(\mathcal{S})+\mathbf{W}(\bar{\mathcal{S}})+\mathbf{Z}(\bar{\mathcal{S}}))^H\right]\right)
\end{aligned} \tag{5.77}$$

where $\mathbf{S}(\mathcal{S}), \mathbf{W}(\bar{\mathcal{S}}), \mathbf{Z}(\bar{\mathcal{S}})$ are correspondingly punctured vectors. The covariance matrix is

$$\mathrm{E}\left[\left(\mathbf{H}(\mathcal{S})\mathbf{S}(\mathcal{S}) + \mathbf{W}(\bar{\mathcal{S}}) + \mathbf{Z}(\bar{\mathcal{S}})\right)\left(\mathbf{H}(\mathcal{S})\mathbf{S}(\mathcal{S}) + \mathbf{W}(\bar{\mathcal{S}}) + \mathbf{Z}(\bar{\mathcal{S}})\right)^{\mathrm{H}}\right]$$
$$= P\mathbf{H}(\mathcal{S})\mathbf{H}^{\mathrm{H}}(\mathcal{S}) + \sigma_w^2\mathbf{I} + \sigma_z^2\mathbf{I}. \tag{5.78}$$

Then, we get for the first term in the total flow mutual information

$$\mathcal{H}\left[\hat{X}(\bar{\mathcal{S}})|S(\bar{\mathcal{S}})\right] = \lg\left((\pi\,\mathrm{e})^{|\bar{\mathcal{S}}|}\det\left((\sigma_w^2 + \sigma_z^2)\mathbf{I} + P\mathbf{H}(\mathcal{S})\mathbf{H}^{\mathrm{H}}(\mathcal{S})\right)\right) \tag{5.79}$$

and for the second term

$$\mathcal{H}\left[\hat{X}(\bar{\mathcal{S}})|S^K\right] = |\bar{\mathcal{S}}|\lg\left(\pi\,\mathrm{e}(\sigma_w^2 + \sigma_z^2)\right). \tag{5.80}$$

Finally, we get

$$I\left(S(\mathcal{S}); \hat{X}(\bar{\mathcal{S}})|S(\bar{\mathcal{S}})\right) = \lg\det\left(\mathbf{I} + \frac{P}{\sigma_w^2 + \sigma_z^2}\mathbf{H}(\mathcal{S})\mathbf{H}^{\mathrm{H}}(\mathcal{S})\right) \tag{5.81}$$

where we used $\det(\alpha\mathbf{A}) = \alpha^n\det\mathbf{A}$ for $\mathbf{A} \in \mathbb{C}^{n \times n}$.

THEOREM 5.5 (Noisy Network Coding Achievable Rates for Gaussian Network) *Assume memoryless Gaussian network* $\mathbf{x} = \mathbf{Hs} + \mathbf{w}$ *with independent Gaussian sources* s_i *each having power P and noise variance* σ_w^2 *per node. The data message rates* $(R_1, \ldots, R_K)$ *are achievable by noisy network coding (NNC) strategy with Gaussian compression test channels* $\hat{X}_k = X_k + Z_k$ *with variance* σ_z^2, *if for all* $\mathcal{S} \subset [1 : K]$ *such that the destination is in the complement set* $\bar{\mathcal{S}} = [1 : K] \setminus \mathcal{S}$, *it holds that*

$$\sum_{i \in \mathcal{S}, \bar{\mathcal{S}} \cap \mathcal{S}_D \neq \emptyset} R_i < \lg\det\left(\mathbf{I} + \frac{P}{\sigma_w^2\left(1 + \frac{\sigma_z^2}{\sigma_w^2}\right)}\mathbf{H}(\mathcal{S})\mathbf{H}^{\mathrm{H}}(\mathcal{S})\right) - |\mathcal{S}|\lg\left(1 + \frac{\sigma_w^2}{\sigma_z^2}\right). \tag{5.82}$$

Proof　See the derivation preceding the theorem.　　□

5.5.4　Examples

Numerical examples for NNC achievable rates and cut-set bound in a Gaussian network are shown now. It is worth noticing that the rates depend only on *relative ratios* of second-order moments of transmitted signals P, observation noise σ_w^2, and compression test channel noise σ_z^2. The cut-set bound rates depend on the ratio P/σ_w^2 and the second-order characteristics of the transfer matrix coefficients. The achievable rates on top of this also depend on the ratio σ_z^2/σ_w^2.

Example 5.6　Assume a full duplex butterfly network (Figure 3.2b) with sources $\mathcal{S}_s = \{1, 2\}$ and destinations $\mathcal{S}_D = \{4, 5\}$, and symmetric channels w.r.t. both sources. The channel transfer matrix is

$$
\mathbf{H} = \begin{bmatrix}
0 & 0 & 0 & 0 & 0 \\
0 & 0 & 0 & 0 & 0 \\
h_{\mathrm{SR}} & h_{\mathrm{SR}} & 0 & 0 & 0 \\
h_{\mathrm{SD}} & 0 & h_{\mathrm{RD}} & 0 & 0 \\
0 & h_{\mathrm{SD}} & h_{\mathrm{RD}} & 0 & 0
\end{bmatrix}. \tag{5.83}
$$

We define SNRs $\gamma_{\mathrm{SR}} = |h_{\mathrm{SR}}|^2 P/\sigma_w^2$, $\gamma_{\mathrm{SD}} = |h_{\mathrm{SD}}|^2 P/\sigma_w^2$, $\gamma_{\mathrm{RD}} = |h_{\mathrm{RD}}|^2 P/\sigma_w^2$ and relative compression test channel variance $\gamma_{zw} = \sigma_z^2/\sigma_w^2$.

Since the system is symmetric, we can evaluate only the rates for source 1. The second-order bound thus gives the bound on $R_1/2$. All first-order cut-sets are $\mathcal{S}_1 = \{\{1\},\{1,3\},\{1,3,4\},\{1,3,5\},\{1,4\},\{1,5\}\}$ and second-order cut-sets are $\mathcal{S}_2 = \{\{1,2\},\{1,2,3\},\{1,2,3,4\},\{1,2,3,5\},\{1,2,4\},\{1,2,5\}\}$. The notation for cut-sets is defined as a set of sets where the inner sets are the sets of node indices belonging to the "source" side of the cut. The resulting graphs for various settings are shown in Figure 5.8. Graph (a) has high γ_{SD} and thus models an almost perfect S–D link, i.e. it corresponds to 2WRC. Graphs (b), (c), and (d) show the impact of the compression test channel variance.

In all setups, we can see that the NNC achievable rates (solid lines) are quite far from the cut-set bound (dashed lines). In all cases, the SNR on the R–D link (shown as multiple lines parametrized by γ_{RD}) essentially hard-limits the rate, and the R–D link clearly becomes a fully saturated bottleneck. A comparison of perfect vs. non-perfect

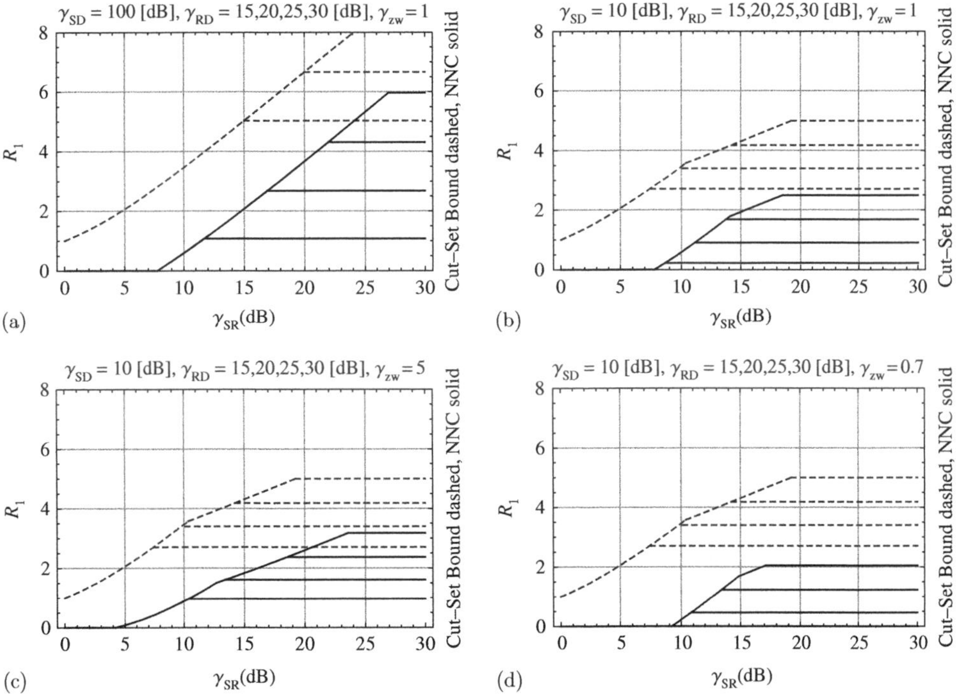

Figure 5.8 NNC in a Gaussian network – example of a butterfly network.

HSI (graph (a) vs. (b)) nicely shows how the limitation of the HSI starts to soft-limit the rates for high SNR on the S–R link (γ_{SR} on the horizontal axis). The S–R link thus carries most of the rate and the side-link S–D does not help much resolving of the self-interference of the source. The effect of the compression model noise (graphs (b), (c), and (d)) shows that high fidelity of the compression model (small γ_{zw}) consumes some rate in the S–R link (in order to reach the same rate we need more SNR). It also emphasizes the hard-limiting effect of R-D link. Both effects can be explained by a high compression message rate occupying the link. On the other hand, the low fidelity (graph (c)) clearly limits the performance on high S–R SNRs where the added compression codebook rate does not matter much but its low fidelity degrades the performance.

5.6   Compute and Forward

5.6.1   Core Principle

Compute and Forward (CF)[6] [45], [46] is a specific form of NCM based on *nested lattice codes* (Section A.5) with *decode and forward* relay strategy. Essentially, CF extends the lattice codes [63] to H-MAC fading channels and allows us to decode a linear HNC message map on the relay. Before going into more rigorous details, we explain the core principle.

The core principle of CF stands on realizing that a linear combination of nested lattice codewords is again a nested lattice codeword. They simply share a common fine lattice and the linear combination again lies on that lattice. The fundamental phenomenon is the modulo shaping (coarse) lattice operation which is applied at the receiver. When it is applied to superposed codewords as a modulo lattice operation it equivalently appears on data messages as a GF-based linear superposition with modulo coefficients. Simply speaking, the modulo lattice operation turns into modulo GF-based operation, clearly owing to a linearity of lattice code construction and distributiveness of the modulo operation. This creates an isomorphism between the H-message and the modulo lattice processed codeword superposition. Then we apply the standard principles of lattice decoding (Section A.5) essentially without any major modification.

In a slightly more detailed description, it is as follows. All component H-MAC nodes transmit lattice codes based on common fine and shaping lattices. These codes are superposed in H-MAC with some channel scalar fading coefficients. Owing to fundamental properties of the lattice code, an integer multiple of codeword modulo shaping lattice, is again a lattice codeword. A sum of codewords taken from the common fine lattice with a subsequently applied modulo shaping lattice operation also produces a valid codeword. If the fading coefficients were integers, then the scaled superposed signal with modulo shaping lattice operation applied is a valid codeword. Owing to the linearity of the lattice code, this resulting codeword would correspond to a linear combination of

---

[6] Sometimes it is also called Lattice Network Coding or Lattice Compute and Forward.

the component messages. Thus the scheme is *isomorphic* NCM and we can decode the desired H-message as if we had a single-user system.

However, the H-MAC channel fading coefficients are *not* integers. The key idea of the CF receiver is to approximate the real coefficients by the integers; or, from another viewpoint, behave as if they were integers and then minimize the after-effects of this mismatch. For that purpose, CF uses a linear single tap equalizer, which scales the received signal to minimize the *lattice mismatch* between the real received signal and the integer-scaling approximation. The mismatch is minimized in the sense of MMSE. This also allows a relatively straightforward interpretation of the residual lattice mismatch as an almost-Gaussian interference, which in turn allows determining a simple expression for the achievable rates.

In the following text, we first show a simplified motivation example following the main lines of the core principle. Then we explain the CF strategy in more detail, however for a full depth rigorous details and proofs, the reader should refer to [44], [45], [46], [63].

5.6.2 Simplified Motivation Example

This motivation example demonstrates the core principles of CF in a simplified form and with an omission of some details. Particularly, we assume only two-component H-MAC with real-valued codebooks and AWGN channels. We omit dithering and we also neglect any issues related to the shaping of the lattices.

Assume two-component H-MAC using the same nested lattice code based on the fine lattice Λ_c and transmitting with the same power P_s and at the same rate. The received signal at the relay is

$$\mathbf{x} = h_A \mathbf{c}_A + h_B \mathbf{c}_B + \mathbf{w} \tag{5.84}$$

where h_A, h_B are real-valued channel coefficients and $\mathbf{w}$ is real-valued AWGN with power P_w.

At the receiver, we apply scaling by α, quantize to lattice Λ_c by the nearest-neighbor quantizer Q_{Λ_c}, and take the result modulo coarse lattice Λ_s

$$\left(Q_{\Lambda_c}(\alpha \mathbf{x})\right) \bmod \Lambda_s. \tag{5.85}$$

The key idea is that $\mathbf{y} = \alpha \mathbf{x}$ can be approximated to the sum of integer multiples of $\mathbf{c}_A$ and $\mathbf{c}_B$, which are on the fine lattice

$$\alpha \mathbf{x} = \alpha h_A \mathbf{c}_A + \alpha h_B \mathbf{c}_B + \alpha \mathbf{w}$$
$$\approx a_A \mathbf{c}_A + a_B \mathbf{c}_B \tag{5.86}$$

where $a_A, a_B \in \mathbb{Z}$. The error of the approximation is

$$\mathbf{e} = (\alpha h_A - a_A)\mathbf{c}_A + (\alpha h_B - a_B)\mathbf{c}_B + \alpha \mathbf{w} \tag{5.87}$$

and we choose α to minimize the mean square approximation error. At the same time we can also choose a_A, a_B. These coefficients, apart from affecting the approximation fidelity, also form HNC map and must guarantee the end-to-end solvability of the network.

In a simplistic solution, we could choose them such that $a_B/a_A = h_B/h_A$ and $\alpha = a_A/h_A$ (assuming h_B/h_A is a rational number and $h_B > h_A$), but this is likely to make α large and cause the noise enhancement $\alpha\mathbf{w}$. On the other side, if we keep α small, the rounding effect of the integer approximation becomes dominant. It is clear that the optimization of the mean square error jointly depends on the coefficients a_A, a_B.

A true MMSE solution minimizes $E[\|\mathbf{e}\|^2]$ by finding such $\hat{\alpha}$ that minimizes a mismatch between the desired integer map $a_A\mathbf{c}_A + a_B\mathbf{c}_B$ and the true scaled observation including the scaled noise

$$\hat{\alpha} = \arg\min_{\alpha} E\left[\|(\alpha h_A - a_A)\mathbf{c}_A + (\alpha h_B - a_B)\mathbf{c}_B + \alpha\mathbf{w}\|^2\right]. \qquad (5.88)$$

In fact, it minimizes the lattice misalignment interference mean power in the noisy observation. The MMSE solution (postponing the details for later) is

$$\hat{\alpha} = \frac{P_s(h_A a_A + h_B a_B)}{P_s(h_A^2 + h_B^2) + P_w} \qquad (5.89)$$

and the residual lattice *misalignment-only* interference power is

$$P_i = P_s\left((\hat{\alpha} h_A - a_A)^2 + (\hat{\alpha} h_B - a_B)^2\right). \qquad (5.90)$$

Figure 5.9 shows an example of the lattice misalignment.

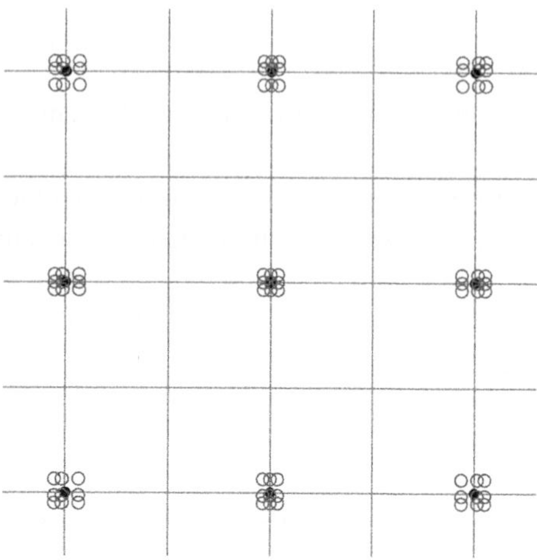

Figure 5.9 Compute and Forward – a simple example of two-source real-valued $\mathbb{Z}^2$ lattices with square shaping lattice (black points) superposed by the channel $h_A\mathbf{c}_A + h_B\mathbf{c}_B$, scaled $\alpha h_A\mathbf{c}_A + \alpha h_B\mathbf{c}_B$ and modulo shaping lattice processed by the receiver (gray circles). Numerical evaluation is shown for $h_A = 0.52$, $h_B = 1$, and $\alpha = 1.95$.

5.6.3   Nested Lattice Codebooks for H-MAC

Now we expose the CF strategy in deeper detail. We start with defining NCM codebooks used by the component nodes. All K H-MAC component sources use a common N-dimensional Λ_c-code Λ_s-shaped nested lattice $\Lambda_s \subset \Lambda_c$ codebook $\mathcal{C}_0$, $\mathbf{c}_k = \mathcal{C}_0(\mathbf{b}_k)$, $k \in [1 : K]$, $\mathbf{b}_k \in \mathbb{F}_{M_b^{N_b}}$, $M_b^{N_b} = M$, with equal powers $P_k = \frac{1}{N} \mathrm{E}[\|\mathbf{c}_k\|^2] = P_s = P(\Lambda_s)$, where we assume sufficiently dense Λ_c. A common codebook rate is defined as $R_0 = \frac{1}{N} \lg |\Lambda_c/\Lambda_s|$ where $M = |\Lambda_c/\Lambda_s|$ is the size of the quotient group, i.e. the number of cosset shifts available in a fine lattice *inside* the coarse lattice fundamental cell. In the case of some node having unequal message rates demands, we must (as a consequence of having a common lattice codebook) choose the highest rate for the common codebook and other less-demanding sources are zero-padded.[7]

In order to keep the compatibility with the assumptions used throughout the text, we allow *complex-valued* lattice codebooks with *identical properties in real and imaginary parts*. Essentially, at the information-theoretic level (not for finite small cardinality alphabets, see Section 5.3.2), we can treat it as two (real and imaginary) subspaces and it effectively doubles the rates. The arguments of Section 5.3.2 in a particular context of lattice codes mean that a complex channel coefficient rotation of the complex lattice again creates a lattice in the projection into real and imaginary axes.

The transmission strategy of individual nodes is exactly the same as for a standard nested lattice code (Section A.5). Before transmitting, the encoded lattice codewords are dithered by a continuously valued random vector $\mathbf{u}_k$ uniformly distributed over the fundamental Voronoi cell of the shaping lattice $\mathbf{u}_k \in \mathcal{V}_0(\Lambda_s)$. The transmitted signal is

$$\mathbf{s}_k = (\mathbf{c}_k + \mathbf{u}_k) \bmod \Lambda_s. \tag{5.91}$$

The dither vectors are assumed to be known to all receivers. The purpose of using dithering in lattice coding is a randomization of the data that appear in the equivalent noise of the equivalent channel, and this makes the equivalent noise independent of codewords (see Section A.5).

In order to simplify the notation, we will assume only one relay at the moment. At the end, we extend that for multiple relays. The received signal in H-MAC channel at the relay is

$$\mathbf{x} = \sum_{k=1}^{K} h_k \mathbf{s}_k + \mathbf{w} \tag{5.92}$$

---

[7] We can slightly generalize the treatment for the lattice codebooks of unequal rates. The idea stands on ordering the sources according to the rates. We create multiple nested lattice codes $\Lambda_s \subset \Lambda_{c_1} \subset \cdots \subset \Lambda_{c_K}$ with rates $R_1 \leq R_2 \leq \cdots \leq R_K$. Each source is then assigned one of these lattice codes according to its rate. The receiver might want to decode an HNC map that does *not* involve *all* sources. The sources that are not participating in the map are assigned *zero* coefficients. Then we can use the lattice quantizer $Q_{\Lambda_{c'}}$ corresponding to the finest code lattice $\Lambda_{c'}$ for which the HNC map has a *non-zero* coefficient. See [45] for details. In our treatment, we will generally assume *all non-zero coefficients*. The usage of multiply nested lattices then does not have any advantage. The quantizer will need to be the finest lattice anyway. The underlying message linear HNC map must be over the common GF, i.e. having the largest cardinality among all component messages. In this situation, the zero-padding will have the same effect.

where $\mathbf{w}$ is IID N-dimensional complex-valued Gaussian zero mean noise with variance $P_w = \frac{1}{N} \mathrm{E}\left[\|\mathbf{w}\|^2\right]$ per dimension, and $h_k \in \mathbb{C}$ are complex-valued channel coefficients and we also define $\mathbf{h} = [h_1, \ldots, h_K]^{\mathrm{T}}$.

5.6.4   H-Codeword with Complex Integer Linear HNC Map

Assume that our target desired H-codeword is a *complex integer linear* HNC map

$$\mathbf{c} = \left(\sum_{k=1}^{K} a_k \mathbf{c}_k\right) \bmod \Lambda_s \tag{5.93}$$

where $a_k \in \mathbb{Z}_\mathrm{j}$ are *complex integers*, and we also denote $\mathbf{a} = [a_1, \ldots, a_K]^{\mathrm{T}}$. The codeword $\mathbf{c} \in \mathcal{C}_0$ belongs to the same nested lattice codebook as the component codes since the complex integers are closed under multiplication and addition.

If the underlying structure of the nested lattice code is such that the NCM is *isomorphic layered NCM*, i.e. there exists a GF-based linear H-message HNC map

$$\mathbf{b} = \sum_{k=1}^{K} q_k \mathbf{b}_k \tag{5.94}$$

where $q_k \in \mathbb{F}_{M_b}$ are GF coefficients such that $\mathbf{c} = \mathcal{C}_0\,(\mathbf{b})$ is one-to-one mapping, then the codeword HNC map $\mathbf{c}$ can be used to decode the H-message $\mathbf{b}$. See Section 4.7.1 for details.

Using the properties of modulo lattice operation, we get

$$\left(\sum_{k=1}^{K} a_k \mathbf{s}_k\right) \bmod \Lambda_s = \left(\sum_{k=1}^{K} a_k \left((\mathbf{c}_k + \mathbf{u}_k) \bmod \Lambda_s\right)\right) \bmod \Lambda_s$$
$$= \left(\left(\sum_{k=1}^{K} a_k \mathbf{c}_k\right) \bmod \Lambda_s + \left(\sum_{k=1}^{K} a_k \mathbf{u}_k\right) \bmod \Lambda_s\right) \bmod \Lambda_s. \tag{5.95}$$

Clearly, the integer-linear modulo combination of transmitted signals

$$\mathbf{s} = \left(\sum_{k=1}^{K} a_k \mathbf{s}_k\right) \bmod \Lambda_s \tag{5.96}$$

appears to be

$$\mathbf{s} = (\mathbf{c} + \mathbf{u}) \bmod \Lambda_s \tag{5.97}$$

where the corresponding dither is

$$\mathbf{u} = \left(\sum_{k=1}^{K} a_k \mathbf{u}_k\right) \bmod \Lambda_s \tag{5.98}$$

and $\mathbf{u} \sim \mathcal{U}(\Lambda_s)$. It means that observing the channel combined signal with complex integer coefficients is equivalent to observing a hypothetical transmitted signal $\mathbf{s}$ carrying H-codeword $\mathbf{c}$ with the dither $\mathbf{u}$. It becomes the same as if a single user transmitted the codeword **c**. If our targets are not the individual component messages but only the H-message **b**, we can thus construct the lattice decoder as for a standard single-user case.

5.6.5 Hierarchical Euclidean Lattice Decoding

The decoding strategy, the same as for single-user lattice coding (Section A.5): (1) linearly preprocesses (by a scaling equalizer) the received signal, (2) removes dither, (3) quantizes by fine lattice quantizer, and (4) performs modulo shaping lattice operation. The only difference is that, instead of one individual node's codeword, we decode the HNC map of the codewords.

If the channel coefficients were complex integers (as in the previous section), then the received signal would look like a single-user system with an equivalent transmitted signal **s** carrying the lattice codeword **c** with equivalent dither **u**, and we could apply the standard lattice decoding to obtain the H-message **b**. However, the channel coefficients are not complex integers. The preprocessor equalizer scaling now serves a two-fold goal. It tries (a) to minimize the impact of the lattice mismatch (misalignment) among the individual component lattices, and (b) to minimize the combined second moment of the additive noise and the misalignment against the fine lattice quantizer. The first is a consequence of the fact that the channel coefficients are not integers and strongly depend on the values **a**. The second is the same as in the standard lattice decoder. Notice that the equalizing preprocessor has only one degree of freedom (one scalar scaling coefficient) for a compensation of a complex received signal imperfection structure. Section 9.3 introduces some advanced processing options.

The decision metric[8]

$$\mathbf{y} = \alpha\mathbf{x} - \mathbf{u} \tag{5.99}$$

serves for a decision on the H-codeword

$$\hat{\mathbf{c}} = \left(Q_{\Lambda_c}(\mathbf{y}) \right) \bmod \Lambda_s \tag{5.100}$$

which, under the *isomorphic* assumption, corresponds one-to-one to the H-message **b**. Since we use a complex-valued system model, the scaling coefficient is allowed to be complex $\alpha \in \mathbb{C}$.

5.6.6 Equivalent Hierarchical Modulo Lattice Channel

The receiver processing from the perspective of the H-codeword is identical with a standard lattice decoding. The only difference is in the actual observation model. Therefore

---

[8] The variable **y** is called decision metric since this is the *only* input needed to make the "hard" decision in the decoder by quantization operation $\arg\max_{\lambda \in \Lambda_c} \|\mathbf{y} - \lambda\|$ and, at the same time, it is not directly the received signal but rather its preprocessed form. In some sense, it is a lattice form of matched filter. Some authors also call the value the $\alpha\mathbf{x}$ estimator (Wiener) because it provides the preprocessing optimizing the mean square error.

the equivalent hierarchical modulo lattice channel derivation follows almost identical lines as in Section A.5. We realize that

$$\left(Q_{\Lambda_c}(\mathbf{y})\right) \bmod \Lambda_s = \left(Q_{\Lambda_c}(\mathbf{y} \bmod \Lambda_s)\right) \bmod \Lambda_s \qquad (5.101)$$

and define

$$\mathbf{y}' = \mathbf{y} \bmod \Lambda_s = (\alpha\mathbf{x} - \mathbf{u}) \bmod \Lambda_s. \qquad (5.102)$$

A substitution of the received signal and subsequent manipulations using properties of $\bmod\,\Lambda_s$ give

$$
\begin{aligned}
\mathbf{y}' &= \left(\alpha \left(\sum_{k=1}^{K} h_k(\mathbf{c}_k + \mathbf{u}_k) \bmod \Lambda_s + \mathbf{w} \right) - \mathbf{u} \right) \bmod \Lambda_s \\[2mm]
&= \left(\mathbf{c} - \mathbf{c} + \alpha \left(\sum_{k=1}^{K} h_k(\mathbf{c}_k + \mathbf{u}_k) \bmod \Lambda_s + \mathbf{w} \right) - \mathbf{u} \right) \bmod \Lambda_s \\[2mm]
&= \left(\mathbf{c} - \left(\sum_{k=1}^{K} a_k\mathbf{c}_k \right) \bmod \Lambda_s + \alpha \sum_{k=1}^{K} h_k(\mathbf{c}_k + \mathbf{u}_k) \bmod \Lambda_s \right. \\[2mm]
&\qquad\left. - \left(\sum_{k=1}^{K} a_k\mathbf{u}_k \right) \bmod \Lambda_s + \alpha\mathbf{w} \right) \bmod \Lambda_s \\[2mm]
&= \left(\mathbf{c} + \sum_{k=1}^{K} (\alpha h_k - a_k)(\mathbf{c}_k + \mathbf{u}_k) \bmod \Lambda_s + \alpha\mathbf{w} \right) \bmod \Lambda_s. \qquad (5.103)
\end{aligned}
$$

Since $\mathbf{u}_k \sim \mathcal{U}(\mathcal{V}(\Lambda_s))$ then also $(\mathbf{c}_k + \mathbf{u}_k) \bmod \Lambda_s \sim \mathcal{U}(\mathcal{V}(\Lambda_s))$ for arbitrary $\mathbf{c}_k$, and we substitute the actual dither by the equivalent one $\mathbf{u}_{k,\mathrm{eq}} = (\mathbf{c}_k + \mathbf{u}_k) \bmod \Lambda_s$ that has the same stochastic properties $\mathbf{u}_{k,\mathrm{eq}} \sim \mathcal{U}(\mathcal{V}(\Lambda_s))$ and is *independent* of $\mathbf{c}_k$. Equivalent dither is zero mean and has the same power as the transmitted signal (assuming sufficiently dense Λ_c), $\frac{1}{N}\mathrm{E}\left[\|\mathbf{u}_{k,\mathrm{eq}}\|^2\right] = P(\Lambda_s) = P_s$.

The equivalent hierarchical modulo lattice channel will have the same stochastic properties as the original one and it is expressed as

$$\mathbf{y}_{\mathrm{eq}} = \left(\mathbf{c} + \mathbf{w}_{\mathrm{eq}}\right) \bmod \Lambda_s \qquad (5.104)$$

where the equivalent noise is

$$\mathbf{w}_{\mathrm{eq}} = \sum_{k=1}^{K} (\alpha h_k - a_k)\mathbf{u}_{k,\mathrm{eq}} + \alpha\mathbf{w}. \qquad (5.105)$$

The equivalent hierarchical channel is an additive noise channel with a modulo lattice operation. The important observation is that, from the *perspective of H-codeword* $\mathbf{c}$, the channel looks like a *standard lattice coding equivalent channel*, and thus all the standard lattice coding theorems will hold unchanged. The only, minor, modification will reflect that now we have a *complex-valued* system model while the treatment in Section A.5 was done for simplicity for the real-valued one. All other facts about the importance of the *uniform random dither* for making the equivalent noise *independent* on the transmitted signal for *arbitrary* scaling α and the discussion about equivalent noise distribution hold the same as for the standard lattice decoder in Section A.5.

The variance per dimension (power) of the equivalent noise is

$$
\begin{aligned}
P_{w_{\text{eq}}} &= \frac{1}{N} \, \mathrm{E}\left[\|\mathbf{w}_{\text{eq}}\|^2 \right] \\
&= \frac{1}{N} \sum_{k=1}^{K} |\alpha h_k - a_k|^2 \, \mathrm{E}\left[\|\mathbf{u}_{k,\text{eq}}\|^2 \right] + \frac{|\alpha|^2}{N} \, \mathrm{E}\left[\|\mathbf{w}\|^2 \right] \\
&= P_s \|\alpha \mathbf{h} - \mathbf{a}\|^2 + |\alpha|^2 P_w.
\end{aligned}
\tag{5.106}
$$

The expression nicely demonstrates a two-fold α scaling impact. First, it provides the degree of freedom to match *all involved* lattices at once (by $\alpha \mathbf{h}$) to look as close as possible to the set of complex integers $\mathbf{a}$. Second, at the same time, it balances the noise power contribution from the lattices misalignment and AWGN noise.

5.6.7 Optimized Single-Tap Linear MMSE Equalizer

The optimization of the receiver is performed by minimizing the equivalent noise power $P_{w_{\text{eq}}}$. As in standard lattice decoding, this corresponds to an MMSE single-tap filter minimizing the mean square error between the desired $\mathbf{c}$ and the real observed ($\mathbf{c} + \mathbf{w}_{\text{eq}}$). Notice that the MMSE optimization is done on the equivalent noise, which is still inside the modulo operation in (5.104). See also the discussion in Section 4.4.4. Apart from optimizing α we can also choose the set of coefficients $\mathbf{a}$. However, the set of coefficients must (1) be consistent with the NCM *isomorphism* assumption, and (2) be such that it guarantees end-to-end solvability of WPNC network (to be discussed later).

We evaluate the MMSE solution for a given set of coefficients $\mathbf{a}$. We first manipulate the expression for equivalent noise power (5.106)

$$
P_{w_{\text{eq}}} = P_s \left(\alpha^* \mathbf{h}^H - \mathbf{a}^H \right) (\alpha \mathbf{h} - \mathbf{a}) + \alpha^* \alpha P_w.
\tag{5.107}
$$

Now we find the stationary point w.r.t. α. But since the noise power is a real-valued function of a complex-valued parameter, we *must* use a generalized derivative (Section A.3.5)

$$
\frac{\tilde{\partial} P_{w_{\text{eq}}}}{\tilde{\partial} \alpha} = P_s \left(\alpha^* \mathbf{h}^H - \mathbf{a}^H \right) \mathbf{h} + \alpha^* P_w.
\tag{5.108}
$$

Finding a solution of $\dfrac{\tilde{\partial} P_{w_{\text{eq}}}(\hat{\alpha})}{\tilde{\partial} \hat{\alpha}} = 0$ gives the MMSE coefficient

$$
\hat{\alpha} = \frac{P_s \mathbf{h}^H \mathbf{a}}{P_s \|\mathbf{h}\|^2 + P_w}.
\tag{5.109}
$$

The resulting minimized equivalent noise power is obtained by substituting the MMSE solution into (5.106) and after some manipulations we get

$$
\begin{aligned}
P_{w_{\text{eq}}}(\hat{\alpha}) &= P_s \left\| \frac{P_s \mathbf{h}^H \mathbf{a}}{P_s \|\mathbf{h}\|^2 + P_w} \mathbf{h} - \mathbf{a} \right\|^2 + \left| \frac{P_s \mathbf{h}^H \mathbf{a}}{P_s \|\mathbf{h}\|^2 + P_w} \right|^2 P_w \\
&= P_s \|\mathbf{a}\|^2 - \frac{P_s^2 |\mathbf{h}^H \mathbf{a}|^2}{P_s \|\mathbf{h}\|^2 + P_w}.
\end{aligned}
\tag{5.110}
$$

5.6.8   Achievable Computation Rate

Now we evaluate the achievable rate of the presented CF strategy for one given receiving relay and a given set of coefficients $\mathbf{a}$. Since the equivalent channel, from the perspective of the H-message, has exactly the same form as in standard lattice coding, we can use modulo lattice channel rate achievability (Theorem A.4) with the modification for complex codebooks (doubling the rate). The achievable hierarchical modulo lattice rate is, in the context of CF, often called a *computation rate*. For a given α and $\mathbf{a}$, it is

$$
\begin{aligned}
R_c(\alpha, \mathbf{a}) &= \lg^+ \frac{P_s}{P_{w_{\mathrm{eq}}}} \\
&= \lg^+ \frac{P_s}{P_s \|\alpha \mathbf{h} - \mathbf{a}\|^2 + |\alpha|^2 P_w}.
\end{aligned}
\tag{5.111}
$$

The computation rate is maximized by setting the scaling coefficient to MMSE $\hat{\alpha}$. It is obtained by substituting $P_{w_{\mathrm{eq}}}(\hat{\alpha})$ into the hierarchical modulo lattice rate

$$
\begin{aligned}
R_c(\hat{\alpha}, \mathbf{a}) &= \max_{\alpha} R_c(\alpha, \mathbf{a}) \\
&= \lg^+ \frac{P_s}{P_{w_{\mathrm{eq}}}(\hat{\alpha})} \\
&= \lg^+ \frac{P_s \|\mathbf{h}\|^2 + P_w}{P_s \left(\|\mathbf{h}\|^2 \|\mathbf{a}\|^2 - \left|\mathbf{h}^{\mathrm{H}}\mathbf{a}\right|^2 \right) + P_w \|\mathbf{a}\|^2}.
\end{aligned}
\tag{5.112}
$$

The MMSE optimized computation rate is still a function of HNC map coefficients $\mathbf{a}$ and it is an object of further optimization. We can, however, quite elegantly constrain the range of the search. The condition is defined by requiring $R_c(\hat{\alpha}, \mathbf{a}) \geq 0$, which in turn requires $P_s \geq P_{w_{\mathrm{eq}}}(\hat{\alpha})$. A substitution of $P_{w_{\mathrm{eq}}}(\hat{\alpha})$ from (5.110) gives

$$
\|\mathbf{a}\|^2 \leq 1 + \frac{P_s \left|\mathbf{h}^{\mathrm{H}}\mathbf{a}\right|^2}{P_s \|\mathbf{h}\|^2 + P_w}.
\tag{5.113}
$$

Using Cauchy–Schwarz inequality $\left|\mathbf{h}^{\mathrm{H}}\mathbf{a}\right|^2 \leq \|\mathbf{a}\|^2 \|\mathbf{h}\|^2$ we get

$$
\|\mathbf{a}\|^2 \leq 1 + \frac{P_s \left|\mathbf{h}^{\mathrm{H}}\mathbf{a}\right|^2}{P_s \|\mathbf{h}\|^2 + P_w} \leq 1 + \frac{P_s \|\mathbf{a}\|^2 \|\mathbf{h}\|^2}{P_s \|\mathbf{h}\|^2 + P_w}
\tag{5.114}
$$

and finally

$$
\|\mathbf{a}\|^2 \leq 1 + \frac{P_s}{P_w} \|\mathbf{h}\|^2.
\tag{5.115}
$$

5.6.9   Special Cases

We analyze the MMSE optimized computation rate under several special cases that give a nice insight into the CF performance limits.

Single-User Case and No Fading $a = h = 1, K = 1$

The simplest possible case is shown for demonstrating the coherence with a standard single-user lattice coding case $a = h = 1, K = 1$. Then

$$R_c(\hat{\alpha}, 1) = \lg\left(1 + \frac{P_s}{P_w}\right) \tag{5.116}$$

and

$$\hat{\alpha} = \frac{P_s}{P_s + P_w} \tag{5.117}$$

and, as expected, the results exactly correspond to standard MMSE optimized single user lattice coding.

Perfect Lattice Alignment and No Fading a $= $ h $= 1$

A slightly more general case is the one where we have K components and the channel coefficients are all unity $\mathbf{a} = \mathbf{h} = \mathbf{1}$. The computation rate is

$$R_c(\hat{\alpha}, 1) = \lg^+\left(\frac{1}{K} + \frac{P_s}{P_w}\right) \tag{5.118}$$

and the corresponding MMSE scaling is

$$\hat{\alpha} = \frac{KP_s}{KP_s + P_w} \tag{5.119}$$

where we nicely see the penalty for having K components with total power KP_s but with underlying lattice code designed for mean power P_s.

Perfect Lattice Alignment a $= $ h $\in \mathbb{Z}_j$

The channel coefficients do not need to be unity. In this case, the channel coefficients are complex integers and we set the HNC map coefficients as $\mathbf{a} = \mathbf{h} \in \mathbb{Z}_j$. Under this condition, all superposed component code lattices will be perfectly aligned. The computation rate is

$$R_c(\hat{\alpha}, \mathbf{h}) = \lg^+ \frac{P_s\|\mathbf{h}\|^2 + P_w}{P_s\left(\|\mathbf{h}\|^2\|\mathbf{h}\|^2 - \left|\mathbf{h}^H\mathbf{h}\right|^2\right) + P_w\|\mathbf{h}\|^2}$$
$$= \lg^+\left(\frac{1}{\|\mathbf{h}\|^2} + \frac{P_s}{P_w}\right). \tag{5.120}$$

As we see, even when removing all lattice misalignment, we still suffer a loss against the AWGN rate $\lg(1 + P_s/P_w)$ for a hypothetical single-user H-message equivalent channel. The reason for this can be seen when we evaluate the corresponding MMSE scaling coefficient

$$\hat{\alpha} = \frac{P_s\|\mathbf{h}\|^2}{P_s\|\mathbf{h}\|^2 + P_w}, \tag{5.121}$$

which differs from the one for the single-user system (A.157) $\hat{\alpha}_1 = P_s/(P_s + P_w)$. The MMSE equalizer in the lattice decoder behaves as if the total H-codeword power was $P_s\|\mathbf{h}\|^2$, i.e. the total power of all components; however, the underlying nested lattice has power P_s. This is clearly a consequence of the MMSE equalization being performed on *unconstrained (i.e. without modulo-lattice operation)* equivalent noise power (5.106).

Lattice Combination Coefficients with Common Scaling

A further generalization even does not assume integer channel coefficients but only a *common scaling* by $\eta \in \mathbb{C}$ such that $\eta\mathbf{h} = \mathbf{a} \in \mathbb{Z}_j$. Then

$$
\begin{aligned}
R_c(\hat{\alpha}, \eta\mathbf{h}) &= \lg^+ \frac{P_s\|\mathbf{h}\|^2 + P_w}{P_s\left(|\eta|^2\|\mathbf{h}\|^2\|\mathbf{h}\|^2 - |\eta|^2\left|\mathbf{h}^H\mathbf{h}\right|^2\right) + P_w|\eta|^2\|\mathbf{h}\|^2} \\
&= \lg^+ \left(\frac{1}{|\eta|^2}\left(\frac{1}{\|\mathbf{h}\|^2} + \frac{P_s}{P_w}\right)\right) \\
&= \lg^+ \left(\underbrace{\frac{1}{|\eta|^2\|\mathbf{h}\|^2}}_{=\|\mathbf{a}\|^2}\left(1 + \frac{P_s}{P_w}\|\mathbf{h}\|^2\right)\right).
\end{aligned}
\tag{5.122}
$$

Again, the lattice alignment is perfect and we pay the penalty for higher total power additionally emphasized by the scaling η.

5.6.10   Multiple Relays

Now we extend the situation for multiple receiving relays. The received signal at the ith relay is

$$
\mathbf{x}_i = \sum_{k=1}^{K} h_{ik}\mathbf{s}_k + \mathbf{w}_i
\tag{5.123}
$$

where $\mathbf{w}_i$ is IID N-dimensional complex-valued Gaussian zero mean noise with variance $P_w = \frac{1}{N}\mathrm{E}\left[\|\mathbf{w}_i\|^2\right]$ per dimension (all receivers have identical noise variance), and $h_{ik} \in \mathbb{C}$ are complex-valued channel coefficients, and we also define $\mathbf{h}_i = [h_{i1}, \ldots, h_{iK}]^\mathrm{T}$.

Each of the receiving relays performs CF strategy and calculates the HNC map for some HNC lattice combination coefficients $\mathbf{a}_i$ that are mapped to message combination coefficients $\mathbf{q}_i = [q_{i1}, \ldots, q_{iK}]^\mathrm{T}$. The scaling coefficient α_i is MMSE optimized at each relay for the local HNC map. Since the HNC maps are *linear*, we must have at least K *independent* maps to be able to solve for all component codewords. The overall coefficient matrix is row-wise composed of vectors $\mathbf{a}_i$, $\mathbf{A} = [\mathbf{a}_1, \ldots, \mathbf{a}_K]^\mathrm{T}$. It must be *full-rank* rank $\mathbf{A} = K$. Clearly, the relays must coordinate the choice of coefficients to fulfill that.

The full-rank matrix $\mathbf{A}$ condition guarantees a solvability at the level of lattice codewords and their unconstrained integer linear combinations. The information messages themselves are, however, combined by q coefficients mapped to a coefficients by the modulo operation (see Section 4.7.2). Dense lattices with high-cardinality alphabets might somewhat relax the consequences, but proper care of this phenomenon is generally needed.

The resulting achievable rate for each individual component R_0 is then given by the minimum of computation rates over all involved relays, where each relay chooses the best HNC map $\mathbf{a}_i$, such that the matrix $\mathbf{A}$ is full-rank, and individually MMSE optimizes its scaling α_i

$$
R_0 = \min_i \ \max_{\mathbf{a}_i:\text{rank}\,\mathbf{A}=K} \ \max_{\alpha_i} R_c(\alpha_i, \mathbf{a}_i)
$$

$$
= \min_i \ \max_{\mathbf{a}_i:\text{rank}\,\mathbf{A}=K} \ \lg^+ \frac{\left(P_s\|\mathbf{h}_i\|^2 + P_w\right)}{P_s\left(\|\mathbf{h}_i\|^2\|\mathbf{a}_i\|^2 - \left|\mathbf{h}_i^{\mathrm{H}}\mathbf{a}_i\right|^2\right) + P_w\|\mathbf{a}_i\|^2}. \tag{5.124}
$$

We have generally assumed that all relays actively participate in CF for all component sources, i.e. all coefficients a_{ik} are non-zero. If this is not the case, and some sources are not participating in some relay maps, we can make more specific and perhaps a somewhat relaxing evaluation of the minimum over the relays, and also possibly be able to allow different rates from the selected subgroup of sources; see [45] for details.

5.6.11 Compute and Forward in the Perspective of WPNC

(1) CF is a *linear isomorphic layered* NCM with self-folded H-codebook (Sections 4.5 and 4.5.4) *enforced* by modulo lattice receiver preprocessing. Properties of the modulo lattice operation used in the receiver and the usage of component nested lattice codes turn the system straightforwardly into an equivalent *single-user* system processing the H-message using a standard lattice coding technique.

(2) CF is constrained to *linear* HNC message maps over a *common* GF shared by all involved components in the H-MAC stage. As a consequence, the number of relays evaluating the CF linear maps in all network cuts must be the same as the number of sources – we need to have K linear independent equations on GF to solve K unknowns.

(3) CF does not solve the coding for a *general* topology of WPNC. It requires homogenous stages activated in a sequence. Particularly, it does not solve the topologies with cycles and with mixed stages.

(4) CF is fundamentally an information-theoretic strategy with the results relying on the usage of "perfect" lattices. However, recent results on low-density lattice coding [53] give some promising boost towards practical usability of the lattice coding.

(5) The HNC message maps $\mathbf{q}_i$ are linked to HNC lattice maps $\mathbf{a}_i$, which are in turn linked to channel coefficients $\mathbf{h}_i$. Therefore the HNC message maps depend on random channel coefficient realizations and may differ realization from realization. At the same time, they need to guarantee the end-to-end solvability (full-rank $\mathbf{A}$ and/or $\mathbf{Q}$). This requires continuous negotiation among relays. The performance penalty of not using maps adapted to channel coefficients can be quite high since the sensitive lattice structure is quite prone to misalignment. In this sense, the CF is the strategy with an adaptation of maps that can be hardly avoided. Compare this to the predetermined maps of non-CF-based NCM, which can use one map, with a modest performance penalty, for a wide range of channel states, or even regardless of channel state.

(6) CF on its own, i.e. the utilization of the lattice coding principle, focuses on one-stage H-MAC only. It essentially assumes that the CF equations of one-stage relays are reliably conveyed to the final destination. All other aspects, as for example a combination of the individual H-MAC stages and the end-to-end solvability, are *not* specific to CF and are common to other strategies.

(7) The evaluation of achievable rates and the receiver processing is constrained to *block-constant* (over the whole lattice codeword) fading in *AWGN* channel.

(8) A major problem of CF is the single degree of freedom in a *scalar* α equalizer that is used jointly for matching lattices observed through the channel $\alpha\mathbf{h}$ to be as close as possible to the set of complex integers $\mathbf{a}$, and, at the same time, to balance a contribution between lattice misalignment and AWGN noise. Of course, we could decrease the granularity of the problem by having large values of $\mathbf{a}$ but this, in turn, increases the contribution of the AWGN noise.

(9) All received H-MAC stage signals are used to make a hard decision on the HNC combined messages. This is performed by fine lattice quantizer. There is no straightforward way of incorporating (in the sense discussed e.g. in Section 4.6) or obtaining imperfect (or partial) information (or soft decision) of some participating received components.

5.6.12   Examples

Example 5.7   The example shows a 2WRC ($K = 2$) scenario with the setup compatible with Example 5.6 where we assume only that the HSI link is perfect (i.e. $\gamma_{SD} \to \infty$) and we concentrate only on H-MAC. Unlike the NNC case, the CF strategy requires a bit more specific definition of the channel coefficients. We will assume a special case according to (5.122), even in the more specific setup $\eta\mathbf{h}_{SR} = \mathbf{a} \in \mathbb{Z}_j$ where $\mathbf{h}_{SR} = h_{SR}\mathbf{1}$, $h_{SR} \in \mathbb{C}$. The combination coefficient is then $\mathbf{a} = \eta h_{SR}\mathbf{1} \in \mathbb{Z}_j$. We define SNR $\gamma_{SR} = |h_{SR}|^2 P_s / P_w$.

The computation rate (5.122) is

$$R_c(\hat{\alpha}, \eta h_{SR}\mathbf{1}) = \lg^+ \left(\frac{1}{\underbrace{|\eta h_{SR}|^2 K}_{=\|\mathbf{a}\|^2}} (1 + \gamma_{SR}K) \right). \tag{5.125}$$

Any complex rotation angle of channel coefficient is clearly compensated by a complex angle of η. The rate maximizing $\mathbf{a}$ must have the smallest norm $\|\mathbf{a}\|^2$ such that its components are non-zero, i.e. $\hat{\mathbf{a}} = [1, 1]^{\mathrm{T}}$, and it corresponds to $\eta = 1/h_{SR}$. This map is end-to-end solvable considering a perfect HSI provided by destinations of the 2WRC. The maximized H-MAC rate is

$$R_c(\hat{\alpha}, \hat{\mathbf{a}}) = \lg^+ \left(\frac{1}{2} + \gamma_{SR} \right). \tag{5.126}$$

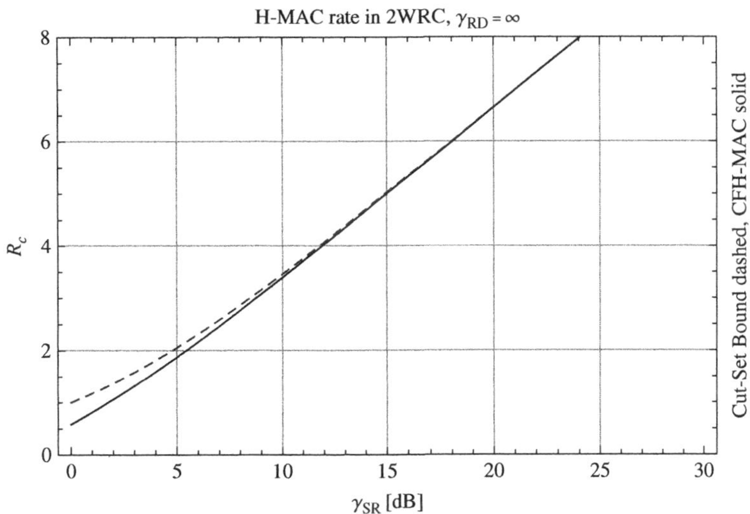

Figure 5.10 Example: computation rate of CF strategy in H-MAC of 2WRC.

CF does not solve the H-BC stage; however, if H-BC has a rate higher than the H-MAC, then $R_c(\hat{\alpha}, \hat{\mathbf{a}})$ is the end-to-end rate for both sources. Since the H-BC is a simple single-user channel transmitting one H-message, the corresponding H-BC rate evaluation is trivial. Figure 5.10 shows the results numerically in comparison to a single-user cut-set bound, i.e. as if there were only *one* user in H-MAC. With no surprise, the CF rate approaches the cut-set bound for medium to high SNR where the coefficient $1/2$ inside the logarithm becomes negligible in comparison with γ_{SR}.

5.7 Hierarchical Decode and Forward in Single-Stage H-MAC

This section will concentrate on the evaluation of the information-theoretic properties of a *single-stage H-MAC* processing. We will allow quite generic settings of the NCM and HNC maps but concentrate only on a single stage and H-MAC processing. This is useful in WPNC for the class of processing strategies where the node *decodes* the H-message (e.g. HDF). This is called hierarchical decoding (H-decoding) (Section 4.3). Each hierarchical stage is required to obtain a reliable data decision, and these are then hierarchically encapsulated in the subsequent stages. Each encapsulation stage is thus a decision bottleneck of the global end-to-end flow. Compare this with the NNC technique, which deals with the whole network but assumes QF type of node processing and requires a large number of transmission blocks, or with the CF technique, which assumes quite a specific form of lattice NCM encoding and specific front-end preprocessing using modulo lattice operation.

In developing an information-theoretic assessment of H-MAC hierarchical decoding, one might be tempted simply to evaluate some form of the mutual information and "extrapolate" the classical coding results into this new situation without paying attention to the details. However, it could lead to some misinterpretations and improper conclusions. Any form of mutual information should have a clear and rigorously defined connection to the coding rates. It might be a more complicated relationship than it is in single-user scenarios. Essentially, we need to develop coding theorems for hierarchical H-MAC decoding under a variety of specific scenarios.

The coding theorems will be given in the form of the achievable and converse rates for H-messages. The H-rate is called an *achievable H-rate* if there exist some given H-codebook construction and some given H-decoding strategy such that the probability of the H-message decoding error approaches zero.[9] For the achievability theorems (Theorems 5.7 and 5.10), we use a common methodology (Section A.4) of random codebook construction and joint typicality decoding. It allows a relatively straightforward mean error probability analysis utilizing a joint typicality lemma. In both theorems, we use random component codebooks that are randomly generated according to $S_k^N \sim \prod_{n=1}^N p(s_{k,n})$ for component messages $b_k \in [1 : 2^{NR_k}]$. The HNC message map definition and its relation to the H-code depends on the relay decoding strategy. In joint-metric H-decoding on the product codebook (Theorem 5.7), we directly decode the H-message and we do not need the H-codebook itself but we only need to know the number of H-messages 2^{NR}. The H-message map is defined also as a random one in order to facilitate the joint typicality decoding approach. On the other hand, the layered H-decoding strategy for layered isomorphic NCM (Theorem 5.10) requires a random regular isomorphic H-codebook. The details are specified directly in the theorems.

The *converse H-rate* (Theorem 5.13) is defined as a bound on the H-rate that is implied by having the decoding error approaching zero *without* any particular specifications of the encoding and decoding technique – they can be arbitrary. The arbitrary code construction means that we specify neither how the codebook is created nor what the codeword's properties are. However, it provides some insight to separately treat two generic categories of random and structured codes. The random codebooks are characterized as in the achievability theorems. The structured codebook is a deterministic codebook construction where we generally allow a coupling among the code symbols, i.e. generally c_n and c^{n-1} are *not* independent.

If the achievable and converse rates meet and can at the same time be maximized over the codesymbol distribution, we get the capacity. Under the HDF strategy, we still have many possible coding and decoding subcategories and constraints. The structure of these subcategories and constraints is, however, different for the achievability and for the converse theorems. We will (Section 5.7.7) be able to find one common subcategory (self-folded isomorphic layered NCM) where the achievable rate and the converse rate meet and are superior to other subcategories for given fixed input distribution. However,

---

[9] Sometimes, we cannot state this rate/region exactly and we only approximate it by upper bounding it. It is then called an achievable rate/region *upper bound*. But still, it is obtained for some particular coding and decoding strategy and it does not mean that there could not be some other better encoding/decoding strategy. It should not be confused with coding *converse* theorems. The achievable rate region means that "something exists everywhere inside" the rate region, while the converse rate region means "nothing exists anywhere outside."

their common maximization over the input distribution is still an open problem and therefore it allows only conjecture of the capacity.

5.7.1 System Model

We assume a single H-MAC stage with component signals s_k^N indexed by $k \in [1 : K]$, where K is the number of nodes. The component signals can be arbitrary nodes within the encapsulation hierarchy, both sources or relays already transmitting hierarchical information. We solve the single H-MAC stage regardless of the actual contents of the component signals. Each component signal is a signal space codeword with the dimension N. The signal space nth symbol $s_{k,n}$ is one-to-one mapped to the code symbol $s_{k,n} = \mathcal{A}_k(c_{k,n})$. This allows us to use GF-based codebooks for codewords c_k^N. The codebook $\mathcal{C}_k$ maps the message $b_k \in [1 : 2^{NR_k}]$ to the codeword $c_k^N(b_k)$. We will also use the "tilde" notation for the complete set of components as defined earlier in the text.

Our target is to decode the hierarchical message $b = \chi(\tilde{b})$. At this moment we do not solve the problem of whether this hierarchical function allows end-to-end solvability at final destinations. We simply assume that it is given and has a cardinality $b \in [1 : 2^{NR}]$ where R will be called the *hierarchical coderate* of the H-codebook (see Section 3.5).

It is important to distinguish three different levels on which the hierarchical function is applied. The first one is defined at the level of messages and it is denoted by χ. The second one is defined on the level of hierarchical codesymbols $c_n = \chi_c(\tilde{c})$ in the case of layered NCM (see Section 4.2.3). The third level is the H-constellation of the signal space code directly related either to the H-message $\mathcal{U}(b)$ or to the H-code symbols $\mathcal{U}_c(c_n)$ in the case of layered NCM. In principle, the only important one is the map χ defined at the *H-message* level since this is the only one that is related to the flow of information data in the WPNC network. It is also directly related to the hierarchical rate. All the others, $\chi_c, \mathcal{U}(b), \mathcal{U}_c(c_n)$, are only secondary in the sense that they should provide *adequate support* for the NCM with the desired hierarchical rate.

5.7.2 HDF Decoding

Our ultimate goal is to obtain reliable decoding of hierarchical messages $b = \chi(\tilde{b})$. There are essentially two approaches (see Section 4.3) relevant to WPNC with HDF relay node strategy (Section 3.3.6). The first one utilizes the *product code* structure of the observed signal $p(x^N|\tilde{c}^N(\tilde{b})) = p(x^N|\tilde{s}^N(\tilde{b}))$ and tries to directly (using product complexity decoder) obtain an estimation of b *without*[10] first explicitly decoding individual estimates of $\tilde{b}$. It is called *joint-metric* hierarchical decoding (Section 4.3.2).[11]

The second HDF decoding approach is the *layered* hierarchical decoding (Section 4.3.3). It assumes that we can describe the system by *layered NCM* (Section 4.2.3).

---

[10] The variant with first obtaining $\tilde{b}$ estimates (hard decisions) and then applying the GF NC map to obtain b is a concatenation of classical multi-user decoding and classical discrete NC (NC-JDF; see Section 3.3.6).

[11] There exists a related concept of non-unique decoding [7] where the decoder in a multi-terminal setup is interested only in some subset of the component messages. Notice that our scenario differs by considering many-to-one codebook functions in contrast to simply dividing codebooks into wanted and unwanted subsets.

Then the observation is expressed in terms of a hierarchical *channel* symbol $p(x^N|c^N)$. If the layered NCM is also *isomorphic*, a one-to-one H-codebook exists (see Section 4.2.4) and we can express the observation as $p(x^N|c^N(b))$.

In the following, we state the coding theorems for both joint-metric and layered H-decoder. Coding theorems provide achievable rate and converse proofs under various conditions.

5.7.3 Joint-Metric Hierarchical Decoding on Product Codebook

Joint-metric H-decoding exploits the product codebook $\tilde{C} = C_1 \times \cdots \times C_K$ to obtain the hierarchical message $b = \chi(\tilde{b})$ directly *without explicitly decoding component messages*. We use joint typicality decoding and randomly generated component codebooks. The fact that we use the product codebook but do not decode component messages needs explanation. The use of a product codebook means that we reveal all component codebooks to the decoder and let it use them. No other derived code structure is revealed to the decoder. Particularly, we do *not* reveal any form of H-codebook. So the decoder can utilize only the parallel structure of all C_k to obtain the hierarchical message. The fact that the decoder does *not* explicitly decode individual $\tilde{b}$ messages is reflected by an adequate way of counting errors. The decoding error is declared only if $\hat{b} \neq b$ regardless of the component messages provided that they are consistent with the H-message.

We want to allow the HNC message maps χ to be as generic as possible. Realizing that the information-theoretic statements assume *randomly* generated codebooks, we can abstract to a large extent in defining the HNC map. Random generation of component codebooks makes the particular assignment of message indices to the actual codeword irrelevant. It means that the only important elements are the HNC map codomain cardinality and its distribution over the component messages subspace cuts. We do not need to care about the exact input–output mapping of the indices. For this purpose, we define a uniform random HNC map that uniformly divides all component subspaces.

DEFINITION 5.6 (Uniformly Random HNC Map)   Assume component messages $\tilde{b} = \{b_1, \ldots, b_K\}$ of cardinalities $M_k = |\{b_k\}|$. Denote the total product cardinality of all component messages $\tilde{M} = \prod_{k=1}^{K} M_k$. A uniformly random HNC map $b = \chi(\tilde{b})$ is defined by having the cardinality $M = |\{b\}| \leq \tilde{M}$ and uniform number of elements in subspace cuts.

Subspace cuts are defined by fixing the messages $b_{k_1}, \ldots, b_{k_L}$ on a subset of L indices $\{k_1, \ldots, k_L\}$ to some particular values. The HNC map bin in a given subspace cut is a set of all $\tilde{b}$ inside that subspace that are consistent with the given b value. The number of all different $\tilde{b}$ indices belonging to one particular b HNC map bin is

$$M(k_1, \ldots, k_L) = \frac{\tilde{M}}{M \prod_{\ell=1}^{L} M_{k_\ell}}. \tag{5.127}$$

We also denote $M(\emptyset) = M(0)$. All cardinalities can be also expressed in terms of the code rates $M = 2^{NR}$, and similarly for others.

The map output is random for any fixed input message $\tilde{b}$ and the distribution is such that the number of elements in all subspace cuts is uniform. This will later allow us to fix the actual transmitted H-message when evaluating the mean H-message decoding error in the proof of Theorem 5.7. The size of the bin describes the amount of uncertainty in the set $\tilde{b}$ consistent with particular b and further constrained by fixing some of the $\tilde{b}$ components.

In a special case of $K = 2$, the definition means the following. The total size of all component messages is $\tilde{M} = M_1 M_2$. Globally without fixing any subset, i.e. $L = 0$, the total number of component messages in one HNC map bin is $M(0) = \tilde{M}/M$. If we fix some particular value b_1, then one HNC map bin contains $M(1) = \tilde{M}/(MM_1) = M_2/M$ component messages, and similarly for fixing b_2 it is $M(2) = \tilde{M}/(MM_2) = M_1/M$.

A special case of $K = 3$ gives $\tilde{M} = M_1 M_2 M_3$, $M(0) = \tilde{M}/M$, $M(1) = M_2 M_3/M$ and similarly for $M(2), M(3)$. Second-order cuts are $M(1,2) = M_3/M$ and similarly for $M(2,3), M(1,3)$.

In the following we establish a theorem for achievable hierarchical rate for *joint-metric H-decoding* with uniformly random HNC map.

THEOREM 5.7 (Achievable Rate for Joint-Metric H-Decoding on Product Codebook) *Assume H-MAC with K components. NCM has each component signal for $k \in [1 : K]$ encoded by randomly generated codebook C_k drawn according to $S_k^N \sim \prod_{n=1}^N p(s_{k,n})$ and with component message $b_k \in [1 : 2^{NR_k}]$. H-messages are defined by uniformly random HNC map $b = \chi(\tilde{b})$, $b \in [1 : 2^{NR}]$. The observation model is a joint-metric of the memoryless channel $p(x^N|\tilde{s}^N(\tilde{b})) = \prod_{n=1}^N p(x_n|\tilde{s}_n(\tilde{b}))$. The decoder decodes H-message on a product codebook but without explicitly decoding (i.e. making decisions on) individual component messages.*

The H-message rate R for joint-metric H-decoding on product codebook is achievable if all component rates R_k are achievable by a classical multi-user MAC joint decoding. The achievable hierarchical rate is the same as for NC-JDF strategy.

Proof We prove the theorem for two-component H-MAC $K = 2$. The generalization is straightforward.

The joint typicality H-message joint-metric and product codebook decoder (see Section A.4, and some notation remarks at joint typicality lemma) declare the decoded H-message $\hat{b}$ if

$$\left(S_1^N(b_1), S_2^N(b_2), X^N\right) \in \mathcal{T} \tag{5.128}$$

for some b_1, b_2 such that $\chi(b_1, b_2) = \hat{b}$. We declare as decoded only the first set, searched in the order of their indices, of messages that are jointly typical and ignore any possible other cases. If no b_1, b_2 fulfills this, the decoder declares $\hat{b} = \emptyset$. The fact that we do *not* decode individual messages is reflected by allowing b_1, b_2 to be "some" values consistent with $\hat{b}$, which is the only important one. We do not pay attention to individual b_1, b_2.

Assume that messages b_1, b_2 were sent and the corresponding H-message is $b = \chi(b_1, b_2)$. The decoding error event $\hat{b} \neq b$ appears in two forms, a decoding failure $\hat{b} = \emptyset$ and an erroneous decision. The decoding failure means that the decoder does

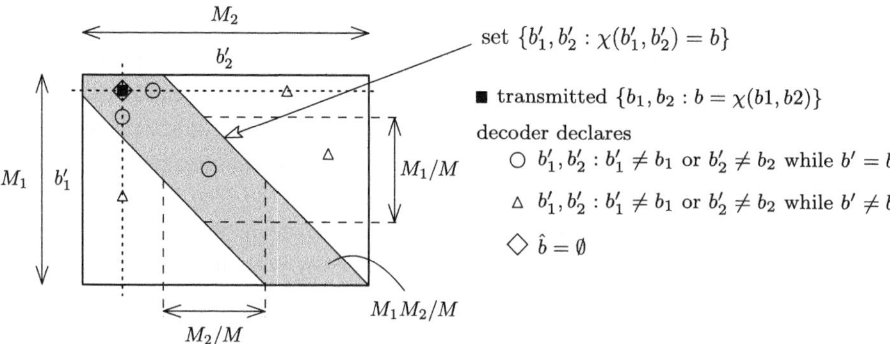

Figure 5.11 Error events in the proof of the achievable H-rate for joint-metric H-decoding on a product codebook.

not recognize as a typical set the situation where $b_1' = b_1$ and $b_2' = b_2$. The erroneous decision is the case when the decoder declares as the estimate a different H-message than it was sent, i.e. b_1', b_2' lie outside the correct H-message set (Figure 5.11). A correct decision does not necessarily mean that both components are correct. Figure 5.11 shows various error event options. It is particularly important to realize that each component individually, or even both components, can be in error while the resulting H-message is decoded correctly.

The overall probability of the error P_e is the probability of the union of these events and it is upper bounded by the sum of all individual error event probabilities. In order to make $P_e \to 0$ for $N \to \infty$, we must make sure that all of them have the probability approaching zero. In the following, we investigate probabilities of these individual cases.

In all error events, we assume that we send b_1, b_2 such that $b = \chi(b_1, b_2)$. Under a standard assumption of random ensemble of symbol-wise IID codebooks (see Section A.4) and the symmetry of their random generation, the average error probability over all codebooks and all random transmitted messages can be evaluated by assuming some fixed transmitted message $b = \chi(b_1, b_2)$.

A complete set of joint typicality decoding variables, i.e. the transmitted codewords $S_1^N(b_1), S_2^N(b_2)$, their corresponding observation X^N, and the test codewords $S_1^N(b_1'), S_2^N(b_2')$ of the decoding process are jointly described by

$$\left(S_1^N(b_1'), S_2^N(b_2'), X^N, S_1^N(b_1), S_2^N(b_2)\right) \sim$$

$$\sim \prod_{n=1}^{N} p(x_n|s_{1,n}(b_1), s_{2,n}(b_2)) p(s_{1,n}(b_1), s_{1,n}(b_1')) p(s_{2,n}(b_2), s_{2,n}(b_2')) \quad (5.129)$$

where the codeword joint PDFs depend on a mutual relation of messages

$$p(s_{k,n}(b_k), s_{k,n}(b_k')) = \begin{cases} p(s_{k,n}(b_k)) \delta(s_{k,n}(b_k) - s_{k,n}(b_k')); & b_k = b_k' \\ p(s_{k,n}(b_k)) p(s_{k,n}(b_k')); & b_k \neq b_k' \end{cases} \quad (5.130)$$

All required derived PDFs are then simply obtained by the marginalization properly respecting the relation of the messages. For example, the received signal for a given (b_1, b_2) is drawn according to $X^N \sim \prod_{n=1}^{N} p(x_n(b_1, b_2))$ where the PDF $p(x_n(b_1, b_2)) = \sum_{s_{1,n}(b_1), s_{2,n}(b_2)} p(x_n | s_{1,n}(b_1), s_{2,n}(b_2)) p(s_{1,n}(b_1)) p(s_{2,n}(b_2))$ is the observation corresponding to transmitted messages with marginalized test codewords. Assuming the error event involving decoder test messages $b' = \chi(b_1', b_2')$, the joint typicality decoder behavior is determined by the behavior of random variables drawn according to

$$\left(S_1^N(b_1'), S_2^N(b_2'), X^N\right) \sim$$

$$\sim \begin{cases} \prod_{n=1}^{N} p(x_n, s_{1,n}(b_1'), s_{2,n}(b_2')); & b_1' = b_1, b_2' = b_2 \\ \prod_{n=1}^{N} p(x_n, s_{2,n}(b_2')) p(s_{1,n}(b_1')); & b_1' \neq b_1, b_2' = b_2 \\ \prod_{n=1}^{N} p(x_n, s_{1,n}(b_1')) p(s_{2,n}(b_2')); & b_1' = b_1, b_2' \neq b_2 \\ \prod_{n=1}^{N} p(x_n) p(s_{1,n}(b_1')) p(s_{2,n}(b_2')); & b_1' \neq b_1, b_2' \neq b_2 \end{cases} \quad (5.131)$$

If the test message, e.g. $b_1' = b_1$, is the same as the one used for transmission, then the observation and the message have joint distribution, e.g. $p(x_n, s_{1,n}(b_1'))$. If some of the test messages and the actually sent messages are different, e.g. $b_1' \neq b_1$, and some are equal, $b_2' = b_2$, then the variables are conditionally independent, $S_1^N(b_1') \perp X^N | \{S_2^N(b_2')\}$, where we also realized that $S_1^N \perp S_2^N$.[12] This conditional independence is later used in the evaluation of the probability of being jointly typical according the joint typicality lemma.

The decoding failure is the event when $\left(S_1^N(b_1'), S_2^N(b_2'), X^N\right) \notin \mathcal{T}$ while $b_1' = b_1$ and $b_2' = b_2$. If this holds, then the distribution in (5.131) is simply the joint PDF $\prod_{n=1}^{N} p(x_n, s_{1,n}(b_1), s_{2,n}(b_2))$ and the random variables $\left(S_1^N(b_1), S_2^N(b_2), X^N\right)$ drawn according to it are simply jointly typical for $N \to \infty$ and thus the probability of this error event approaches zero.

The erroneous decision events can be further classified into first- and second-order events depending on whether both or just one of the components lie outside of H-messages set. In all cases, we first find the probability for a particular set of erroneous (b_1', b_2'), and the union of events for all possible (b_1', b_2') is then upper bounded by the sum of probabilities.

Error events of the first order where b_2' is correct and b_1' makes an error, the vice versa case, and the event of the second-order are defined as

$$\Omega_{e1}(b_1') = \left\{ \left(S_1^N(b_1'), S_2^N(b_2'), X^N\right) \in \mathcal{T}, \, b_1' \neq b_1, b_2' = b_2, \chi(b_1', b_2') \neq b \right\},$$
$$(5.132)$$

$$\Omega_{e2}(b_2') = \left\{ \left(S_1^N(b_1'), S_2^N(b_2'), X^N\right) \in \mathcal{T}, \, b_1' = b_1, b_2' \neq b_2, \chi(b_1', b_2') \neq b \right\},$$
$$(5.133)$$

$$\Omega_{e12}(b_1', b_2') = \left\{ \left(S_1^N(b_1'), S_2^N(b_2'), X^N\right) \in \mathcal{T}, \, b_1' \neq b_1, b_2' \neq b_2, \chi(b_1', b_2') \neq b \right\}.$$
$$(5.134)$$

---

[12] We can also realize that the *grouped pair* of variables is independent $S_1^N(b_1') \perp (X^N, S_2^N(b_2'))$. Then we could use a special two-variable form of the joint typicality lemma and due to a specific property of the variables, namely $S_1^N(b_1') \perp S_2^N(b_2')$, we would get exactly the same result as by using the full joint typicality lemma.

Probability $\Pr\{\Omega_{e12}(b_1', b_2')\}$ is given by a special form of joint typicality lemma (A.112) since $(S_1^N(b_1'), S_2^N(b_2'))$ and X^N are independent as a consequence of codebook generation, and thus

$$\Pr\{\Omega_{e12}(b_1', b_2')\} \approx 2^{-NI(S_1, S_2; X)}. \tag{5.135}$$

Assuming uniformly random HNC map, one H-message bin with two degrees of freedom has $M(0) = 2^{N(R_1 + R_2 - R)}$ components out of the total number $2^{N(R_1 + R_2)}$ of component pairs. The decoder decision on the pair (b_1', b_2') falling into the bin means a correct H-message decision. The number of pairs falling outside this bin determines the number of error events $\Omega_{e12}(b_1', b_2')$. The probability of the union of all those events can be then upper bounded by

$$\begin{aligned}
\Pr\{\Omega_{e12}\} &= \Pr\left\{\cup_{(b_1', b_2')}\Omega_{e12}(b_1', b_2')\right\} \\
&\lesssim \left(2^{N(R_1 + R_2)} - 2^{N(R_1 + R_2 - R)}\right)\Pr\{\Omega_{e12}(b_1', b_2')\} \\
&= \left(2^{N(R_1 + R_2)} - 2^{N(R_1 + R_2 - R)}\right) 2^{-NI(S_1, S_2; X)}.
\end{aligned} \tag{5.136}$$

This probability will go to zero for $N \to \infty$ if $(R_1 + R_2) < I(S_1, S_2; X)$ and $(R_1 + R_2 - R) < I(S_1, S_2; X)$. Realizing that $R \geq 0$ we get the second-order condition

$$R_1 + R_2 < I(S_1, S_2; X). \tag{5.137}$$

First-order error event probability $\Pr\{\Omega_{e1}\}$ for some fixed b_2' is given by realizing that $S_1^N(b_1')$ and $X^N|\{S_2^N(b_2')\}$ are independent, and using (A.108) it gives

$$\Pr\{\Omega_{e1}(b_1')\} \approx 2^{-NI(S_1; X|S_2)}. \tag{5.138}$$

The H-message bin size for fixed b_2' is $M(2) = 2^{N(R_1 - R)}$ out of the total number of components 2^{NR_1}. The number of possible H-message errors with fixed b_2' is $(2^{NR_1} - 2^{N(R_1 - R)})$ and events union probability upper bound is

$$\Pr\{\Omega_{e1}\} = \Pr\left\{\cup_{b_1'}\Omega_{e1}(b_1')\right\} \lesssim \left(2^{NR_1} - 2^{N(R_1 - R)}\right) 2^{-NI(S_1; X|S_2)}. \tag{5.139}$$

Again, realizing that $R \geq 0$, the condition for $\Pr\{\Omega_{e1}\} \approx 0$ as $N \to \infty$ is

$$R_1 < I(S_1; X|S_2). \tag{5.140}$$

The condition for $\Pr\{\Omega_{e2}\} \approx 0$ is obtained similarly as

$$R_2 < I(S_2; X|S_1). \tag{5.141}$$

$\square$

REMARK 5.1   Theorem 5.7 states that the achievable H-rate for joint-metric product codebook H-decoding is the same as for NC-JDF, i.e. the strategy which first explicitly decodes individual component messages using a classical multi-user MAC decoding and then it builds the discrete NC H-message. This result has a wider context (see Footnote 11) but it showed itself to be useful to rigorously try to see whether the more generous definition of H-decoding error events would help to increase the rate. The H-decoding error events are more relaxed than in the classical JDF because making an error in one or both components does not necessarily mean H-message error.

However, the underlying structure of the achievability proof shows that this does not give any advantage in the resulting H-rate. Simply speaking, the error events are more relaxed but still both decoders would have to make the errors only in some coordinated way to keep the H-message correct. And, obviously, there is no binding structure[13] provided to the decoder that would additionally complement the product codebook and that would directly reduce the codeword set inside the decoder. The only connection is the outer one combining the messages. It does not matter how the messages are combined, whether by iterative sharing of soft-message information or any other way.

REMARK 5.2   There is no advantage provided by the joint-metric product codebook H-decoder over the NC-JDF. But on the other hand, there is also absolutely no constraint on the size and the type of HNC map either on the structure of the H-code or the alphabet of channel symbols. Particularly, the H-codebook does *not* need to be the isomorphic layered NCM.

5.7.4 Layered Hierarchical Decoding for Isomorphic Layered NCM

We now focus on the evaluation of information-theoretic properties for a specific NCM case – the isomorphic layered NCM (Section 4.2.4). This more specific setup enables to use the layered hierarchical decoding (Section 4.3.3). On one hand, the structure of NCM is rather constrained by this assumption, but on the other hand, it allows us to develop a useful equivalent model of the system and in turn to evaluate the achievable hierarchical rate in a relatively straightforward manner.

Most importantly, although being somewhat constrained in assumptions, this encoding and decoding strategy potentially provides *an advantage in terms of achievable hierarchical rate* over the joint-metric hierarchical decoding on a product codebook or an NC-JDF. As such, it is one of the WPNC strategies that shows the performance benefits w.r.t. classical techniques.

Equivalent Hierarchical Channel Model

The equivalent hierarchical channel model defines the input–output behavior of the system from the perspective of the H-message using the isomorphic layered property. We assume an isomorphic layered NCM (Section 4.2.4) with a one-to-one symbol-wise $s_{k,n} = \mathcal{A}_k(c_{k,n})$ channel symbol mapper, component codebooks $\mathcal{C}_k$

$$c_k^N = \mathcal{C}_k(b_k), \; s_{k,n} = \mathcal{A}_k(c_{k,n}), \;\; k \in [1:K], \, n \in [1:N], \tag{5.142}$$

H-message map χ, and symbol-wise H-code symbol HNC map χ_c

$$b = \chi(\tilde{b}), \;\; c_n = \chi_c(\tilde{c}_n), \;\; c^N = \mathcal{C}(b), \tag{5.143}$$

---

[13]   This binding structure is the H-codebook, but, as assumed under the joint-metric decoder strategy, it is not revealed to the decoder, and in fact it even does not need to exist.

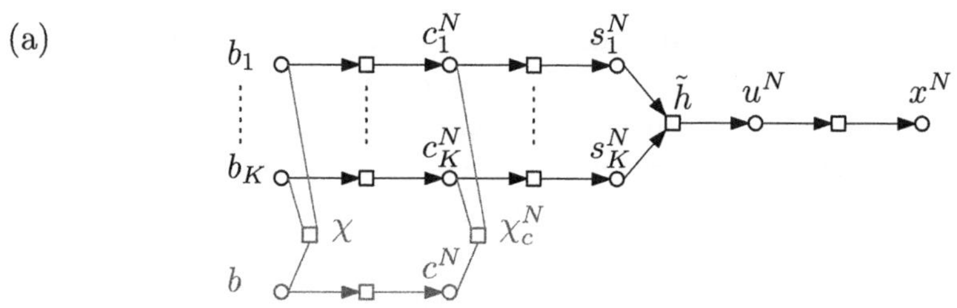

(a)

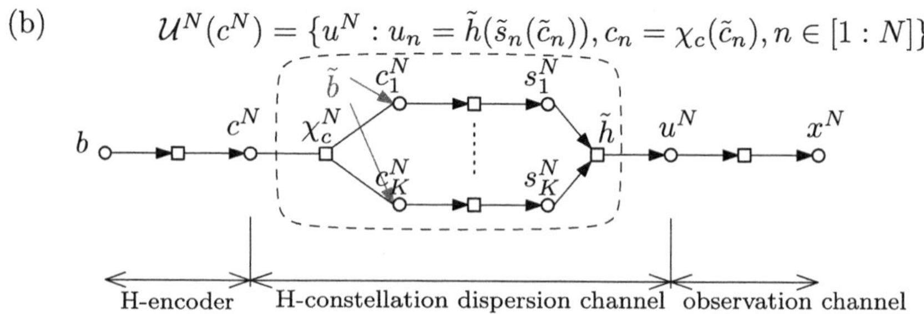

(b)

$$\mathcal{U}^N(c^N) = \{u^N : u_n = \tilde{h}(\tilde{s}_n(\tilde{c}_n)), c_n = \chi_c(\tilde{c}_n), n \in [1:N]\}$$

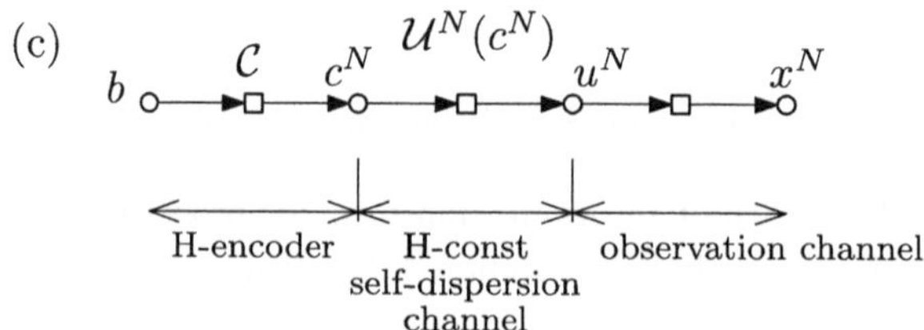

(c)

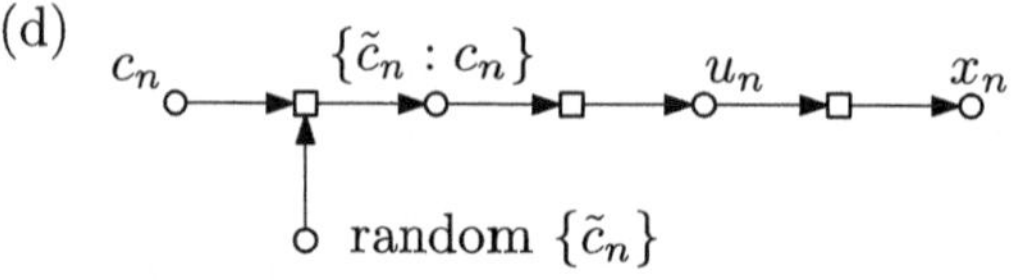

(d)

Figure 5.12 Equivalent hierarchical channel model for layered hierarchical decoding of isomorphic layered NCM: (a) real system model, (b) equivalent model in a basic form, (c) equivalent model in a compact form, (d) symbol-wise model with explicit randomness of H-constellation self-dispersion.

where $\mathcal{C}$ is an H-codebook. A memoryless channel observation model forms Markov chain $\tilde{C}_n \mapsto U_n \mapsto X_n$ (Section 3.5.1)

$$u_n = \tilde{h}\left(\tilde{s}_n(\tilde{c}_n)\right), \quad x_n = x(u_n) \tag{5.144}$$

where u_n is a channel-combined symbol. The channel function $\tilde{h}(\cdot)$ inherently includes all channel parametrization. The real system is described by the model in Figure 5.12a shown together with the relationships to the hierarchical entities. Figure 5.12b shows the input–output relationship from the perspective of the H-message. It is clear that the middle part of the model contains a degree of freedom over all $\tilde{c}^N$ consistent with c^N. We can model the system by a serial concatenation of a two-part channel. The first part $c_n \mapsto u_n$ will be called the *H-constellation dispersion channel*. It captures the degrees of freedom in the H-constellation and its self-dispersion. The sources of the randomness are the component messages $\tilde{b}$ and subsequently the codewords $\tilde{c}^N$. Owing to hierarchical many-to-one mapping, there are multiple $\tilde{b}$ and multiple $\tilde{c}^N$ that are consistent with a particular c^N (see Section 4.4.3 for a discussion of some aspects of the distribution). The H-constellation dispersion means that there remains a randomness (excitation by $\tilde{b}$) that is, however, constrained by allowing only such $\tilde{c}^N$ that are consistent with c^N. This consistency relationship between c^N and $\tilde{c}^N$ is captured by χ_c^N (Figure 5.12b). In other words, for a given c^N, all $\{\tilde{b} : c^N\} = \{\tilde{b} : \chi_c^N\left(\tilde{c}^N(\tilde{b})\right) = c^N\}$ are allowed and this makes the u^N random for a fixed c^N.

The second part $u_n \mapsto x_n$ is the *observation channel*. Figure 5.12c shows a compact form of the model where the self-dispersion part is represented by the H-constellation operator. It inherently represents the channel parametrization $\tilde{h}$ and channel symbol maps $\mathcal{A}_k$.

The equivalent H-channel model is thus described by the *H-constellation self-dispersion* part

$$u_n(c_n) \in \mathcal{U}_n(c_n) = \left\{ u_n : u_n = \tilde{h}\left(\tilde{s}_n(\tilde{c}_n)\right), c_n = \chi_c(\tilde{c}_n)\right\}. \tag{5.145}$$

The input–output relationship $u_n(c_n)$ is stochastic and the randomness comes from the random component messages $\tilde{b}$ and the corresponding multiple random $\tilde{c}_n$ that are consistent with c_n. Assuming that the component channel symbols are symbol-wise one-to-one mappers $s_{k,n} = \mathcal{A}_k(c_{k,n})$, the channel function $u_n = \tilde{h}(\tilde{s}_n)$ is symbol-wise and memoryless, and assuming that the component codebooks are randomly generated with IID symbols according to $S_k^N \sim \prod_{n=1}^N p(s_{k,n})$ then the H-constellation symbols U_n are also symbol-wise IID and drawn according to $U^N \sim \prod_{n=1}^N p(u_n|c_n)p(c_n)$. Any symbol-wise function of symbol-wise IID sequences remains symbol-wise IID.

The likelihood $p(u_n|c_n)$ can be easily obtained from

$$p(u_n|c_n) = \frac{\sum_{\tilde{c}_n : \chi_c(\tilde{c}_n)=c_n} p(u_n|\tilde{c}_n)p(\tilde{c}_n)}{\sum_{\tilde{c}_n : \chi_c(\tilde{c}_n)=c_n} p(\tilde{c}_n)} \tag{5.146}$$

where $p(u_n|\tilde{c}_n)$ is uniquely given by $u_n = \tilde{h}\left(\tilde{s}_n(\tilde{c}_n)\right)$. The marginalization is performed over all $\tilde{c}_n$ consistent with c_n, i.e. $\{\tilde{c} : c_n\} = \{\tilde{c} : \tilde{c}_n : \chi_c(\tilde{c}_n) = c_n\}$. This set defines the randomness in the self-dispersion (Figure 5.12d). However, the overall effect of this randomness greatly depends on the way in which it is mapped to $u_n = u_n(\tilde{c}_n)$. Different H-constellations will thus have different dispersion impact on the channel. In a special case of *self-folded NCM* (Section 4.5), there will be *no randomness* in the H-constellation self-dispersion channel.

The second part of the channel model is a standard memoryless observation channel described by the likelihood $\prod_{n=1}^N p(x_n|u_n)$, e.g. AWGN channel. The overall channel likelihood is

$$p(x_n|c_n) = \sum_{u_n} p(x_n|u_n, c_n)p(u_n|c_n)$$

$$= \sum_{u_n} p(x_n|u_n)p(u_n|c_n) \tag{5.147}$$

where we used the fact that $c_n \mapsto u_n \mapsto x_n$ also form a Markov chain. Then by substituting $p(u_n|c_n)$ we get

$$p(x_n|c_n) = \frac{1}{p(c_n)} \sum_{u_n} p(x_n|u_n) \left(\sum_{\tilde{c}_n : \chi_c(\tilde{c}_n)=c_n} p(u_n|\tilde{c}_n) p(\tilde{c}_n) \right)$$

$$= \frac{1}{p(c_n)} \sum_{\tilde{c}_n : \chi_c(\tilde{c}_n)=c_n} \left(\sum_{u_n} p(x_n|u_n) p(u_n|\tilde{c}_n) \right) p(\tilde{c}_n) \qquad (5.148)$$

where $p(c_n) = \sum_{\tilde{c}_n : \chi_c(\tilde{c}_n)=c_n} p(\tilde{c}_n)$. Since $\tilde{c}_n \mapsto u_n \mapsto x_n$, it leads us to the final expression

$$p(x_n|c_n) = \frac{1}{\sum_{\tilde{c}_n : \chi_c(\tilde{c}_n)=c_n} p(\tilde{c}_n)} \sum_{\tilde{c}_n : \chi_c(\tilde{c}_n)=c_n} p(x_n|\tilde{c}_n) p(\tilde{c}_n). \qquad (5.149)$$

This is an expected result directly related to H-SODEM (Section 4.4).

The equivalent channel transforms the system into an equivalent single user system where the input is the H-message b, which is encoded by isomorphic layered NCM with H-codebook C into H-code c^N. The code passes through a serial concatenation of two memoryless random channels: (1) the hierarchical self-dispersion channel $c_n \mapsto u_n$, and (2) the observation channel $u_n \mapsto x_n$. For the two-step equivalent channel, viewed from the perspective of H-messages, we define the following forms of mutual information.

Example 5.8   Equivalent hierarchical memoryless channel model

Assume H-MAC with two components. Component codes are on $\mathbb{F}_2$ with constellation mapper $s_{k,n} = \mathcal{A}_k(c_{k,n})$, $k \in \{A, B\}$, $\mathcal{A}_k(0) = -1$, and $\mathcal{A}_k(1) = +1$ (the BPSK alphabet). The HNC map is a symbol-wise XOR $c_n = c_{A,n} + c_{B,n}$. The H-MAC is, for a simplicity, a unity gain memoryless AWGN channel $x_n = s_{A,n} + s_{B,n} + w_n$. The H-constellation becomes $\mathcal{U}(0) = \{\pm 2\}$ and $\mathcal{U}(1) = \{0\}$.

The H-constellation self-dispersion channel model part is as follows (see Figure 5.12). Component code symbols consistent with a given H-symbol $\{\tilde{c}_n : c_n\} = \{(c_{A,n}, c_{B,n}) : c_n\}$ are $\{(c_{A,n}, c_{B,n})\} = \{(0,0), (1,1)\}$ for $c_n = 0$ and $\{(c_{A,n}, c_{B,n})\} = \{(0,1), (1,0)\}$ for $c_n = 1$. For the first case, the corresponding H-alphabet points are $u_n \in \mathcal{U}_n(0) = \{-2, +2\}$, and for the second case it is one second-order point $u_n \in \mathcal{U}(1) = \{0\}$. We see that this example has non-zero self-dispersion, which demonstrates a random "scrambling" of $u_n \in \mathcal{U}_n(0) = \{-2, +2\}$. If the component codes are symbol-wise IID with uniformly distributed symbols then the H-constellation self-dispersion is also symbol-wise IID (memoryless).

The self-dispersion channel equivalent model is described by the conditional PDF $p(u_n|c_n)$, where $p(u_n|0) = (\delta(u_n + 2) + \delta(u_n - 2))/2$ and $p(u_n|1) = \delta(u_n)$. The observation part of the channel is a standard single-input single-output model described by $p(x_n|u_n)$ for the memoryless observation. For the AWGN observation it is $p(x_n|u_n) = p_w(x_n - u_n)$ where $p_w(w)$ is Gaussian PDF.

DEFINITION 5.8 (Hierarchical Mutual Information)   Assume a memoryless equivalent hierarchical channel model forming Markov chain $C_n \mapsto U_n \mapsto X_n$ and described by $p(u_n|c_n)$ and $p(x_n|u_n)$. We define $I(C;X)$ as *hierarchical mutual information*, $I(C;U)$ as *H-constellation mutual information*, and $I(U;X)$ as *hierarchical observation mutual information*.

Achievable Hierarchical Rate

The equivalent model allows us to view the real system as a single-user equivalent system. We might be tempted to extrapolate the classical results related to achievable rates without paying attention to details. However, there are some important aspects in achievability theorem assumptions that are trivially fulfilled in the single-user system although they do *not* need to be valid in our H-coding case. Therefore, we state formally the hierarchical achievable rate theorem and go through its proof. Although it proceeds along traditional lines, it will allow us to identify the *critical differences*. A critical factor is the *independence* of the H-codewords. For that purpose, we need to extend the definition of isomorphic hierarchical codebook (Definition 4.4) by a regularity condition.

DEFINITION 5.9 (Regular Hierarchical Isomorphic Codebook)   Isomorphic hierarchical codebook $c^N = \mathcal{C}(b)$ is called *regular* if all H-codewords are independent, i.e.

$$\forall b' \neq b: \quad C^N(b') \!\perp\!\!\!\perp C^N(b) \tag{5.150}$$

where b, b' are particular distinct messages.

The definition essentially states that the H-codebook *behaves* as if its codewords were drawn randomly according to some distribution, i.e. exactly the same way as for a standard single-user random codebook (see Section A.4). The word "behaves" stresses that, in reality, the H-codebook is not directly generated as this, but its codewords are a result of the interaction of the component codewords in the channel. The result of this interaction should, however, look as if it was generated by a random generator. Now, we can state the theorem.

THEOREM 5.10 (Achievable H-rate for Layered H-decoding and Isomorphic Layered NCM)   *Assume H-MAC with K components and isomorphic layered NCM defined by component codes $c_k^N = \mathcal{C}_k(b_k)$ with one-to-one symbol-wise channel symbol mappers $s_{k,n} = \mathcal{A}_k(c_{k,n})$ for all $k \in [1:K]$, $n \in [1:N]$. Component codebooks are randomly generated according to $S_k^N \sim \prod_{n=1}^{N} p(s_{k,n})$ for component messages $b_k \in [1:2^{NR_k}]$.*

The isomorphic H-code $c^N = \mathcal{C}(b)$ with H-rate R has H-message map $b = \chi(\tilde{b})$, $b \in [1:2^{NR}]$, and symbol-wise H-code symbol HNC map $c_n = \chi_c(\tilde{c}_n)$. The memoryless equivalent hierarchical channel observation model forms Markov chain $C_n \mapsto U_n \mapsto X_n$ and it is described by $p(u_n|c_n)$ and $p(x_n|u_n)$.

If the isomorphic H-codebook is regular *then the hierarchical rate R is achievable if $R < I(C;X)$ and $R < I(C;U)$ and $R < I(U;X)$.*

Proof   The decoder is a joint typicality decoder. It declares decoded H-message $\hat{b}$ if

$$\left(C^N(\hat{b}), X^N \right) \in \mathcal{T} \tag{5.151}$$

at the first occurrence of suitable $\hat{b}$ ignoring the other ones. If no such message exists, it declares $\hat{b} = \emptyset$. Under the assumption of randomly generated codebooks symmetry, we can assume that the fixed H-message b was sent, particularly some component messages $\tilde{b}$ were sent, such that $b = \chi(\tilde{b})$. The decoding error event $\hat{b} \neq b$ has two forms, the decoding failure $\hat{b} = \emptyset$ and erroneous decision $\hat{b} \neq b$.

Assuming an equivalent hierarchical memoryless channel, a complete description of all variables involved in the decoding, i.e. the transmitted codeword $C^N(b)$, the received signal X^N, and the test codewords $C^N(b')$, is described by

$$\left(C^N(b'), X^N, C^N(b)\right) \sim \prod_{n=1}^{N} p(x_n|c_n(b))p(c_n(b), c_n(b')). \tag{5.152}$$

Utilizing the *regularity* assumption for the H-codebook, we have[14]

$$p(c_n(b), c_n(b')) = \begin{cases} p(c_n(b))\delta(c_n(b) - c_n(b')); & b' = b \\ p(c_n(b))p(c_n(b')); & b' \neq b \end{cases}. \tag{5.153}$$

All other derived PDFs can be obtained by a marginalization. For a test message b' used in the decoder typicality test, the decoder behavior is dictated by the distribution

$$\left(C^N(b'), X^N\right) \sim \begin{cases} \prod_{n=1}^{N} p(x_n|c_n(b'))p(c_n(b')); & b' = b \\ \prod_{n=1}^{N} p(x_n)p(c_n(b')); & b' \neq b \end{cases}. \tag{5.154}$$

The test codeword $C^N(b')$ for $b' \neq b$ is independent with X^N.

The decoding failure event Ω_{e0} means that the decoder does not recognize $(c^N(b'), x^N)$ as a typical set member even if the test message is $b' = b$. The probability of decoding failure is

$$P_{e0} = \Pr\{\Omega_{e0}\} = \Pr\left\{\left(C^N(b), X^N\right) \notin T\right\}. \tag{5.155}$$

If $b' = b$, then the PDF in (5.154) becomes a *joint* PDF $\prod_{n=1}^{N} p(x_n|c_n(b))p(c_n(b))$ and by the properties of the typical set $P_{e0} \approx 0$ as $N \to \infty$.

The erroneous decision event is

$$\Omega_{e1}(b') = \left\{\left(C^N(b'), X^N\right) \in T, b' \neq b\right\}. \tag{5.156}$$

If $C^N(b')$ and $X^N(b)$ are *independent*, then we can use joint typicality lemma (A.112) and

$$\Pr\{\Omega_{e1}(b')\} \approx 2^{-NI(C;X)}. \tag{5.157}$$

There are all together $(2^{NR} - 1)$ messages b' and the probability of the union of events is upper bounded by

$$\begin{aligned} P_{e1} &= \Pr\left\{\cup_{b' \neq b}\Omega_{e1}(b')\right\} \\ &\lesssim 2^{NR}2^{-NI(C;X)}. \end{aligned} \tag{5.158}$$

If $R < I(C;X)$ then $P_{e1} \approx 0$ for $N \to \infty$.

---

[14] We must keep in mind the *random* codebook generation principle. Therefore, an explicit notation using $\delta(c_n(b) - c_n(b'))$ is needed to allow random c_n for some fixed b.

The assumption of the regularity that makes $C^N(b')$ for $b' \neq b$ independent with X^N is *critical*. While the independence of the decoder test codeword and the observation of some other codeword is trivially fulfilled for a classical single-user case, where it directly follows from the random codebook generation, it is *not* automatically fulfilled for NCM. $C^N(b')$ and $X^N(b)$ are independent if and only if $C^N(b')$ and $C^N(b)$ are independent for $b' \neq b$. This is guaranteed for a *regular* isomorphic H-codebook.

The remaining statements of the theorem are consequences of $C_n \mapsto U_n \mapsto X_n$ being Markov chain. Then, the data processing inequality holds and

$$I(C;X) \leq I(C;U), \tag{5.159}$$

$$I(C;X) \leq I(U;X). \tag{5.160}$$

The achievable rate is constrained by the complete hierarchical equivalent channel properties $I(C;X)$ and also individually by the self-dispersion part $I(C;U)$ and the observation part $I(U;X)$ properties. Which of these becomes a dominant bound depends on the particular properties of the H-code and H-constellation/alphabet. $\square$

REMARK 5.3   The theorem explicitly required a *regular* H-codebook. The reasons for this follow. Component codewords themselves are independent $C_1^N(b_1) \perp\!\!\!\perp C_2^N(b_2)$ for any b_1, b_2. But it does not extend to H-codewords. The situation where the independence of $C^N(b')$ and $C^N(b)$ is not automatically satisfied is the one where two different H-codewords have one or more common component codewords. In the case $K = 2$, consider $b' \neq b$, such that $b = \chi(b_1, b_2)$ and $b' = \chi(b_1, b_2')$, $b_2' \neq b_2$, i.e. the components have one common message. Then we have

$$C^N(b) = \chi_c^N \left(C_1^N(b_1), C_2^N(b_2) \right), \tag{5.161}$$

$$C^N(b') = \chi_c^N \left(C_1^N(b_1), C_2^N(b_2') \right). \tag{5.162}$$

Depending on a particular form of χ_c^N, $C^N(b')$ and $C^N(b)$ may (but do not have to) be independent. If both differ, $b_1' \neq b_2$, $b_2' \neq b_2$, the H-codewords are independent by the independence of all involved components.

REMARK 5.4   Notice that the size and the form of the channel symbol HNC map χ_c affect only inherently the numerical value of $I(C;U)$ but do not play any explicit role in the theorem.

REMARK 5.5   The component codebooks $\mathcal{C}_k$ must have codewords generated as IID sequences and the code symbol HNC map must be *symbol-wise*. This guarantees that the H-codebook will also have IID codewords sequences. This assumption is critical for our capability to describe the system by a single-letter characterization.

We saw that the *regularity* of H-codebook was an important condition for the interpretation of the hierarchical mutual information as the upper bound on the achievable hierarchical rate. Without this, we could not interpret this value. In what follows, we focus on the condition under which the H-codebook becomes regular.

THEOREM 5.11 (Regular H-Codebook with Linear GF Map on Uniform Discrete Alphabets)   *If the isomorphic layered NCM has all component codebook symbols $c_{k,n}$,*

$k \in [1 : K]$, *from a common finite alphabet* $c_{k,n} \in \mathcal{A}_c$ *of the size* $|\mathcal{A}_c| = M_c$, *all component codebooks are randomly drawn IID sequences* $C_k^N \sim \prod_{n=1}^N p_C(c_{k,n})$ *with uniform PMF* $p_{C_k}(c_{k,n}) = 1/M_c$, *and the HNC code symbol-wise map* $c_n = \chi_c(\tilde{c}_n)$ *is an arbitrary linear map on GF* $\mathbb{F}_{M_c} = (\mathcal{A}_c, +, \times)$, *then the H-codebook is* regular.

Proof   For the isomorphic layered NCM, if $b' \neq b$ then there exists a non-empty set of indices $\mathcal{K} \subset [1 : K]$ such that $b_k' \neq b_k$, for $k \in \mathcal{K}$. Different H-messages must differ in at least one component message. By the random generation of component codebooks, the corresponding component codewords must be independent and IID symbol sequences, i.e. $C_k^N(b_k') \perp\!\!\!\perp C_k^N(b_k)$ and $C_{k,n}(b_k') \perp\!\!\!\perp C_{k,n}(b_k)$, for $k \in \mathcal{K}$. We denote $C_k = C_{k,n}(b_k)$, $C_k' = C_{k,n}(b_k')$, and similarly for H-codesymbols $C = C_n(b)$, $C' = C_{k,n}(b')$.

A linear symbol-wise HNC χ_c map on GF is defined

$$c_n = \chi_c(\tilde{c}_n) = \sum_{k \in [1:K]} a_k c_{k,n} \tag{5.163}$$

where $a_k \in \mathbb{F}_{M_c}$. The map for C and C' has a common part and the part that differs

$$C = \sum_{k \notin \mathcal{K}} a_k C_k + \sum_{k \in \mathcal{K}} a_k C_k, \tag{5.164}$$

$$C' = \sum_{k \notin \mathcal{K}} a_k C_k + \sum_{k \in \mathcal{K}} a_k C_k'. \tag{5.165}$$

We denote the common part $C_0 = \sum_{k \notin \mathcal{K}} a_k C_k$ and clearly it is some value $C_0 \in \mathcal{A}_c$ regardless of the coefficients a_k. All random variables C_k, C_k' for $k \in \mathcal{K}$ are mutually independent. If C_k is uniformly distributed on $\mathcal{A}_c$ then, by the properties of multiplication on GF (the existence of the unique inversion), $a_k C_k$ is also uniformly distributed on $\mathcal{A}_c$. Similarly, by the properties of addition on GF, $\sum_{k \in \mathcal{K}} a_k C_k$ and $\sum_{k \in \mathcal{K}} a_k C_k'$ are uniformly distributed and, by $C_k \perp\!\!\!\perp C_k', k \in \mathcal{K}$, they are independent $\left(\sum_{k \in \mathcal{K}} a_k C_k\right) \perp\!\!\!\perp \left(\sum_{k \in \mathcal{K}} a_k C_k'\right)$.

An addition of an arbitrary constant C_0 to the uniformly distributed random variable on GF is again a uniformly distributed random variable. This comes from the properties of addition on GF. As a consequence, C and C' are uniformly distributed on GF and do not depend on a common constant C_0, thus $C \perp\!\!\!\perp C'$. Since component codebooks are symbol-wise IID, then also $C^N \perp\!\!\!\perp C'^N$, which proves that the H-codebook is regular. $\square$

REMARK 5.6   Notice that both (1) the uniform component codesymbol distribution, and (2) the HNC being defined as symbol-wise linear on a common GF were needed in the proof. The extensions of the claim to continuous-valued alphabets, even with uniform distribution, or non-uniform distributions, or nonlinear mappings, are unclear.

REMARK 5.7 (Minimal HNC Map)   The theorem implies, as a consequence of GF operation properties, that the linear map must be the *minimal* cardinality HNC map over a *common* component alphabet.

5.7.5 Properties of Hierarchical Mutual Information

The equivalent H-channel is a serial concatenation of $c_n \mapsto u_n \mapsto x_n$ and the hierarchical mutual information $I(C;X)$ is upper bounded by the H-constellation mutual information $I(C;X) \le I(C;U)$ and the hierarchical observation mutual information $I(C;X) \le I(U;X)$. In order to maximize the $I(C;X)$ we must jointly match the PDF/PMF $p_C(c_n)$ and $p_U(u_n)$ to achieve the maximum. The degrees of freedom are the distribution of component code symbols $p_{S_k}(s_{k,n})$ and HNC map χ_c. For a given channel $\tilde{h}$, these two fully define the properties of $p_U(u_n)$. The hierarchical mutual information is maximized by

$$I_{\max}(C;X) = \max_{p_{\tilde{S}}(\tilde{s}_n), \chi_c(\tilde{c}_n)} I(C;X). \tag{5.166}$$

LEMMA 5.12 (H-Constellation Mutual Information for Self-Folded H-Constellation)
The self-folded H-constellation maximizes the H-constellation mutual information $I(C;U)$, and the hierarchical mutual information is equal to the hierarchical observation mutual information $I(C;X) = I(U;X) \le I(C;U)$.

Proof The self-folded H-constellation is a *one-to-one* mapping $c_n \mapsto u_n$. Then $\mathcal{H}[U|C] = 0$. The H-constellation mutual information is thus the maximal one

$$I(C;U) = \mathcal{H}[U] - \mathcal{H}[U|C] = \mathcal{H}[U]. \tag{5.167}$$

For the second part of the lemma, we realize that due to $c_n \mapsto u_n$ being a one-to-one mapping, both $C_n \mapsto U_n \mapsto X_n$ and $U_n \mapsto C_n \mapsto X_n$ are Markov chains. Then we use the chain rule for the mutual information in two forms

$$I(X; U, C) = I(X;C) + I(X;U|C)$$
$$= I(X;U) + I(X;C|U). \tag{5.168}$$

Owing to the Markov property $I(X;U|C) = 0$ and $I(X;C|U) = 0$ and thus $I(X;C) = I(X;U)$. The inequality $I(U;X) \le I(C;U)$ is a consequence of $I(C;U) = \mathcal{H}[U]$ and $I(U;X) = \mathcal{H}[U] - \mathcal{H}[U|X]$ where $\mathcal{H}[U|X] \ge 0$. $\square$

REMARK 5.8 (Self-Folded NCM Optimality) The achievable rate for a general isomorphic layered NCM (Theorem 5.10) is limited by both capabilities of the hierarchical self-dispersion and the observation

$$R \le I(C;X) \le I(C;U), \tag{5.169}$$
$$R \le I(C;X) \le I(U;X). \tag{5.170}$$

The self-folded constellation removes the first bottleneck since in that case it is $I(U;X) \le I(C;U)$ and the system is only observation-limited and *not* self-dispersion limited.

In a general (non-self-folded) NCM case the values $I(C;U)$ and $I(U;X)$ are in a general relationship ($\lessgtr$) and the one that becomes dominant depends on the amount of the self-dispersion and the quality of the observation channel. However, in the case of highly reliable observation,[15] i.e. when the observation equivocation $\mathcal{H}[U|X] \to 0$ and

---

[15] For example in a special case of an additive Gaussian noise observation channel, it would correspond to the high signal-to-noise ratio.

thus $I(U;X) \to \mathcal{H}[U]$, then the self-dispersion becomes dominant because $I(C;U) = \mathcal{H}[U] - \mathcal{H}[U|C] \leq \mathcal{H}[U]$. On the other hand, the self-folded NCM will be superior regardless of the observation channel properties.

5.7.6   HDF Coding Converse Rate

The coding converse theorem identifies what conditions on the rate are implied by the errorless decoding requirement while not being constrained to *any particular* encoding/decoding strategy. Essentially it means that there does *not exist any code* with errorless decoding with a higher rate.

The analysis technique for the coding converse rate needs to use only generic information-theoretic properties applicable to any processing. Frequently, it uses Fano's inequality. In the following, we state the HDF coding converse rate theorem. The HDF coding is understood as any possible strategy having the *H-message* as the target of the information transfer. The generic form of the converse rate theorem holds for an arbitrary (structured or random) codebook construction. However, we also show its special forms for some code subclasses that allow us to find some common relations to achievability theorems. The equivalent system model showing the impact of the structured codebooks is in Figure 5.13.

THEOREM 5.13 (HDF Coding Converse Rate)   *Assume H-MAC with K components. NCM has each component signal for $k \in [1 : K]$ encoded by a codebook C_k with*

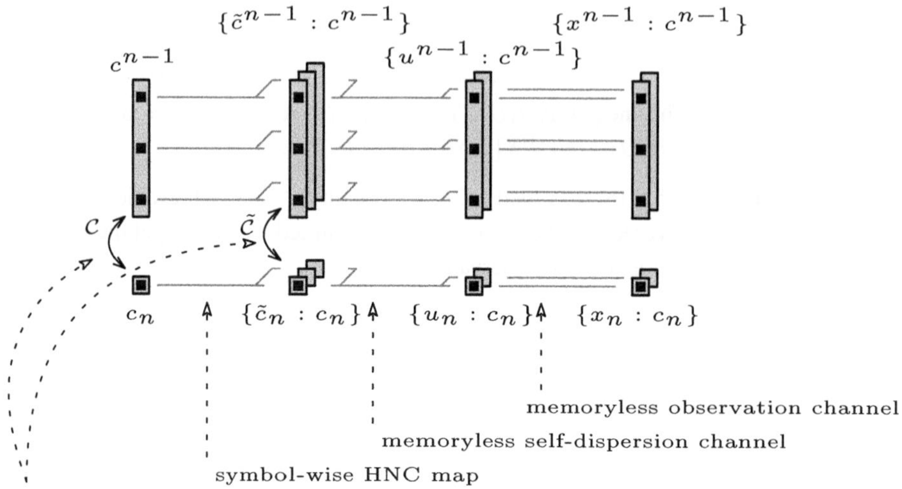

Figure 5.13 Equivalent system model for structured and random codebook constructions. A structured code C introduces the dependence between c_n and c^{n-1} (and similarly for $\tilde{C}$). Multiple blocks or circles represent the fact that there are multiple possible values consistent with given H-symbol c_n.

component message $b_k \in [1 : 2^{NR_k}]$. H-messages are defined by HNC map $b = \chi(\tilde{b})$, $b \in [1 : 2^{NR}]$. H-messages are assumed to have uniform distribution. The observation model is a joint-metric of the memoryless channel forming Markov chain $\tilde{S}_n \mapsto U_n \mapsto X_n$ and it is described by $p(u_n|\tilde{s}_n(\tilde{b}))$ and $p(x_n|u_n)$.

Any H-coding and H-decoding strategy with error probability $P_e = \Pr\{\hat{b} \neq b\}$ approaching zero $P_e \approx 0$ as $N \to \infty$ must have

$$R \leq I(U;X) - I_{\mathrm{SD}} \tag{5.171}$$

where the self-dispersion penalty is $I_{\mathrm{SD}} = \lim_{N \to \infty} \frac{1}{N} I(U^N; X^N|B) \geq 0$ for arbitrary code construction.

For special codebook subclasses, it is (a) $I_{\mathrm{SD}} = 0$ for self-folded isomorphic NCM, and (b) $I_{\mathrm{SD}} = I(U;X|B)$ for random component codes with IID symbols $S_k^N \sim \prod_{n=1}^{N} p(s_{k,n})$.

Proof The proof will be performed in the following detailed steps.

(1) Fano's inequality gives

$$\mathcal{H}[B|\hat{B}] \leq 1 + P_e NR = N(1/N + P_e R). \tag{5.172}$$

For $N \to \infty$, the term $\epsilon = (1/N + P_e R)$ goes to zero $\epsilon \to 0$.

(2) Since $B \mapsto X^N \mapsto \hat{B}$ form the Markov chain, we get from the data processing inequality that $I(B; X^N) \geq I(B; \hat{B})$ and in turn $\mathcal{H}[B|X^N] \leq \mathcal{H}[B|\hat{B}]$.

(3) Combining this and Fano's inequality result, we get

$$\mathcal{H}[B|X^N] \leq N\epsilon. \tag{5.173}$$

(4) From the definition of mutual information between random H-message B and the observation X^N, i.e. $I(B; X^N) = \mathcal{H}[B] - \mathcal{H}[B|X^N]$ and the entropy of uniform H-messages, we get

$$\begin{aligned}
NR &= \mathcal{H}[B] \\
&= I(B; X^N) + \mathcal{H}[B|X^N] \\
&\leq I(B; X^N) + N\epsilon. \tag{5.174}
\end{aligned}$$

(5) We use the chain rule for the mutual information to get

$$I(B; X^N) = \sum_{n=1}^{N} I(B; X_n|X^{n-1}). \tag{5.175}$$

(6) Individual terms can be further manipulated again using the chain rule with non-negativity of mutual information

$$\begin{aligned}
I(B; X_n|X^{n-1}) &= I(B, X^{n-1}; X_n) - I(X_n; X^{n-1}) \\
&\leq I(B, X^{n-1}; X_n). \tag{5.176}
\end{aligned}$$

(7) Then

$$I(B;X^N) \leq \sum_{n=1}^{N} I(B,X^{n-1};X_n).$$ (5.177)

(8) In the next step, we introduce a new variable U_n into some mutual information expression in such a way that the original expression $I(B,X^{n-1};X_n)$ will be part of it. The newly introduced variable can be essentially anything. There is formally no constraint on what the variable should look like. But it makes sense to introduce it such that we can interpret it in some usable information-theoretic link with the H-message b. Of course, for our intentions, the H-constellation variable U_n has the useful interpretation. Consider the expression with the new variable

$$I(B,U_n,X^{n-1};X_n) = \mathcal{H}[B,U_n,X^{n-1}] - \mathcal{H}[B,U_n,X^{n-1}|X_n].$$ (5.178)

The first entropy is

$$\mathcal{H}[B,U_n,X^{n-1}] = \mathcal{H}[U_n|B,X^{n-1}] + \mathcal{H}[B,X^{n-1}]$$ (5.179)

and the second one is

$$\mathcal{H}[B,U_n,X^{n-1}|X_n] = \mathcal{H}[U_n|B,X^{n-1},X_n] + \mathcal{H}[B,X^{n-1}|X_n].$$ (5.180)

Substituting these expressions back into the original one gives

$$\begin{aligned}
I(B,U_n,X^{n-1};X_n) &= \left(\mathcal{H}[U_n|B,X^{n-1}] - \mathcal{H}[U_n|B,X^{n-1},X_n]\right) \\
&\quad + \left(\mathcal{H}[B,X^{n-1}] - \mathcal{H}[B,X^{n-1}|X_n]\right) \\
&= I(U_n;X_n|B,X^{n-1}) + I(B,X^{n-1};X_n),
\end{aligned}$$ (5.181)

which contains the original expression $I(B,X^{n-1};X_n)$ and thus

$$I(B,X^{n-1};X_n) = I(B,U_n,X^{n-1};X_n) - I(U_n;X_n|B,X^{n-1}).$$ (5.182)

Notice again that we did not need to assume any special properties or relations for U_n for this step.

(9) We now focus on the first mutual information $I(B,U_n,X^{n-1};X_n)$ in (5.182). Since $\tilde{s}_n \mapsto U_n \mapsto X_n$ and $\tilde{s}_n = \tilde{s}_n(b)$ then also $(B,X^{n-1}) \mapsto U_n \mapsto X_n$ is a Markov chain. We use the chain rule

$$I(B,U_n,X^{n-1};X_n) = I(U_n;X_n) + I(B,X^{n-1};X_n|U_n)$$ (5.183)

and the Markov property $I(B,X^{n-1};X_n|U_n) = 0$. Notice that the fact that H-message B can produce multiple values (many-to-one function of $\tilde{B}$) of U_n does not matter at all. Conditioning by U_n and the Markov property $(B,X^{n-1}) \mapsto U_n \mapsto X_n$ is enough to have $I(B,X^{n-1};X_n|U_n) = 0$. Then we get

$$I(B,U_n,X^{n-1};X_n) = I(U_n;X_n).$$ (5.184)

(10) The second mutual information $I(U_n; X_n|B, X^{n-1})$ in (5.182) is

$$
\begin{aligned}
I(U_n; X_n|B, X^{n-1}) &= \\
&= I(U_n; X_n|B) + I(U_n; X_n|B, X^{n-1}) - I(U_n; X_n|B) \\
&= I(U_n; X_n|B) + \mathcal{H}[X_n|B, X^{n-1}] - \mathcal{H}[X_n|B, U_n, X^{n-1}] \\
&\quad - \mathcal{H}[X_n|B] + \mathcal{H}[X_n|B, U_n] \\
&= I(U_n; X_n|B) + I(X_n; X^{n-1}|B, U_n) - I(X_n; X^{n-1}|B).
\end{aligned}
\tag{5.185}
$$

Since the observation channel is memoryless, then for any B it holds $X_n \mapsto U_n \mapsto X^{n-1}|B$ and thus $I(X_n; X^{n-1}|B, U_n) = 0$.

(11) Combining the results of all previous steps gives

$$
\begin{aligned}
NR &\leq I(B; X^N) + N\epsilon \\
&\leq \sum_{n=1}^{N} I(B, X^{n-1}; X_n) + N\epsilon \\
&= \sum_{n=1}^{N} I(B, U_n, X^{n-1}; X_n) - I(U_n; X_n|B, X^{n-1}) + N\epsilon \\
&= \sum_{n=1}^{N} I(U_n; X_n) - I(U_n; X_n|B) + I(X_n; X^{n-1}|B) + N\epsilon \\
&= NI(U; X) - NI(U; X|B) + \sum_{n=1}^{N} I(X_n; X^{n-1}|B) + N\epsilon.
\end{aligned}
\tag{5.186}
$$

Thus, for $N \to \infty$, $R \leq I(U; X) - I_{\mathrm{SD}}$, where

$$
I_{\mathrm{SD}} = I(U; X|B) - \lim_{N \to \infty} \frac{1}{N} \sum_{n=1}^{N} I(X_n; X^{n-1}|B).
\tag{5.187}
$$

(12) In a general case of arbitrary (structured or random) codebooks, we get for the penalty term

$$
\begin{aligned}
I_{\mathrm{SD}} &= \mathcal{H}[X|B] - \mathcal{H}[X|B, U] \\
&\quad - \lim_{N \to \infty} \frac{1}{N} \sum_{n=1}^{N} \mathcal{H}[X_n|B] - \mathcal{H}[X_n|B, X^{n-1}].
\end{aligned}
\tag{5.188}
$$

For stationary code and observation, it holds that $\sum_{n=1}^{N} \mathcal{H}[X_n|B] = N\mathcal{H}[X|B]$. The chain rule gives $\sum_{n=1}^{N} \mathcal{H}[X_n|B, X^{n-1}] = \mathcal{H}[X^N|B]$. Then

$$
I_{\mathrm{SD}} = \lim_{N \to \infty} \frac{1}{N} \mathcal{H}[X^N|B] - \mathcal{H}[X|B, U].
\tag{5.189}
$$

The memoryless observation implies $B \mapsto U_n \mapsto X_n$ and thus

$$
I_{\mathrm{SD}} = \lim_{N \to \infty} \frac{1}{N} \mathcal{H}[X^N|B] - \mathcal{H}[X|U].
\tag{5.190}
$$

A chain rule implies

$$\mathcal{H}[X^N, U^N | B] = \mathcal{H}[X^N | U^N, B] + \mathcal{H}[U^N | B], \tag{5.191}$$

$$\mathcal{H}[X^N, U^N | B] = \mathcal{H}[U^N | X^N, B] + \mathcal{H}[X^N | B], \tag{5.192}$$

and thus

$$\mathcal{H}[X^N | B] = \mathcal{H}[X^N | U^N, B] + \mathcal{H}[U^N | B] - \mathcal{H}[U^N | X^N, B]. \tag{5.193}$$

Since $B \mapsto U^N \mapsto X^N$ is a Markov chain and the observation is memoryless, we get

$$\mathcal{H}[X^N | U^N, B] = \mathcal{H}[X^N | U^N] = N\mathcal{H}[X | U]. \tag{5.194}$$

The penalty term of self-dispersion is then

$$I_{\mathrm{SD}} = \lim_{N \to \infty} \frac{1}{N} \left(\mathcal{H}[U^N | B] - \mathcal{H}[U^N | X^N, B] \right)$$

$$= \lim_{N \to \infty} \frac{1}{N} I(U^N; X^N | B). \tag{5.195}$$

(13) In a *special case* of random symbol-wise IID component codewords, i.e. the PDF is $\prod_{n=1}^{N} p(x_n | u_n) p(u_n | \tilde{s}_n) p(\tilde{s}_n | \tilde{b})$, then $X_n \perp\!\!\!\perp X^{n-1}$ for any fixed $\tilde{b}$ and thus also for any $\tilde{b}$ consistent with b. Then $I(X_n; X^{n-1} | B) = 0$ and (5.187) gives $I_{\mathrm{SD}} = I(U; X | B)$.

(14) In a *special case* of self-folded and isomorphic NCM, U^N is one-to-one mapped to C^N and also to B. Since the observation channel is memoryless, then for any B it holds $X_n \mapsto U_n \mapsto X^{n-1} | B$. Then both parts of the penalty term are zero

$$I(X_n; X^{n-1} | B) = 0, \tag{5.196}$$

$$I(U; X | B) = 0, \tag{5.197}$$

and $I_{\mathrm{SD}} = 0$. This holds regardless of whether the code is structured or random.

$$\square$$

REMARK 5.9 (Alphabet U)   We recognize $I(U; X)$ as the hierarchical observation mutual information. It describes the capabilities of the channel $u_n \mapsto x_n$ in terms of how many bits per symbol can be reliably recognized. Notice that it simply describes the capability of the U alphabet *regardless* of the internal structure the H-constellation, i.e. a particular grouping of symbols to some HNC map. For example, in the constellation according to the example in Figure 3.12b, we would simply consider the alphabet of cardinality 3 with properly set probabilities of the symbols; here it would be $\{0.25, 0.5, 0.25\}$. The HNC map does not need to be defined at all. This is fully in line with the "converse" proof, which should abstract from any particular coding strategy. Notice that the mutual information $I(C; X)$ captures, contrary to $I(U; X)$, the HNC map and also inherently the U alphabet properties.

REMARK 5.10 (Hierarchical Self-Dispersion Mutual Information)   The term $I_{\mathrm{SD}} = \lim_{N \to \infty} \frac{1}{N} I(U^N; X^N | B)$, or $I_{\mathrm{SD}} = I(U; X | B)$ for a random IID codebook, describes the amount of the *H-constellation self-dispersion* and will be called *Hierarchical Self-Dispersion Mutual Information*. It captures residual degrees of freedom in $u_n \mapsto x_n$

when we fix the H-message b. Unlike for $I(U; X)$, I_{SD} captures the internal structure of the H-code/constellation. It has the form of the rate penalty for having self-dispersed H-constellation. The self-folded NCM does not have any residual dispersion $I_{SD} = 0$ because u^N is uniquely given by b in one-to-one mapping. Therefore, the self-folded NCM is *optimal*.

REMARK 5.11 (Structured vs. Symbol-Wise IID Random Component Codes)   The self-folded isomorphic NCM has $I_{SD} = 0$ regardless of whether the component codes $\tilde{C}$ are structured or symbol-wise IID random.

REMARK 5.12 (Hierarchical and Hierarchical Self-Dispersion Flow)   The subtractive form of the H-rate condition also suggests that the observation channel alphabet U must support two information flows, the hierarchical one and the self-dispersion one. Having an H-constellation with multiple points that could correspond to one H-message only increases demands on their resolving in $I(U; X)$, the same way as if they are all distinct codewords. At the same time, it does not help towards the H-rate. The "superfluous" H-constellation points, which belong to the same H-message, just cause the interference flow. But unlike the "classical" interference, which simply "blurs" the observation, this self-dispersion must be decodable, i.e. recognizable at the receiver. This is implied by the subtractive form of the associated information flow.

REMARK 5.13 (Converse for Isomorphic NCM)   In a special case of isomorphic NCM, the converse can be further manipulated. As a consequence of isomorphism, the H-codeword and the H-message are one-to-one mapping $b \mapsto c^N$. The conditioning by the message B thus uniquely determines the H-codeword C^N and thus $I(U^N; X^N | B) = I(U^N; X^N | C^N)$. Markov chain $C^N \mapsto U^N \mapsto X^N$ implies $\mathcal{H}[X^N | U^N, C^N] = \mathcal{H}[X^N | U^N]$ and the memoryless observation channel implies $\mathcal{H}[X^N | U^N] = N\mathcal{H}[X | U]$. The converse rate becomes

$$R \leq I(U; X) - \lim_{N \to \infty} \frac{1}{N} I(U^N; X^N | C^N)$$

$$= (\mathcal{H}[X] - \mathcal{H}[X | U]) - \lim_{N \to \infty} \frac{1}{N} \left(\mathcal{H}[X^N | C^N] - \mathcal{H}[X^N | U^N, C^N] \right)$$

$$= (\mathcal{H}[X] - \mathcal{H}[X | U]) - \lim_{N \to \infty} \frac{1}{N} \left(\mathcal{H}[X^N | C^N] - N\mathcal{H}[X | U] \right)$$

$$= \mathcal{H}[X] - \lim_{N \to \infty} \frac{1}{N} \mathcal{H}[X^N | C^N]. \tag{5.198}$$

REMARK 5.14 (Converse for Isomorphic Layered NCM with Symbol-Wise IID Component Codes)   Building on the previous case of isomorphic NCM we can proceed even further by adding the layered NCM property and assuming symbol-wise IID competent codes. As a consequence of the layered NCM, the H-codewords are mapped symbol-wise and together with memoryless property of the channel and symbol-wise IID $\tilde{C}^N$, $C_n \mapsto U_n \mapsto X_n$ forms a Markov chain and $I_{SD} = I(U; X | B)$. A conditioning by the message B uniquely determines the H-symbol C. Moreover, U and X depend on C symbol-wise and then $I(U; X | B) = I(U; X | C)$.

The converse rate is then

$$R \leq I(U;X) - I(U;X|C)$$
$$= (\mathcal{H}[X] - \mathcal{H}[X|U]) - (\mathcal{H}[X|C] - \mathcal{H}[X|U,C]). \tag{5.199}$$

Because $C_n \mapsto U_n \mapsto X_n$ form a Markov chain, X and C are conditionally independent given U, and thus $\mathcal{H}[X|U,C] = \mathcal{H}[X|U]$. Finally, for the isomorphic layered NCM, the converse rate is

$$R \leq I(U;X) - I(U;X|C) = I(C;X). \tag{5.200}$$

As a side-effect, it is also worth noting that the self-dispersion mutual information is the difference of equivocations

$$I(U;X|C) = \mathcal{H}[X|C] - \mathcal{H}[X|U]. \tag{5.201}$$

Also, notice that "end-to-end" hierarchical mutual information $I(C;X)$ hides the actual contributions from the hierarchical observation $I(U;X)$ and the H-constellation self-dispersion penalty $I(U;X|C)$. *Self-folded* isomorphic layered NCM, however, has this penalty zero $I(U;X|C) = 0$ and $c_n \mapsto u_n$ is one-to-one mapping, thus $I(C;X) = I(U;X)$.

REMARK 5.15 (Self-Dispersion Rate Gap)   Because $\lim_{N\to\infty} \frac{1}{N} I(U^N;X^N|B)$, or $I(U;X|B)$ for IID symbol-wise random codes, or $I(U;X|C)$ for IID symbol-wise random isomorphic layered NCM, describe the rate drop w.r.t. to the optimal self-folded NCM case, we also call them *Hierarchical Self-Dispersion Rate Gap*.

5.7.7   Hierarchical Capacity

The summary of the previous results is in Table 5.1. The achievable and converse H-rates results table suggests that $I(U;X)$ is achievable by isomorphic layered regular self-folded NCM with symbol-wise IID random component codebooks. The converse for self-folded isomorphic NCM is $R \leq I(U;X)$ and it is the *maximal* one for any coding scheme with a given fixed $p_U(u)$. This means that the achievable and converse rates for *self-folded isomorphic layered regular NCM with symbol-wise IID random component codebooks* are identical and maximal for a *given distribution* of $p_U(u)$. Such an NCM is thus an optimal one and its H-rate cannot be exceeded by any other scheme for a given fixed $p_U(u)$.

The previous statement identifies the optimal solution for a fixed $p_U(u)$. In order to proceed in identifying the true capacity, we need to allow optimization over the input distribution $p_U(u)$. However, the self-folding requirement strongly constrains the possible PDFs/PMFs $p_U(u_n)$. The constraint on possible alphabet U might affect the maximization of converse rate. Both $I(U;X)$ and $I(U;X|B)$ can be coupled by the self-folding constraint. The minimization of $I(U;X|B)$ to zero might not necessarily mean maximization of $I(U;X)$.

A particular codebook construction fulfilling the self-folded property in a natural way for the given channel $\tilde{s}_n \mapsto u_n \mapsto x_n$ is still an open problem. For example, in the case of the linear Gaussian observation channel, it can be obtained by *enforcing* it by

Table 5.1 Achievable and converse H-rates of various forms of NCM with HDF strategy in H-MAC and the corresponding assumptions.

NCM with HDF strategy	Achievable H-rate	Converse H-rate
General NCM structured $\tilde{C}$		$I(U;X) - \lim_{N\to\infty} \frac{1}{N}I(U^N;X^N\|B)$ uniform H-message
General NCM symbol-wise IID random $\tilde{C}$		$I(U;X) - I(U;X\|B)$ uniform H-message
NCM Self-folded Isomorphic		$I(U;X)$ uniform H-message
Layered NCM Isomorphic symbol-wise IID random $\tilde{C}$	$I(C;X)$ regular	$I(C;X) = I(U;X) - I(U;X\|C)$ uniform H-message
Layered NCM Self-Folded Isomorphic symbol-wise IID random $\tilde{C}$	$I(C;X) = I(U;X)$ regular	$I(C;X) = I(U;X)$ uniform H-message

specific nonlinear receiver preprocessing, namely $\mod \Lambda_c$ for CF strategy as shown by the following example.

Example 5.9 Modulo-lattice enforced self-folding

Assume uncoded components $s_A = \mathcal{A}(c_A)$, $s_B = \mathcal{A}(c_B)$, where GF data symbols are $c_A, c_B \in \mathbb{F}_2$ and the constellation ASK alphabet mapper is $\mathcal{A}(0) = 0$, $\mathcal{A}(1) = +1$. Assume a linear (and for simplicity) real-valued noiseless H-MAC channel $x = h_A s_A + h_B s_B$, $h_A = h_B = 1$. The HNC many-to-one map of data symbols is $c = \chi(c_A, c_B) = c_A + c_B$ (in $\mathbb{F}_2$ GF). The H-constellation $\mathcal{U}(c)$, i.e. the constellation space received signal viewed as a function of the H-data, observed by the receiver is $\mathcal{U}(0) = \{0, +2\}$, $\mathcal{U}(1) = \{+1\}$. This H-constellation is not self-folded since one H-data symbol $c = 0$ (which corresponds to $(c_A, c_B) \in \{(0,0),(1,1)\}$) is mapped to two distinct constellation points $\{0, +2\}$. It is only partially self-folded since the second point $c = 1$ corresponding to $(c_A, c_B) \in \{(0,1),(1,0)\}$ maps to one common constellation point $\{+1\}$. The CF modulo-lattice operation in real-valued one-dimensional space can be modeled (ignoring for simplicity the lattice inflation) as $y = x \mod 2$, and thus the y observation has H-constellation $\mathcal{U}'(c)$ such that $\mathcal{U}'(0) = 0$ and $\mathcal{U}'(1) = +1$, which is now self-folded.

The following hierarchical capacity conjecture ignores *natural* realizability of the self-folding in the radio channel and only *conjectures* the possibility of decoupling of the maximization of $I(U;X)$ and zeroing of $I(U;X\|B)$ under the self-folding. Therefore it should be understood in a rather relaxed manner.

CONJECTURE 5.14 (Hierarchical Capacity)   *The hierarchical capacity for HDF strategy in H-MAC is*

$$C_H = \max I(U; X) \qquad (5.202)$$

where the maximization is taken over all $p_U(u_n)$ consistent with underlying $p(\tilde{s}_n)$ and $\tilde{s}_n \mapsto u_n$ is such that u_n is a self-folded H-constellation.

5.7.8   Finite Alphabet Regular Layered NCM in Linear Memoryless Gaussian Channel

System Model

The evaluation of the H-rates for a finite alphabet NCM in a linear memoryless Gaussian channel is relatively straightforward. Assume a complex-valued H-MAC symbol combination channel model

$$u_n = u(\tilde{s}_n) = \sum_{k=1}^{K} h_k s_{k,n} \qquad (5.203)$$

in m-dimensional complex constellation space. We will assume a layered NCM and thus component channel symbols are symbol-wise one-to-one mappings $s_{k,n} = \mathcal{A}_k(c_{k,n})$ of code symbols $c_{k,n}$ from a common alphabet of the size M_c. We assume that the discrete component code symbols have a uniform distribution $p(c_{k,n}) = 1/M_c$. The observation channel is a complex Gaussian model

$$x_n = u_n + w_n \qquad (5.204)$$

where the noise variance is σ_w^2 and its PDF for m-dimensional constellation symbols is

$$p_W(w_n) = \frac{1}{\pi^m \sigma_w^{2m}} \exp\left(-\frac{1}{\sigma_w^2}\|w_n\|^2\right). \qquad (5.205)$$

Hierarchical Mutual Information

The simplest is the evaluation of the hierarchical observation mutual information $I(U; X)$

$$\begin{aligned} I(U; X) &= \mathcal{H}[X] - \mathcal{H}[X|U] \\ &= \mathcal{H}[X] - \mathcal{H}[W] \end{aligned} \qquad (5.206)$$

where $\mathcal{H}[W]$ is the noise entropy. The following expressions involve H-code symbols C and thus are applicable on *isomorphic* and *regular* layered NCMs only. The hierarchical self-dispersion mutual information is

$$\begin{aligned} I(U; X|C) &= \mathcal{H}[X|C] - \mathcal{H}[X|C, U] \\ &= \mathcal{H}[X|C] - \mathcal{H}[X|U] \\ &= \mathcal{H}[X|C] - \mathcal{H}[W] \end{aligned} \qquad (5.207)$$

where $\mathcal{H}[X|C, U] = \mathcal{H}[X|U]$ since $C_n \mapsto U_n \mapsto X_n$ form a Markov chain. The last is the overall hierarchical information

$$I(C; X) = \mathcal{H}[X] - \mathcal{H}[X|C]. \qquad (5.208)$$

All mutual information expressions above can be obtained from the knowledge of entropies $\mathcal{H}[W]$, $\mathcal{H}[X]$, $\mathcal{H}[X|U]$, $\mathcal{H}[X|C]$. The simplest is the noise entropy $\mathcal{H}[W] = \lg(\pi e \sigma_w^2)$. For the remaining ones, we get the corresponding PDFs

$$
\begin{aligned}
p(x_n) &= \sum_{\tilde{c}_n} p(x_n|\tilde{c}_n)p(\tilde{c}_n) \\
&= \sum_{\tilde{c}_n} p\left(x_n|u(\tilde{s}_n(\tilde{c}_n))\right) p(\tilde{c}_n) \\
&= \frac{1}{M_c^K} \sum_{\tilde{c}_n} p_W\left(x_n - u(\tilde{s}_n(\tilde{c}_n))\right),
\end{aligned}
\tag{5.209}
$$

$$
p(x_n|u_n) = p_W(x_n - u_n),
\tag{5.210}
$$

$$
p(u_n) = \sum_{\tilde{c}_n:u(\tilde{s}_n(\tilde{c}_n))=u_n} p(\tilde{c}_n),
\tag{5.211}
$$

and using (4.17)

$$
p(x_n|c_n) = \frac{1}{M_c^{K-1}} \sum_{\tilde{c}_n:\chi_c(\tilde{c}_n)=c_n} p_W(x_n - u(\tilde{c}_n)),
\tag{5.212}
$$

where we also realize that the regularity implies the *minimal* HNC map for which the following holds

$$
p(c_n) = \sum_{\tilde{c}_n:\chi_c(\tilde{c}_n)=c_n} p(\tilde{c}_n) = \frac{1}{M_c}.
\tag{5.213}
$$

The entropies are then obtained by m-dimensional integration of the corresponding densities that must be done typically by a numerical evaluation. We also need to keep in mind that, because of the finiteness of the alphabets (see the discussion about complex-valued alphabets in Section 5.3.2), the results are *strongly influenced* by the channel parametrization $\tilde{h}$, which is the phenomenon typical for WPNC.

Monte-Carlo Evaluation of the Integrals

All integrals involved in the entropy evaluations are complicated multi-fold integrals over the m-dimensional complex plane. Their direct evaluation by traditional numerical integration procedures is known to be complex and very slowly converging. Better computational efficiency is achieved by Monte-Carlo numerical evaluation of the integrals. This is not to be confused with Monte-Carlo simulation. We use it for the numerical evaluation of the integrals, not for running the system with some particular coding and signal processing algorithms. The method is usable for evaluation of any summation or integration of the form

$$
I = \int f(x)g(x)\, dx.
\tag{5.214}
$$

We consider $f(x)$ as a PDF, and hence the integral can be evaluated as

$$
I = E_{X \sim f(x)}[g(X)].
\tag{5.215}
$$

Providing that we are able to generate random values with the distribution $f(x)$: $X \sim f(x)$ then the integral is approximated by evaluating the empirical mean $\hat{E}[.]$ of $g(x)$. Generation of the random variable with a given density is a relatively easy task particularly for linear Gaussian mixture densities. We can also always resort to building a full system model producing the desired random variable and feeding that system with its all random excitations – typically data and noise sources.

For example, a numerical evaluation of $\mathcal{H}[X]$ gives

$$\mathcal{H}[X] \approx -\hat{E}_{X \sim p(x_n)}\left[\lg\left(\frac{1}{M_c^K} \sum_{\tilde{c}_n} p_W(X_n - u(\tilde{c}_n)) \right) \right] \qquad (5.216)$$

where X_n is generated with the distribution $p(x_n)$. The conditional entropy approximation is obtained as

$$\mathcal{H}[X|C] = -\hat{E}_{X,C \sim p(x_n,c_n)}\left[\lg p(X|C) \right]. \qquad (5.217)$$

5.8   End-to-End Solvability

The end-to-end solvability of the WPNC cloud is affected by two aspects. First, it is a "contents" provided to the final destination throughout the multiple H-messages formed by a hierarchical encapsulation of the HNC maps. Second, it is a "form" for how this "contents" is presented. It is given by the H-processing operation providing various forms of the metrics μ.

5.8.1   Global Linear HNC Map

Here we analyze how to combine local linear HNC maps into the global one. In a generic case, we encapsulate the maps (Section 3.4). But in a special case of *linear* maps, we can get an elegant and simple solution based on a product of matrices. This situation is particularly useful for HDF strategy, where each encapsulation step provides a *reliable* decision of a *linear hierarchical* message map. Assuming that all linear maps are on a *common* GF, and that all individual steps are cycle-free homogenous stages processed in a sequence, we can get the global map for L encapsulation levels (stages) as a simple product

$$\tilde{\mathbf{b}}' = \tilde{\mathbf{Q}}_L \times \cdots \times \tilde{\mathbf{Q}}_1 \tilde{\mathbf{b}} = \tilde{\mathbf{Q}}\tilde{\mathbf{b}} \qquad (5.218)$$

where the *end-to-end global HNC matrix* is (LHS matrix multiplication)

$$\tilde{\mathbf{Q}} = \prod_{\ell=1}^{L} \tilde{\mathbf{Q}}_\ell. \qquad (5.219)$$

5.8.2   Solvability of Linear HNC Map

Once we have an available end-to-end HNC map description (e.g. a concatenation of linear maps in HDF strategy), we use this information for the ultimate WPNC goal, i.e. for solving the individual messages. In this respect, we need (a) to find criteria for the solvability, and (b) to set up a practical algorithm performing this solution. The solvability criteria are generally described by the generalized exclusive law (Section 3.4) but it is too abstract and we need a more straightforward solution, at least for some special scenarios (linear maps). The solvability and the solution algorithms are applied at the *message* level since this is the only entity we are interested in as a final goal.

The solvability for a one-step (one-stage) linear map is relatively straightforward when the maps are defined on GF. We can use matrix inversions and we simply solve the set of linear equations. The conditions for the solvability are trivial – a full-rank HNC map matrix. This situation is essentially the one covered by a traditional NC.

In WPNC, the situation can frequently be complicated by several facts. First, using a *common* GF over the whole WPNC network might be rather constraining when designing NCM. It is particularly apparent in the isomorphic NCM design variant where all messages, codewords, channel symbols, and codes would have to share a common (e.g. binary) alphabet. To relax this condition, it would be nice to have the capability to "slice" the solution into several sequential steps, each possibly having different properties. Moreover, different stages might use different strategies and the reliable decisions might not be available at all interim stages. Second, in some situations, e.g. when relaxing the size of H-constellation for isomorphic NCM to make it more resistant to channel fading, it might be too constraining even to have GF-based operations. Then we might want to have only *ring-based* operations where we do not have multiplication inverses generally available. Third, the relay H-processing operation and the associated information measure μ_b can have various forms (decision, compression, linear scaling) and we need to know how this influences the solvability.

In this section, however, we investigate the simplest special case of linear common GF HNC maps used in L stages and each stage performing reliable decision on the all H-messages $\tilde{\mathbf{b}}_\ell$. All nodes participating in the ℓth, $\ell \in [1 : L]$, stage transmit reliable decisions on concatenated H-messages

$$\tilde{\mathbf{b}}_\ell = \tilde{\mathbf{Q}}_\ell \tilde{\mathbf{b}}_{\ell-1} \tag{5.220}$$

where the original sources are $\tilde{\mathbf{b}}_0 = \tilde{\mathbf{b}}$. Clearly, $\tilde{\mathbf{Q}} = \prod_{\ell=1}^{L} \tilde{\mathbf{Q}}_\ell$ and all per-stage matrices $\tilde{\mathbf{Q}}_\ell$ must be *full rank*.

5.8.3 Solving Linear Ring-Based HNC Maps

In this section we consider the more general linear case where data symbols b and the coefficients of the linear maps q (the elements of the per-stage matrices $\tilde{\mathbf{Q}}_\ell$) may be from a *ring* rather than from a Galois field (see Section A.2.1). This allows increased flexibility in the mappings available, especially for fading channels. It of course subsumes the case outlined in the previous section. We still assume that a common ring is used over the whole network, but because we will allow that some sources and some relays may use a subset of the ring elements, this effectively allows symbol cardinality to vary over the network. We note that this may have implications for the isomorphic property of the effective mappings, as discussed in Section 4.2.4, but this is not our concern in this section.

In this case the generalized exclusive law (Section 3.4) becomes

$$\forall \tilde{\mathbf{b}}(b_i), \tilde{\mathbf{b}}(b_i') : b_i \neq b_i' \Rightarrow \tilde{\mathbf{Q}}\tilde{\mathbf{b}}\,(b_i) \neq \tilde{\mathbf{Q}}\tilde{\mathbf{b}}\,(b_i') \tag{5.221}$$

where $\tilde{\mathbf{Q}}$ denotes the end-to-end matrix relating the HNC-mapped symbols available at the destination to all source symbols, and (as in Section 3.4), $\tilde{\mathbf{b}}(b_i) = \{\tilde{\mathbf{b}} : b_i\}$ denotes the set of all message vectors $\tilde{\mathbf{b}}$ consistent with b_i, i.e. having b_i in the ith position.

This is clearly fulfilled if symbols and coefficients are defined over a common GF, and $\tilde{\mathbf{Q}}$ is full rank. We will see, however, that it may still be fulfilled if they are from a common ring $\mathbb{S}$ rather than a field, and also in the case addressed by the generalized exclusive law where only one source symbol is of interest at the destination in question. First define $\bar{\mathbf{b}}^{(i)}$ as $\tilde{\mathbf{b}}$ with the ith element omitted, and similarly $\bar{\mathbf{Q}}^{(i)}$ as $\tilde{\mathbf{Q}}$ with the ith column omitted, and also $\mathbf{q}^{(i)}$ as the ith column of $\tilde{\mathbf{Q}}$. Then we may rewrite the generalized exclusive law as

$$\forall \bar{\mathbf{b}}^{(i)}, \bar{\mathbf{b}}'^{(i)}, b_i, b_i' : b_i \neq b_i' \Rightarrow \bar{\mathbf{Q}}^{(i)}\bar{\mathbf{b}}^{(i)} + \mathbf{q}^{(i)}b_i \neq \bar{\mathbf{Q}}^{(i)}\bar{\mathbf{b}}'^{(i)} + \mathbf{q}^{(i)}b_i' \tag{5.222}$$

and thus

$$\mathbf{q}^{(i)}\left(b_i - b_i'\right) \neq \bar{\mathbf{Q}}^{(i)}\left(\bar{\mathbf{b}}'^{(i)} - \bar{\mathbf{b}}^{(i)}\right). \tag{5.223}$$

We consider two cases here: first, when the RHS of this equation is zero. In this case the generalized exclusive law will fail (that is, equality will occur) only if all elements of $\mathbf{q}^{(i)}$ are zero-divisors (see Section A.2.1) in $\mathbb{S}$. Note that this will not occur if the coefficients are from a GF, or if they all have unique inverses in $\mathbb{S}$.

Second, if the RHS is non-zero, equality will occur only if $\mathbf{q}^{(i)}$ is linearly dependent on the columns of $\bar{\mathbf{Q}}^{(i)}$, and conversely the inequality remains true, and thus the generalized exclusive law is valid, provided $\mathbf{q}^{(i)}$ is linearly independent of the columns of $\bar{\mathbf{Q}}^{(i)}$. This of course applies if the algebra is over a field as well as a ring. Note that it does not necessarily require $\tilde{\mathbf{Q}}$ to be full rank, since it does not require that all the columns of $\bar{\mathbf{Q}}^{(i)}$ be linearly independent of one another.

To summarize, the generalized exclusive law holds, and hence the source symbol b_i may be unambiguously decoded at the destination provided both: (1) not all elements of $\mathbf{q}^{(i)}$ are zero-divisors in $\mathbb{S}$, *and* (2) $\mathbf{q}^{(i)}$ is linearly independent of the columns of $\bar{\mathbf{Q}}^{(i)}$. Note that these conditions apply to the end-to-end matrix $\tilde{\mathbf{Q}}$ rather than to the per-stage matrices $\tilde{\mathbf{Q}}_\ell$.

5.8.4   H-Processing Operations

This section explains what conditions must be fulfilled by HNC maps and the H-information measure to allow the end-to-end solvability. They are both defined by the H-processing relay operation (Figure 3.5). The conditions of solvability are necessary but not sufficient from the whole WPNC network information transfer perspective. We can interpret them as a *network source coding* part of WPNC.[16] Here, we will not solve the issues related to the transfer over an unreliable, noisy, channel, i.e. channel coding, but we purely concentrate on the *contents* and its *representation*, i.e. the source coding. Because this is done over multiple nodes, it fits under a network source coding. The contents, i.e. some HNC map b, is represented through the information measure μ_b. Both the contents and its representation measure have an impact on the solvability. Preceding sections concentrated on the simplest cases for both linear HNC maps, as contents, and full reliable message decisions, as the information measure. Here we relax these cases into a more general form.

The WPNC network source coding will be treated independently of the channel coding by simply assuming that node front-end observations and processing/decoding are such that the relay processing provides a *correct measure* $\mu(b)$. Finding the conditions for the end-to-end network solvability can be interpreted as a zero-distortion source coding in the information-theoretic perspective. The problem, although quite similar to the classical NC, will be treated in a more general form than in the traditional discrete NC. It will be done mainly by allowing a variety of forms of the measure μ (DF, SF, QF, CpsF), unlike the classical NC, which assumes only full reliable decisions.

Global H-Message Markov Cut

The WPNC network with some sources, destinations, and relays can be split into a Markov chain of H-message measure processing (Figure 5.14). All front-end and back-end channel coding related processing is considered to be reliable, and the only entities that we treat are H-message measures μ_b that result from the relay processing operation (Figures 3.4 and 3.5). The Markov processing cut is defined for a selected set of sources $\mathcal{S}$ at one end, some "cut" (border) nodes $\mathcal{B}$, and corresponding destinations $\hat{\mathcal{S}}$ at the other end. We will use the notation $\tilde{\mu}_b(\mathcal{S}) = \{\mu_{b_k}\}_{k\in\mathcal{S}}$ for the set of all H-measures processed in the nodes belonging to the set $\mathcal{S}$. It is worth noting that the "cut" is led along the nodes in the set $\mathcal{B}$, not over the wireless links.

DEFINITION 5.15 (Global H-message Markov Cut)   Assume a WPNC network with some subset of source nodes $\mathcal{S}$ and corresponding set of destination nodes $\hat{\mathcal{S}}$, where the map measures in both sets are *identical* $\tilde{\mu}_b(\mathcal{S}) = \tilde{\mu}_b(\hat{\mathcal{S}})$. The set of nodes $\mathcal{B}$ is defined to be a *global H-message Markov cut* for a given $\mathcal{S}, \hat{\mathcal{S}}$ if the H-message measures in the subsets $\mathcal{S}, \mathcal{B}, \hat{\mathcal{S}}$ form a Markov chain

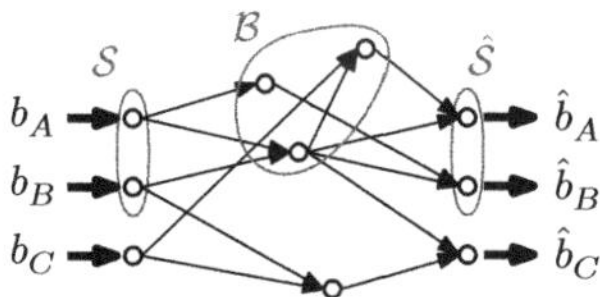

Figure 5.14 Global H-message Markov cut for a given $\mathcal{S}, \hat{\mathcal{S}}$.

[16] Notice that this sets in fact the *source and channel WPNC coding separation*. It differs from the traditional NC used on top of classical PHY by allowing *more general* information measure μ not restricted to common GF-based linear operations.

$$\tilde{\mu}_b(\mathcal{S}) \mapsto \tilde{\mu}_b(\mathcal{B}) \mapsto \tilde{\mu}_b(\hat{\mathcal{S}}). \tag{5.224}$$

REMARK 5.16 (Source and Destination Measures) The source and destination measures are typically directly the data messages, e.g. $\tilde{\mu}_b(\mathcal{S}) = \{b_A, b_B\}$ and the destination estimates $\tilde{\mu}_b(\hat{\mathcal{S}}) = \{\hat{b}_A, \hat{b}_B\}$.

REMARK 5.17 (Destination Node Strategy Including Other Maps) If the destination node with its primary destination target $\hat{b}_k$ also includes evaluation of some HSI $\hat{b}_{k'}$, e.g. as an auxiliary decoding step, it needs to be included in the set $\hat{\mathcal{S}}$.

REMARK 5.18 (Generalization of Source and Destination Sets) The source and destination sets do not need to be the real sources and destinations. The sets $\mathcal{S}, \hat{\mathcal{S}}$ do not need to have the same size and also do not need to include all sources and destinations in the network, but it can be only their subsets we are interested in. The only important aspect is that they have *identical* set of measures $\tilde{\mu}_b(\mathcal{S}) = \tilde{\mu}_b(\hat{\mathcal{S}})$.

H-Message Network Processing Inequality

The following theorem is a simple consequence of the information processing and of Fano's inequalities. However, it allows us to properly understand the role of the H-message information measures inside the network on the end-to-end information transfer.

THEOREM 5.16 (H-Message Network Processing Inequality) *Assume discrete H-message measures $\tilde{\mu}_b(\mathcal{S}) = \tilde{\mu}_b(\hat{\mathcal{S}})$ transmitted by back-ends and decoded by front-ends in the node set $\mathcal{S}$ and $\hat{\mathcal{S}}$ respectively. The end-to-end decoding distortion is defined in terms of the error probability $P_e = \Pr\left\{\tilde{\mu}_b(\mathcal{S}) \neq \tilde{\mu}_b(\hat{\mathcal{S}})\right\}$. In order to achieve this end-to-end goal, the equivocation at arbitrary global H-message Markov cut $\mathcal{B}$ must be*

$$\mathcal{H}\left[\tilde{\mu}_b(\mathcal{S})\big|\tilde{\mu}_b(\mathcal{B})\right] \leq \mathrm{H}(P_e) + P_e \lg |\{\tilde{\mu}_b(\mathcal{S})\}| . \tag{5.225}$$

In a special case of zero distortion errorless deterministic decoding, it must be $\mathcal{H}\left[\tilde{\mu}_b(\mathcal{S})\big|\tilde{\mu}_b(\mathcal{B})\right] = 0$.

Proof For the end-to-end decoding fidelity, Fano's inequality dictates an upper bound on the equivocation

$$\mathcal{H}\left[\tilde{\mu}_b(\mathcal{S})\big|\tilde{\mu}_b(\hat{\mathcal{S}})\right] \leq \mathrm{H}(P_e) + P_e \lg |\{\tilde{\mu}_b(\mathcal{S})\}| . \tag{5.226}$$

Data processing inequality for the Markov chain $\tilde{\mu}_b(\mathcal{S}) \mapsto \tilde{\mu}_b(\mathcal{B}) \mapsto \tilde{\mu}_b(\hat{\mathcal{S}})$ implies

$$I\left(\tilde{\mu}_b(\mathcal{S}); \tilde{\mu}_b(\hat{\mathcal{S}})\right) \leq I\left(\tilde{\mu}_b(\mathcal{S}); \tilde{\mu}_b(\mathcal{B})\right) \tag{5.227}$$

and in turn

$$\mathcal{H}\left[\tilde{\mu}_b(\mathcal{S})\big|\tilde{\mu}_b(\hat{\mathcal{S}})\right] \geq \mathcal{H}\left[\tilde{\mu}_b(\mathcal{S})\big|\tilde{\mu}_b(\mathcal{B})\right] . \tag{5.228}$$

A combination of this with Fano's inequality gives the desired result

$$\mathcal{H}\left[\tilde{\mu}_b(\mathcal{S})\big|\tilde{\mu}_b(\mathcal{B})\right] \leq \mathrm{H}(P_e) + P_e \lg |\{\tilde{\mu}_b(\mathcal{S})\}| . \tag{5.229}$$

The special case result comes simply by realizing that, for $P_e = 0$, we must have $H(P_e) + P_e \lg |\{\tilde{\mu}_b(\mathcal{S})\}| = 0$. $\qquad\qquad\square$

The theorem can be interpreted as a decoding-fidelity quantized sufficient statistic for obtaining the desired estimates at the destination set. It can be used for evaluating trade-offs involving various forms of measures (e.g. full decision vs. compression) at various network nodes. A basic usage is shown in the following examples.

Example 5.10 Assume a two-source two-relay network in Figure 5.15a. Assume that we are interested in the behavior w.r.t. message of source A. Then $\mathcal{S} = \{A\}$, $\hat{\mathcal{S}} = \{\hat{A}\}$ and corresponding measure sets are $\tilde{\mu}_b(\mathcal{S}) = \{b_A\}$ and $\tilde{\mu}_b(\hat{\mathcal{S}}) = \{b_A\}$. Clearly the only available global H-message Markov cut is $\mathcal{B} = \{R_1, R_2\}$. If we want to obtain an errorless decision on $\hat{A}$, the Markov cut equivocation needs to be $\mathcal{H}\left[b_A \middle| \mu_{b_{R1}}, \mu_{b_{R2}}\right] = 0$.

Example 5.11 Assume the butterfly network in Figure 5.15b. Again, we are interested in the behavior w.r.t. the message of source A. Then $\mathcal{S} = \{A\}$, $\hat{\mathcal{S}} = \{\hat{A}\}$ and corresponding measure sets are $\tilde{\mu}_b(\mathcal{S}) = \{b_A\}$ and $\tilde{\mu}_b(\hat{\mathcal{S}}) = \{b_A\}$. The Markov H-message cut is $\mathcal{B} = \{R\}$. However, notice that by letting the target measure be only $\tilde{\mu}_b(\hat{\mathcal{S}}) = \{b_A\}$ we completely ignore message b_B in the decoding strategy of the node $\hat{A}$. It corresponds simply to having an interference channel where channel links $B \to R$ and $B \to \hat{A}$ are treated as an interference. The relay measure μ_{b_R} is then the only source of information for destination $\hat{A}$ and errorless decoding requires $\mathcal{H}\left[b_A \middle| \mu_{b_R}\right] = 0$.

Example 5.12 We now extend the previous butterfly network decoding example and change the strategy to allow the destination $\hat{A}$ to use (and to provide) also data b_B (Figure 5.15c). The destination set is $\hat{\mathcal{S}} = \{\hat{A}\}$ and the corresponding measures are $\tilde{\mu}_b(\hat{\mathcal{S}}) = \{b_A, b_B\}$. In order to find the corresponding source set with the identical

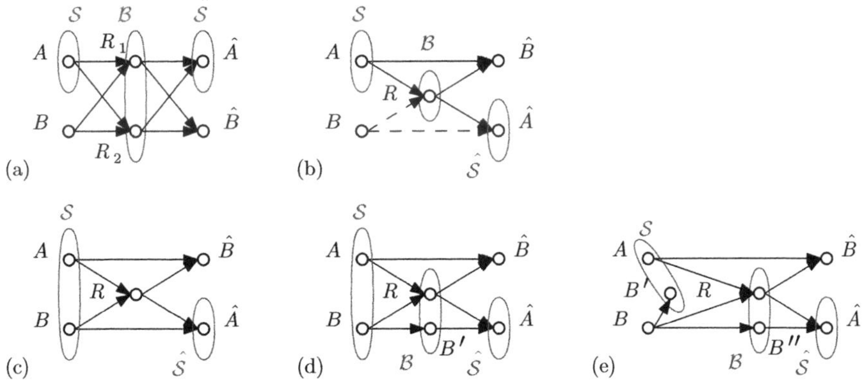

Figure 5.15 Global H-message Markov cut examples.

measures, we would need to include also the source B and thus $\mathcal{S} = \{A, B\}$ and $\tilde{\mu}_b(\mathcal{S}) = \{b_A, b_B\}$. Unfortunately there is no way to define the H-message Markov cut.

We can, however, define a new, virtual, node B' serving purely as a dummy node replicating the b_B message (Figure 5.15d). The H-message Markov cut is then $\mathcal{B} = \{R, B'\}$ and the errorless communication requires $\mathcal{H}\left[b_A, b_B \big| \mu_{b_R}, b_B\right] = \mathcal{H}\left[b_A \big| \mu_{b_R}, b_B\right] = 0$ where we used the chain rule for entropies.

Example 5.13   The previous example can be generalized for the situation where the destination $\hat{A}$ does not have a full decision on the b_B message available but it has available (soft or incomplete) only measure μ_{b_B}. It can be modeled by properly defining the dummy B' relay H-message measure μ_{b_B} and then it must be $\mathcal{H}\left[b_A, b_B \big| \mu_{b_R}, \mu_{b_B}\right] = 0$.

Example 5.14   A further generalization may assume that the measure $\mu_{b_{B'}}$ available to the destination does not need to be the same as the measure used by the destination μ_{b_B} (Figure 5.15e). We use another dummy node B' with measure μ_{b_B} to model the source set $\mathcal{S} = \{A, B'\}$ with measure set $\tilde{\mu}_b(\mathcal{S}) = \{b_A, \mu_{b_B}\}$ and the second dummy node B'' with measure $\mu_{b_{B'}}$ to model the measure available to the destination. Then the errorless decoding requires $\mathcal{H}\left[b_A, \mu_{b_B} \big| \mu_{b_R}, \mu_{b_{B'}}\right] = 0$.

Part III

Design of Source, Relay, and Destination Strategies

6   NCM and Hierarchical Decoding Design for H-MAC

6.1   Introduction

A focus of this chapter is NCM and H-decoding design for a *single-stage H-MAC* channel. It is a basic building block of more complex WPNC networks. We will build on the generic theory background developed in Part II. We show various particular NCM and H-decoding designs for selected scenarios, topologies, and channel models. A more specific setup of the scenarios allows us to deal with more specific aspects of the design, e.g. channel parametrization effects, H-decoding specifically matched to particular NCM design and scenario, numerical evaluation of the performance, etc. Many of these results can be evaluated only under the *particular finite alphabet* choice, in contrast with Part II where we tried to abstract from the particular channel alphabet. It also allows us to show some practical implementation aspects here.

All techniques presented in this chapter fall into the *hierarchical decode and forward* class. All receivers (relays) involved in the single-stage H-MAC are assumed to make a decision on some H-message. From the global WPNC network end-to-end solvability, these H-messages must comply with H-message network processing inequality (see Section 5.8.4). From the perspective of this chapter, we assume that this is fulfilled and we focus only on obtaining the H-messages on the relays.

6.2   NCM with HNC Maps Adapted to Channel Parameters

We start with a simple concept of designing an *uncoded* NCM with a symbol-wise HNC map and the H-constellation optimized for a given channel parametrization.[1]

6.2.1   System Model

Component Symbols

We assume two-component (SA, SB) H-MAC topology (Figure 6.4). Transmitted channel symbols $c_{A,n}, c_{B,n} \in \mathbb{F}_M$ are symbol-wise mapped $s_{A,n} = \mathcal{A}_s(c_{A,n})$, $s_{B,n} = \mathcal{A}_s(c_{B,n})$,

---

[1] This topic appears in many extensions and variants (including also some simple encoded cases) [28], [29], [30], [31] frequently under the name "denoising." The term denoising refers to the fact that the many-to-one HNC map and the corresponding H-constellation lead to the reduction of the number of decision regions and, together with possible merges of the neighboring regions, reduces the ambiguity in the decision process. In this book, we will, however, keep the terminology and notation as was developed in Part II.

$s_{A,n}, s_{B,n} \in \mathcal{A}_s$, on a common channel constellation space *finite* alphabet $\mathcal{A}_s$ of size M. Components are synchronized on the symbol timing level.

Linear AWGN H-MAC Channel

The channel model is a flat-fading linear AWGN channel with relative fading (Section 3.5) coefficient $h \in \mathbb{C}$

$$x_n = \left(s_{A,n} + h s_{B,n}\right) + w_n \tag{6.1}$$

where w_n is complex-valued AWGN with σ_w^2 variance per dimension. A common fading coefficient (Section 3.5.2) is assumed as a unity. The hierarchical channel combined symbol is

$$u_n = s_{A,n} + h s_{B,n} \tag{6.2}$$

and the H-alphabet is $\mathcal{A}_u, u_n \in \mathcal{A}_u$.

H-Constellation

A symbol-wise HNC map is $c_n = \chi_c(c_{A,n}, c_{B,n})$, $c_{A,n}, c_{B,n} \in \mathbb{F}_M$. Values c_n are not necessarily from $\mathbb{F}_M$ as will be shown later. The H-constellation $\mathcal{U}(c_n)$ classifies the H-alphabet symbols according to their corresponding value of HNC map

$$\mathcal{U}(c_n) = \left\{ u_n : u_n = s_{A,n}(c_{A,n}) + h s_{B,n}(c_{B,n}), \ c_n = \chi_c(c_{A,n}, c_{B,n}) \right\}. \tag{6.3}$$

6.2.2   H-Decoding

The H-decoding is the simplest possible one based on uncoded symbol-wise H-symbol decisions

$$\hat{c}_n = \arg \min_{c_{A,n}, c_{B,n} : \chi_c(c_{A,n}, c_{B,n}) = c_n} \left\| x_n - \left(s_{A,n}(c_{A,n}) + h s_{B,n}(c_{B,n})\right) \right\|^2. \tag{6.4}$$

It is the hierarchical minimum distance approximation (4.20), which does not respect potential different multiplicities of H-alphabet points and works only for high SNR (see further details in Section 4.4).

6.2.3   Channel Optimized HNC Maps

Singular Fading

The HNC map χ_c is required to obey the exclusive law (Section 3.4) under all singular fadings (Section 3.5.3) in order to make them resolved. Unfortunately, there is no one common HNC map that would achieve that for all possible relative parametrizations h. This is illustrated for a two-component QPSK ($\mathcal{A}_s(c_{l,n}) = \exp(j\pi(c_{l,n}/2 + 1/4))$) with bit-wise XOR HNC map and various channel parametrizations in Figure 6.1. Figures 6.1a1,a2,a3 show the decision regions in a complex plane of x_n values for the hierarchical minimum distance metric approximation, and Figures 6.1b1,b2,b3 for the true marginalized H-metric (4.17). We used this example scenario to demonstrate the impact of the chosen H-metric. We see that the decision regions can be quite different.

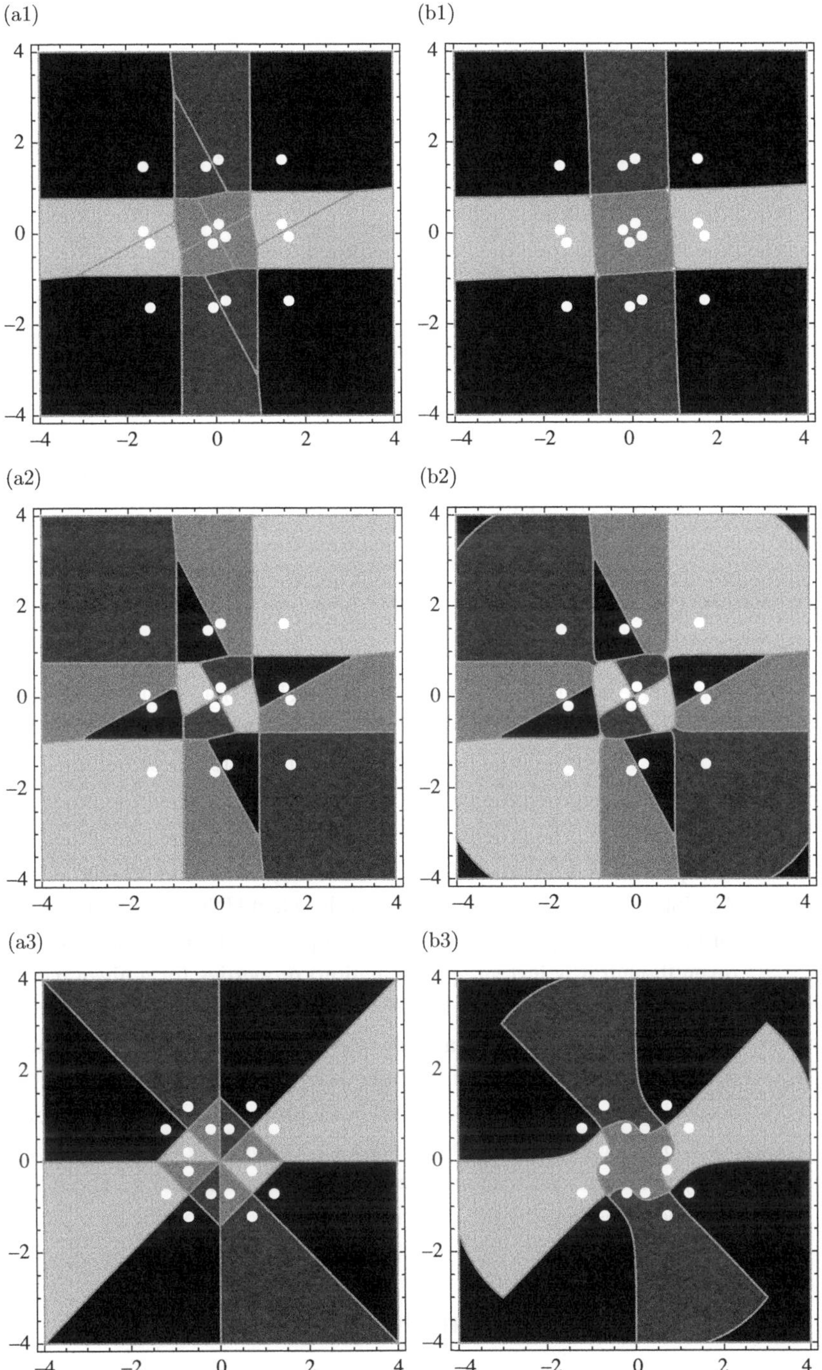

Figure 6.1 Symbol-wise hierarchical decision regions in complex plane for two-component QPSK in H-MAC with bit-wise XOR HNC map. (a) Hierarchical minimum distance metric; (b) true hierarchical marginalized metric for noise variance $\sigma_w^2 = 0.4$. Relative fading is (1) $|h| = 1.2$, $\sphericalangle h = 5°$, (2) $|h| = 1.2$, $\sphericalangle h = 95°$, (3) $|h| = 0.5$, $\sphericalangle h = 45°$.

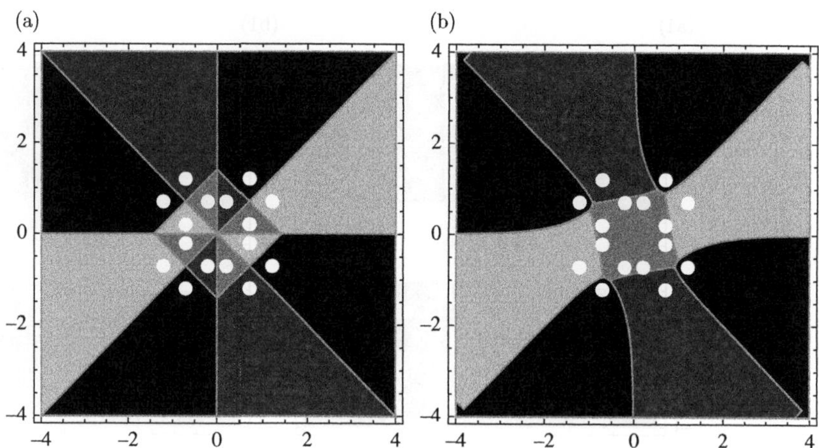

Figure 6.2 Symbol-wise hierarchical decision regions in complex plane for two-component QPSK in H-MAC with bit-wise XOR HNC map. (a) Hierarchical minimum distance metric; (b) true hierarchical marginalized metric for noise variance $\sigma_w^2 = 0.8$. Relative fading is $|h| = 0.5$, $\sphericalangle h = 45°$.

However, in both variants, some of the channel fading values clearly lead to very closely spaced H-symbols that belong to different HNC map values and, in some cases, even to unresolved singular fadings. The behavior of the true metric depends also on the noise variance. At very low SNRs, it can even completely blur the separation of the regions (Figure 6.2).

Adaptive HNC Map with Closest-Neighbor Clustering

A solution to the problem is to adaptively select HNC maps that avoid unresolved singular fadings. On top of this, we can also optimize the H-constellation minimum distance (even though it is just an approximation) properties for a given particular relative fading h. A closest-neighbor clustering [31] selects for *each* relative channel gain h such an HNC map that (1) it guarantees the validity of the exclusive law, and (2) it groups the closest distance points under one H-symbol c_n. The map $\chi_c \in \{\chi_c^{(i)}\}_{i=0}^9$ is selected according to the h value, as shown in Figure 6.3. The maps themselves are defined by Table 6.1. The price paid for this hand-crafted optimization is the fact that maps are generally nonlinear and many of them have even cardinality 5. These properties are major complications in using this principle in layered isomorphic code NCM design.

6.3   Layered NCM and Layered H-Decoding Design

This section shows a particular design of *layered isomorphic NCM* (Section 4.2.4) in simple scenarios. It uses its most straightforward form relying on *linear GF-based HNC maps* (Section 4.7.2), particular *finite* channel constellation alphabets, and *linear GF-based* component codebooks. We focus on a simple scenario

Table 6.1 Adaptive HNC maps $\chi_c^{(i)}$.

$\chi_c^{(i)}$	$c_{A,n}, c_{B,n}$															
	0,0	0,1	0,2	0,3	1,0	1,1	1,2	1,3	2,0	2,1	2,2	2,3	3,0	3,1	3,2	3,3
$\chi_c^{(0)}$	0	1	2	3	1	0	3	2	2	3	0	1	3	2	1	0
$\chi_c^{(1)}$	1	3	0	2	0	2	1	3	3	1	2	0	2	0	3	1
$\chi_c^{(2)}$	0	2	4	1	3	0	2	4	1	3	0	2	4	1	3	0
$\chi_c^{(3)}$	3	2	1	0	0	1	2	4	1	3	4	0	4	0	3	2
$\chi_c^{(4)}$	2	1	0	4	0	4	3	2	3	2	1	0	1	0	4	3
$\chi_c^{(5)}$	2	1	3	0	1	4	0	2	3	0	1	4	0	2	4	3
$\chi_c^{(6)}$	1	4	2	0	4	2	0	3	2	0	3	1	0	3	1	4
$\chi_c^{(7)}$	1	0	2	3	4	2	1	0	0	4	3	1	2	3	0	4
$\chi_c^{(8)}$	4	0	1	2	2	3	4	0	0	1	2	3	3	4	0	1
$\chi_c^{(9)}$	0	3	1	2	2	0	4	1	4	1	0	3	3	4	2	0

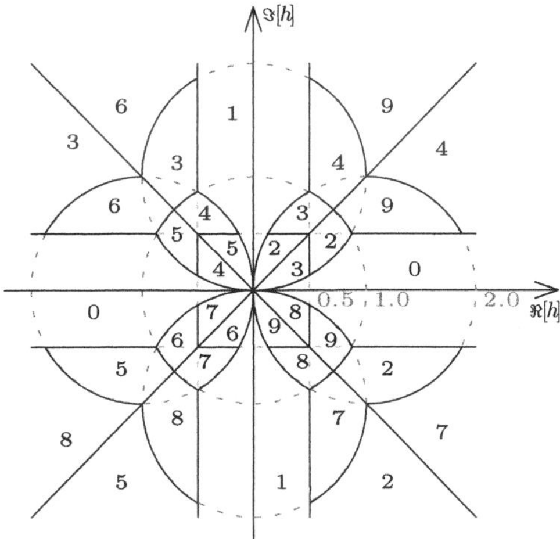

Figure 6.3 Adaptive HNC map selection $\chi_c \in \{\chi_c^{(i)}\}_{i=0}^9$ for a given relative channel gain h. Numbers $i \in [0 : 9]$ indicate the map index $\chi_c^{(i)}$.

of a two-component H-MAC channel with a simple linear AWGN channel model. The simplicity of the scenario allows a straightforward evaluation and interpretation of the performance.

6.3.1 System Model

Component Messages and Codes

We assume a two-component (SA, SB) H-MAC topology (Figure 6.4). Component messages are $\mathbf{b}_A, \mathbf{b}_B \in \mathbb{F}_M^{N_b}$. The relay uses the HDF strategy to decode the H-message

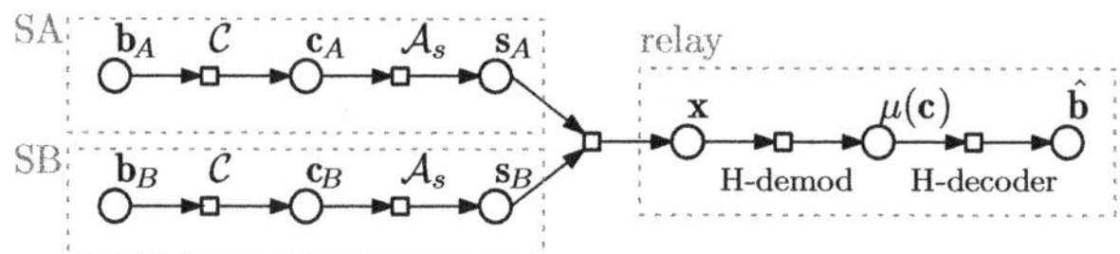

Figure 6.4 System model for layered NCM in a two-component H-MAC.

$\mathbf{b} \in \mathbb{F}_M^{N_b}$ with a minimum cardinality symbol-wise HNC map $b_i = \chi(b_{A,i}, b_{B,i})$, $i \in [1 : N_b]$, $b_i, b_{A,i}, b_{B,i} \in \mathbb{F}_M$. We focus on the performance of H-decoding on one relay and assume that the end-to-end solvability (see Section 5.8.4) is fulfilled by a proper choice of HNC maps and corresponding information measures on all involved relays and paths to the final destination.

The component messages are encoded by a *common* $(2^{NR}, N)$ codebook $\mathcal{C}$ on GF $\mathbb{F}_M$, $\mathbf{c}_A = \mathcal{C}(\mathbf{b}_A)$, $\mathbf{c}_B = \mathcal{C}(\mathbf{b}_B)$, $\mathbf{c}_A, \mathbf{c}_B \in \mathbb{F}_M^N$. The component code rate is $R = \lg(M^{N_b})/N$. Channel symbols transmitted by the component nodes are one-to-one symbol-wise mapped, $s_{A,n} = \mathcal{A}_s(c_{A,n})$, $s_{B,n} = \mathcal{A}_s(c_{B,n})$, $s_{A,n}, s_{B,n} \in \mathcal{A}_s$, on a common channel constellation space *finite* alphabet $\mathcal{A}_s$ of the size M. Later we will use various alphabets for a numerical performance evaluation. The component sources are assumed to be synchronized at the symbol timing level.

Linear AWGN H-MAC Channel

The channel model is a flat-fading linear AWGN channel with relative fading (Section 3.5) coefficient $h \in \mathbb{C}$

$$x_n = \left(s_{A,n} + h s_{B,n}\right) + w_n \tag{6.5}$$

where w_n is complex-valued AWGN with σ_w^2 variance per dimension. We assume a unity common fading coefficient (see Section 3.5.2). Since the common fading cannot influence the performance related to *hierarchical* entities, it is dropped for clarity of presentation. The hierarchical channel combined symbol is

$$u_n = s_{A,n} + h s_{B,n} \tag{6.6}$$

and the H-alphabet is $\mathcal{A}_u$, $u_n \in \mathcal{A}_u$. Clearly, the alphabet depends on the component constellation and the channel parametrization.

H-Constellation

The HNC map on coded symbols is defined as a symbol-wise map $c_n = \chi_c(c_{A,n}, c_{B,n})$, $c_n, c_{A,n}, c_{B,n} \in \mathbb{F}_M$. The H-constellation $\mathcal{U}(c_n)$ classifies the H-alphabet symbols according to their corresponding value of HNC map

$$\mathcal{U}(c_n) = \left\{ u_n : u_n = s_{A,n}(c_{A,n}) + h s_{B,n}(c_{B,n}), \ c_n = \chi_c(c_{A,n}, c_{B,n}) \right\}. \tag{6.7}$$

6.3.2   Linear Isomorphic Layered NCM

A linear isomorphic layered NCM (Section 4.7.2) is a straightforward design option. The component codes are linear GF $\mathbb{F}_M$ codes with generator matrix $\mathbf{G} \in \mathbb{F}_M^{N \times N_b}$

$$\mathbf{c}_k = \mathcal{C}(\mathbf{b}_k) = \mathbf{G}\mathbf{b}_k, \; k \in \{A, B\}. \tag{6.8}$$

Message and code HNC maps are identical linear scalar combinations

$$\mathbf{b} = a_A \mathbf{b}_A + a_B \mathbf{b}_B, \tag{6.9}$$

$$\mathbf{c} = a_A \mathbf{c}_A + a_B \mathbf{c}_B, \tag{6.10}$$

where $a_k \in \mathbb{F}_M$ and all operations are on GF. Using (4.92), we get

$$\mathbf{c} = a_A \mathbf{G}\mathbf{b}_A + a_B \mathbf{G}\mathbf{b}_B = \mathbf{G}\left(a_A \mathbf{b}_A + a_B \mathbf{b}_B\right) \tag{6.11}$$

and the H-code is then $\mathbf{c} = \mathbf{G}\mathbf{b}$.

Next we assume that the component codebook has *IID* code symbols with *uniform* distribution. This is equivalent to assuming that the codebook is a "perfect" finite alphabet single-user codebook.[2] Then, according to Theorem 5.11, the H-codebook is *regular* and *achievability* Theorem 5.10 holds. Any H-rate $R < I(C;X)$ is achievable. Also, we can use the layered decoding and the equivalent hierarchical channel model (Section 5.7.4).

Notice that, in order to be able to interpret the mutual information $I(C;X)$ as the limit on the achievable rate, and to have a possibility of using the layered H-decoding (which requires the isomorphism), all the following conditions have to be fulfilled.[3]

(1) All component codes are *identical GF linear "perfect"* single-user codes. This implies that the weaker component channel becomes a bottleneck. However, we cannot use different codes. The factorization of $\mathbf{G}$ matrix in (4.92) would not be possible.

(2) HNC maps are *linear symbol-wise* maps and they are *identical* for both message and code symbols. This implies a *minimal* cardinality map.

(3) All HNC maps and component codes must be over a *common GF* $\mathbb{F}_M$.

(4) The constellation space map $\mathcal{A}_s$ must be *symbol-wise one-to-one* mapping. Otherwise we could not use a single-letter characterization of the equivalent channel. In Section 6.3.4, we will elaborate more on this aspect.

6.3.3 H-Decoding

Under the assumptions discussed above, we can use layered H-decoding and equivalent hierarchical channel model (Figure 5.12). The H-constellation self-dispersion part (Section 5.7.4) is described by

---

[2] Practically, common capacity approaching single-user codes (e.g. LDPC) closely approach this assumption.

[3] Notice that if the conditions are not fulfilled, i.e. the NCM is *not* isomorphic, we cannot use layered H-decoding and the H-message is not represented by H-code with exploitable structure. Also, currently the GF-based *minimal* map is the only known HNC map enabling the isomorphic solution. Indeed, we have the *generic* option of joint H-decoding on a product codebook, but this strategy does *not* give any rate advantage over classical multi-user MAC (see Section 5.7).

$$p(u_n|c_n) = \frac{\sum_{\tilde{c}_n:\chi_c(\tilde{c}_n)=c_n} p(u_n|\tilde{c}_n)p(\tilde{c}_n)}{\sum_{\tilde{c}_n:\chi_c(\tilde{c}_n)=c_n} p(\tilde{c}_n)} \tag{6.12}$$

where $\tilde{c}_n = (c_{A,n}, c_{B,n})$. The observation part is described by $p(x_n|u_n) = p_w(x_n - u_n)$ where the complex Gaussian PDF for N_u-dimensional constellation symbols is

$$p_w(w_n) = \frac{1}{\pi^{N_u}\sigma_w^{2N_u}} \exp(-\frac{1}{\sigma_w^2}\|w_n\|^2). \tag{6.13}$$

Common constellations, such as MPSK, QAM, have one complex dimension, $N_u = 1$. Complex envelope constellation space Gaussian noise has $\sigma_w^2 = 2N_0$ variance per dimension, where N_0 is a single-sided power spectrum density of the real-valued noise.

Using the properties of the HNC maps (see details in Section 4.4), we get the H-demodulator metric

$$\mu(c_n) = p(x_n|c_n) = \frac{1}{M} \sum_{\tilde{c}_n:\chi_c(\tilde{c}_n)=c_n} p_w(x_n - u(\tilde{c}_n)), \tag{6.14}$$

where the a priori H-code PDF is

$$p(c_n) = \frac{1}{M}. \tag{6.15}$$

The decoding metric $\mu(c_n)$ is fed into the decoder of $\mathcal{C}$ code producing the estimates of the H-message $\hat{\mathbf{b}}$. The decoder is a standard "single-user" decoder. This is a clear benefit of having isomorphic NCM and equivalent H-channel.

6.3.4   Linear HNC Maps on Extended GF

Section 4.7.2 in general, and Section 6.3.2 in particular, showed the design using a linear HNC map on $\mathbb{F}_M$. All HNC map coefficients and also the code itself used the same size of alphabet. In a special case, when this is an *extended* GF, $M = M'^m$, we might, however, be tempted to mix the usage of the operations on $\mathbb{F}_M$ and $\mathbb{F}_{M'}^m$, where the latter is a vector space of m-tuples on $\mathbb{F}_{M'}$. A combination of binary code $M' = 2$ with M'^m sized constellations seems to be particularly attractive.

The formal definitions follow. Assume that $M = M'^m$ where M' is prime. All symbols, codes, and coefficients with elements from $\mathbb{F}_{M'}$ will be denoted by $(.)'$, e.g. c'_n, and the operations on $\mathbb{F}_{M'}$ or the associated vector space (applied element-wise) $\mathbb{F}_{M'}^m$ will be denoted by $\oplus, \otimes$. The symbols and operations on $\mathbb{F}_M$ will keep the notation used in previous sections. The vectors from the vector space on $\mathbb{F}_{M'}$ are segmented into m-tuples. We assume that the size of the vector is an integer multiple of m. We denote m-tuple of $\mathbb{F}_{M'}$ symbols by $\mathbf{c}'_j$, and vector of m-tuples as $\mathbf{c}' = [\ldots, \mathbf{c}'_j, \ldots]^T$, and similarly for other variables. The "conversion" function between extended GF and its m-tuple representation is $f : \mathbb{F}_{M'}^m \mapsto \mathbb{F}_M$, and its form combining the whole vector from m-tuples is $\mathbf{f} : \mathbb{F}_{M'}^{mN} \mapsto \mathbb{F}_M^N$.

We have essentially two possibilities of applying a mixed GF design. The first one tries to keep linear combination coefficients in $\mathbb{F}_M$ and the second one keeps them in $\mathbb{F}_{M'}$. We now analyze the consequences for both.

Linear HNC Map with $\mathbb{F}_M$ Coefficients

We start with a code HNC map with $\mathbb{F}_M$ coefficients

$$\mathbf{c} = a_A \mathbf{c}_A + a_B \mathbf{c}_B. \tag{6.16}$$

Now assume that the component codes are on $\mathbb{F}_{M'}$ GF, $\mathbf{c}'_A = \mathbf{G}' \otimes \mathbf{b}'_A$, $\mathbf{c}'_B = \mathbf{G}' \otimes \mathbf{b}'_B$, where $\mathbf{c}'_A, \mathbf{c}'_B \in \mathbb{F}_{M'}^{mN}$, $\mathbf{b}'_A, \mathbf{b}'_B \in \mathbb{F}_{M'}^{mN_b}$, $\mathbf{G}' \in \mathbb{F}_{M'}^{mN \times mN_b}$. Then the map is

$$\mathbf{c} = a_A \mathbf{f}(\mathbf{c}'_A) + a_B \mathbf{f}(\mathbf{c}'_B) = a_A \mathbf{f}(\mathbf{G}' \otimes \mathbf{b}'_A) + a_B \mathbf{f}(\mathbf{G}' \otimes \mathbf{b}'_B). \tag{6.17}$$

It is clear that any next step that would follow the lines of (6.11) would require that both the multiplication and the addition are isomorphic on $\mathbb{F}_{M'}$ and $\mathbb{F}_{M'}^m$. This means that $f(a')f(b')$ would need to be equal to $f(a' \otimes b')$ and $f(a') + f(b')$ would need to be equal to $f(a' \oplus b')$ where a', b' are m-tuples on $\mathbb{F}_{M'}^m$ and operations $\otimes, \oplus$ are applied *element-wise*. The isomorphism holds for the addition but unfortunately *not for the multiplication*. Thus, we cannot generally get the coefficient multiplication inside the $\mathbf{f}(.)$ operation and the linear HNC map thus cannot propagate to the level of the $\mathbb{F}_{M'}$-based code. A clear underlying reason is that $f^{-1}(a_A)$ and $f^{-1}(a_B)$ are not scalars in $\mathbb{F}_{M'}$.

For example, the element-wise $\otimes$ operations on $\mathbb{F}_2^2$ viewed as the $\mathbb{F}_{2^2}$ operation do not even comply with the GF requirements on multiplication (e.g. missing inverse, and a rather strange mapping for a neutral element for the multiplication $3 = f(11)$) is as follows.

$\otimes$	$0 = f(00)$	$1 = f(01)$	$2 = f(10)$	$3 = f(11)$
$0 = f(00)$	00	00	00	00
$1 = f(01)$	00	01	00	01
$2 = f(10)$	00	00	10	10
$3 = f(11)$	00	01	10	11

The isomorphism for addition, however, works.

$\oplus$	$0 = f(00)$	$1 = f(01)$	$2 = f(10)$	$3 = f(11)$
$0 = f(00)$	00	01	10	11
$1 = f(01)$	01	00	11	10
$2 = f(10)$	10	11	00	01
$3 = f(11)$	11	10	01	00

Linear HNC Map with $\mathbb{F}_{M'}$ Coefficients

The second option is to apply the linear HNC operation on $\mathbb{F}_{M'}$ and map the result on the $\mathbb{F}_M$ using the $f(.)$ function. The HNC map for code is defined as

$$\mathbf{c} = \mathbf{f}(\mathbf{c}') = \mathbf{f}\left(a'_A \otimes \mathbf{G}' \otimes \mathbf{b}'_A \oplus a'_B \otimes \mathbf{G}' \otimes \mathbf{b}'_B\right) \tag{6.18}$$

where $a'_A, a'_B \in \mathbb{F}_{M'}$. This clearly *limits* the range of the coefficients but preserves the linearity. For example, if $M' = 2$ then the only non-singular choice is $a'_A = a'_B = 1$ (i.e. bit-wise XOR) regardless of the size M.

The linear operation on $\mathbb{F}_{M'}$ leads to the desired result

$$
\begin{aligned}
\mathbf{c} &= \mathbf{f}\left(\mathbf{G}' \otimes \left(a'_A \otimes \mathbf{b}'_A \oplus a'_B \otimes \mathbf{b}'_B\right)\right) \\
&= \mathbf{f}\left(\mathbf{G}' \otimes \mathbf{b}'\right)
\end{aligned}
\tag{6.19}
$$

where the H-message HNC map is

$$
\mathbf{b}' = a'_A \otimes \mathbf{b}'_A \oplus a'_B \otimes \mathbf{b}'_B
\tag{6.20}
$$

and the H-code at the level of $\mathbb{F}_{M'}$ GF is $\mathbf{c}' = \mathbf{G}' \otimes \mathbf{b}'$. According to Theorem 5.11, the $\mathbb{F}_{M'}$ H-codebook is regular. The H-codebook on $\mathbb{F}_M$ is then obtained by m-tuple grouping $\mathbf{c} = \mathbf{f}(\mathbf{c}')$. If the $\mathbb{F}_{M'}$ code has uniform and IID symbols then also the m-tuple groups will be uniform and IID on $\mathbb{F}_M$. Thus the regularity condition is fulfilled also for the $\mathbb{F}_M$ H-codebook.

6.3.5   H-Coding Rates

Having an input–output stochastic description of equivalent H-channel (Figure 5.12) $p(x_n|c_n)$ and an a priori PDF $p(c_n)$ now allows us to evaluate the hierarchical mutual information $I(C;X)$ that limits the *achievable* H-coding rates (Section 5.7.4, and Theorem 5.10). We must properly respect the relative channel parametrization. This makes an analytical evaluation rather difficult and we frequently resort to numerical methods (e.g. Monte-Carlo evaluation of integrals, Section 5.7.8). In the following, we provide numerical results for selected component constellations and HNC maps (more details are in [57]).

The H-rates are shown against the reference case of cut-set bounds for the same finite channel symbol alphabet. This setup identifies the limits of a standard multi-user MAC. The first-order cut-set bound I_1 limits the single-user rates. The second-order cut-set bound limits the sum-rate for both users. Since we have a minimal cardinality HNC map which implies a common source rate, we divide the sum-rate bound I_s by 2, to get the interpretable limit $I_2 = I_s/2$.

The received symbol energy $\mathcal{E}_s = \mathrm{E}[|s_{A,n}|^2]/2$ to noise density N_0 ratio, used for the H-rates and also for the cut-set bound reference, is defined as if there was only component SA in the channel, i.e.

$$
\gamma_x = \frac{\mathcal{E}_s}{N_0} = \frac{\mathrm{E}[|s_{A,n}|^2]}{2N_0}.
\tag{6.21}
$$

We recall that the common channel parametrization was assumed to be unity. This definition allows easier comparison to a single-user system having the same power budget.

BPSK

This case is defined by $M = 2$, $c_{A,n}, c_{B,n}, c_n \in \mathbb{F}_2$, $s_{A,n}, s_{B,n} \in \mathcal{A}_s = \{-1, +1\}$, $\mathcal{A}_s(0) = -1$, $\mathcal{A}_s(1) = +1$, and *the only possible* non-singular option for HNC map

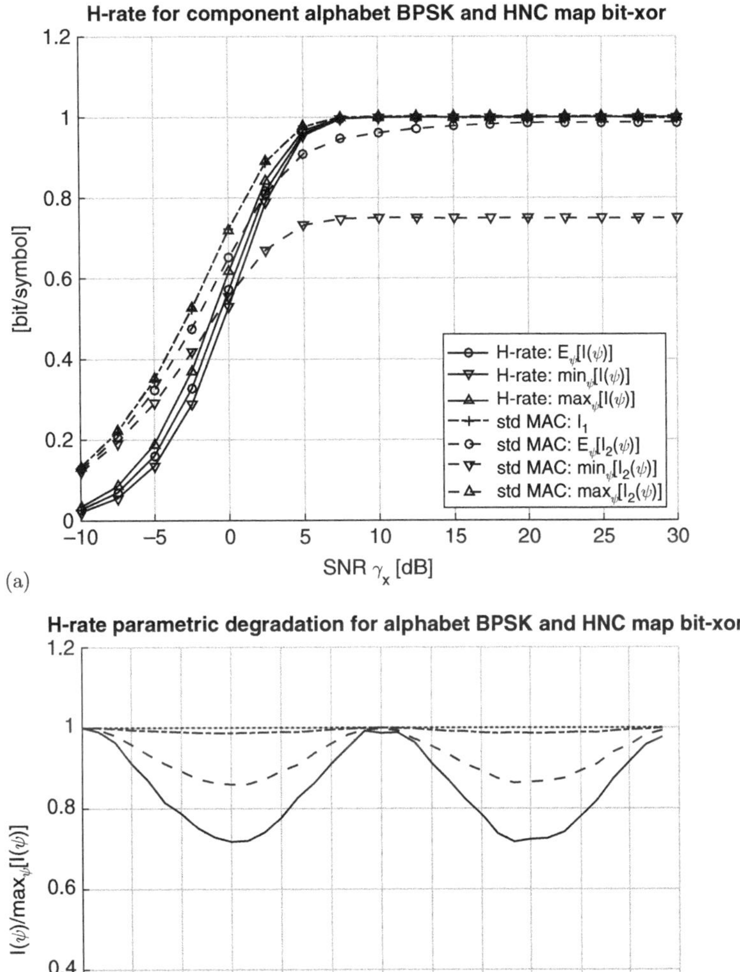

Figure 6.5 H-rate bound $I(C; X)$ and the impact of the channel parametrization for BPSK with XOR HNC map.

$\mathbb{F}_2$ has coefficients $a_A = 1$, $a_B = 1$, i.e. $c_n = c_{A,n} + c_{B,n}$, which is a bit-wise XOR operation. Owing to a 180 degree constellation symmetry, a change of constellation mapper $\mathcal{A}_s(.)$ does not have any effect. Figure 6.5 shows the numerical results of $I(C; X)$ for relative channel parametrization $h = \exp(\mathrm{j}\,\psi)$.

4PSK with Bit-Wise XOR HNC Map

This case is defined by $M = 4$, $s_{A,n}, s_{B,n} \in \mathcal{A}_s = \{1, j, -1, -j\}$, $\mathcal{A}_s(0) = 1$, $\mathcal{A}_s(1) = j$, $\mathcal{A}_s(2) = -1$, $\mathcal{A}_s(3) = -j$, where we have chosen "natural" constellation indexing. There is also an option to use another constellation mapper $\mathcal{A}_s(\cdot)$. Since the size of the component constellation implies codes on the extended GF, we can choose to have either $\mathbb{F}_{2^2}$ or $\mathbb{F}_2$ code alphabet (Section 6.3.4). In the case of the code on $\mathbb{F}_2$ GF, the codesymbols are $c_{A,n}, c_{B,n}, c_n \in \mathbb{F}_2$, and we have again *the only possible* non-singular option for HNC map with $\mathbb{F}_2$ coefficients $a_A = 1$, $a_B = 1$. For $m = 2$, the binary symbols are grouped by pairs and the map $\mathbf{c}'_n = \mathbf{c}'_{A,n} \oplus \mathbf{c}'_{B,n}$ is a bit-wise XOR operation. Figure 6.6 shows the numerical results of $I(C; X)$ for relative channel parametrization $h = \exp(j\,\psi)$.

Performance

All numerical results for H-rate bound $I(C; X)$ show several common performance characteristics that hold generally for any particular H-constellation. The H-rate bounds for linear isomorphic NCM are *achievable*. The system is equivalent to a single-user system (see Section 5.7.4). The practical codes that achieve the H-rate are standard capacity approaching finite alphabet single-user codes. The interpretation of the reference classical MAC case needs, however, some caution. The rate region is theoretically, using joint typicality decoding, achievable for arbitrary alphabets and input distributions (e.g. [18] Section 4.5). Approaching the first-order bound I_1 with practical codes clearly does not present a problem. However, a design of practical codes and decoding strategies for approaching the second-order bound I_2 for *finite* alphabet inputs is still an open issue, including e.g. achievable rates for a practical approach to successive interference cancellation or joint decoding for finite alphabets (see some results in [21]). The rate I_2 also depends on the channel parametrization.

High SNR region In the high SNR region, we can clearly see that the H-rate bound approaches the first-order bound I_1. It means, provided that other paths fulfill the end-to-end solvability, that the WPNC system performs as if there were only one user.[4] The presence of the other one does affect the first one even if they are both sharing the same radio resources. It demonstrates the phenomenon of "friendly interference," which does not hurt the performance. This behavior is observed in the high SNR region, where the classical multi-user technique is interference-limited.

Medium SNR region In the medium SNR region, the HDF has still an advantage over the classical multi-user technique. It is the region where the H-rate bound is higher than the second-order classical MAC limit I_2.

Low SNR region In this region, the classical multi-user technique (both I_1 and I_2 rates) has an advantage over the H-rate. This is a typical situation in the noise-limited system, where any "smart" handling with the interference cannot help.

---

[4] For example in the butterfly network, if the destinations have perfect HSI and the H-BC stage is not a bottleneck, the overall end-to-end performance will be as if there were only one user in the system.

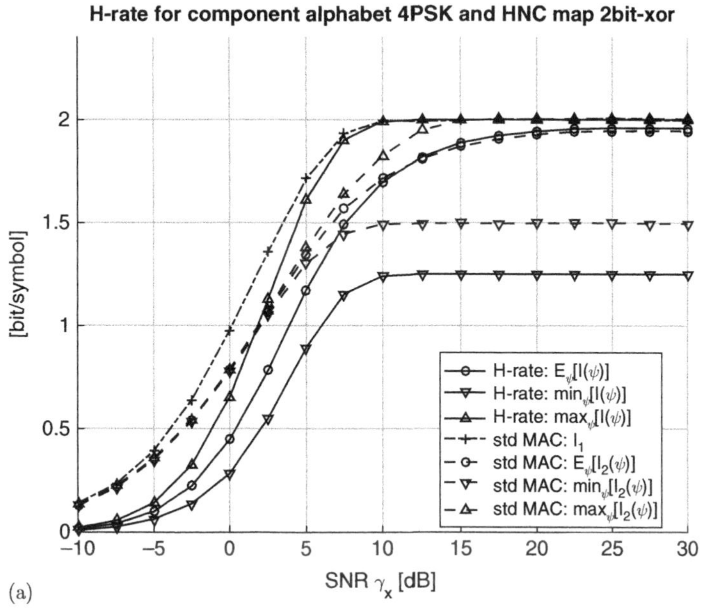

(a)

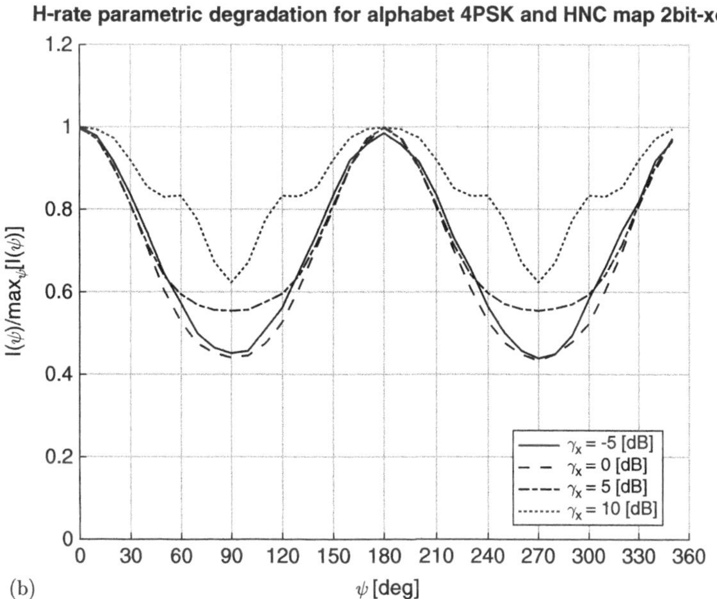

(b)

Figure 6.6 H-rate bound $I(C; X)$ and the impact of the channel parametrization for 4PSK with "natural" indexing and with bit-wise XOR HNC map.

Phase parametrization All H-constellations have H-rate performance depending on the channel phase. Phase rotation substantially changes the shape of the H-constellation. This behavior is pronounced in medium and low SNR regions. In these conditions, the H-metric conditional PDF (see e.g. Figure 4.5) has wide tails and the system H-performance depends on the locations of neighboring constellation points. Their position, in turn, depends on the channel parametrization. Figures 6.5b and 6.6b show an explicit impact of the phase on the H-rate for various SNRs. Figures 6.5a and 6.6a show that impact by plotting the minimum, the maximum, and the average value of the H-rate bound over the whole relative phase range, assuming it has a uniform distribution. We also plot the mean, the minimum, and the maximum values for the reference I_2 rates. In the 4PSK case (Figure 6.6b), we see also the deteriorating effect of unresolved singular fades at 90 and 270 degrees.

7 NCM Design and Processing for Parametric Channels

7.1 Introduction

This chapter is dedicated to the NCM design and receiver processing that we need to use to properly address the effects of the *channel parametrization*. This generally leads to various synchronization and equalization techniques or NCM design resistant (at least to some extent) against the parametrization. However, all these must properly reflect two fundamental WPNC characteristics that dominate to any H-MAC channel. First, the received signals from component nodes interact in a common signal space. Second, the target of our effort at the receiving node is to decode the H-codeword and the receiver node is neither interested nor potentially having access to the individual component codebooks.

Section 7.2 deals with a straightforward task of estimating the channel state in the H-MAC channel. We concentrate on a pilot-based technique. The goal is, if possible, to avoid the orthogonal pilot signals. The usage of orthogonal pilots would require additional resources dedicated purely to pilots that also might require an excessive resource management in large networks. It also somewhat goes against the WPNC paradigm of sharing the signal space among all H-MAC component signals. We will see that a complete sharing of pilot signal space among component nodes creates fundamental problems, which can be somewhat reduced by a partial sharing or under some specific channel state conditions.

Section 7.3 extends the scope for frequency selective channels and also, as a special case, for frequency flat channels with uneven delays among H-MAC components. We show that OFDM type of NCM waveform can easily turn this case, similarly to classical single-user systems, into the set of multiplexed flat fading sub-carrier channels. This appears as an intuitive step. However, rather more important, and less obvious, is the fact that it solves the case of uneven delays. This is specific for WPNC and cannot be ignored. We will show that NCM with OFDM waveform easily accepts arbitrary timing misalignment among the signals received from H-MAC components approximately up to the length of a cyclic prefix.

Finally, in Section 7.4, we analyze the conditions under which the NCM could be made *resistant* to various forms of H-MAC parametrization. We will define various levels of resistance, ranging from complete invariance w.r.t. some channel parameters, to the uniformly most powerful performance. We will also discuss various forms of the performance criteria used to measure the resistance. The H-distance spectrum will be shown to be relatively easy to handle and will provide some easily applicable rules. However, we need to interpret them with high caution, because they are only based on an approximation with a limited validity.

7.2   Synchronization and Pilot Design

7.2.1   Synchronization and Channel State Estimation in WPNC Context

Decoding in Parametric Channel

The synchronization or the channel state estimation (CSE) technique, regardless of whether our system is a point-to-point one or a WPNC complex network, is the way of handling the fact that the received signal is, apart from the desired payload data, parametrized by the *nuisance* channel parameters. They are called nuisance parameters because we are not interested in them; however, all receiver processing must *unavoidably* respect their presence. The synchronization or CSE technique is a very wide topic with a plethora of options, approaches, and implementations. They must precisely match to the particular system scenario, channel model, and implementation capabilities, and they are almost always very individually hand-crafted for this particular situation. The effectiveness and the optimality level of handling the CSE problem also substantially influence the overall performance of the system (throughput, allocated resources, error rate, etc.).

In order to introduce a systematic concept into the problem solution, we can broadly classify the possible approaches. The most generic approach is to devise a parametric decoder that decodes the desired payload while fully respecting and allowing the channel parametrization. This is all achieved by one "monster" joint algorithm for decoding with channel parametrization. This could include a variety of "joint" algorithms (e.g. joint ML parametrization estimator and decoder) with many forms of solvers (e.g. factor-graph-based iterative solvers). Clearly, this approach leads to quite complex solutions.

We can simplify the implementation by separating the data decoding and the processing related to the channel parametrization. It can essentially have two forms – the synchronization and the equalization. In the synchronization approach, we build a separate estimator of the channel nuisance parameters (called CSE) and the parameter estimates are then provided to the data demodulator/decoder that can properly adjust its operation. For example, the CSE estimates the phase rotation and the demodulator/decoder then adequately rotates its decision regions. The equalization approach, possibly using its internal CSE, preprocesses the received signal before entering the demodulator/decoder in such a way that it appears as if there were no channel parametrization at all. For example, it rotates back the phase rotation caused by the channel such that it appears as if it was unrotated and then the demodulator/decoder can operate with decision regions as if the phase shift was zero. The synchronization and equalization approaches are essentially mutually dual and we can choose the one which better suits our implementation needs. The price paid for the separation of the CSE and demodulator/decoding algorithms is, however, a general suboptimality. This can be improved by letting the demodulator/decoder and CSE/equalizer cooperate by iteratively providing each other with their tentative outputs in some form, e.g. full decisions, or some soft information measure. Under some conditions, it can lead to the performance close to the optimal joint type of the algorithm.

A processing for the channel parametrization with many nuisance parameters reduces the fidelity of the estimation. The higher the number of degrees of freedom in the channel parametrization, the worse the estimation precision becomes. Regardless of whether we use a joint or separate synchronization/equalization type of the algorithm, we can try to reduce the number of the degrees of freedom by transmitting specific signals that help the receiver processing. These signals are known as pilots (or pilot signals). Typical pilots are the transmitted signals that do not have any payload data and potentially are also specifically tuned to ease or optimize the performance of the estimation of some given channel parameter. The use of pilots, however, comes at the price of reducing the resources (time, frequency, etc.) used for actual payload data transfer.

The synchronization and CSE in general, and particularly in the context of WPNC, is a very wide and complex area that largely extends the scope of the book. We therefore concentrate only on those aspects that are WPNC specific and we use a very basic form of the algorithms in selected simple scenarios. Particularly, we will concentrate only on a *pilot-based CSE* in a *linear AWGN H-MAC* channel.

CSE in H-MAC Channel

The problem of pilot-based CSE in H-MAC is substantially different from a standard situation in the point-to-point channel. The underlying reason lies in the *interaction* of pilot signals from multiple component nodes participating in the H-MAC. Payload data carrying signals of NCM are designed, in a compliance with the WPNC fundamental paradigm, to *non-orthogonally share* common resources – temporal, frequency, spatial. The same should also apply to pilot signals. Ideally, all pilot signals involved in H-MAC should share the same signal/constellation space among them similarly as the payload NCM. The pilot signals *sharing* the common resources and at the same time capable of serving to CSE of parameters needed for H-decoder (e.g. a relative phase) will be called *H-MAC CSE pilots*.

The problem of decoupling the interaction of signals coming from multiple component nodes in H-MAC channel can of course be trivially solved by using the orthogonal pilots. This turns the problem into multiple independent point-to-point pilot-based CSEs. However, it would require allocating additional dedicated resources; for example, by extending the dimensionality (most typically additional symbol slots) dedicated to H-MAC CSE pilots. The number of possible orthogonal sequences is given by the dimensionality of the space. The resource allocation and the associated signaling would thus have to be kept separate and in parallel for both NCM and pilots. Such an approach obviously contradicts the fundamental paradigm of WPNC and we would like to avoid it or at least to minimize it as much as possible.

7.2.2 Fundamental Limits for Phase and Magnitude Estimators in Linear AWGN H-MAC

Linear Flat-Fading AWGN H-MAC Channel

For the following treatment, we concentrate on a simple linear flat-fading AWGN two-component H-MAC channel. We will assume signals are aligned in time, i.e. we assume zero delay (see Section 7.3 for a generalization). It allows the representation of the system model in N-dimensional signal space

$$\mathbf{x} = h_A \mathbf{s}_A + h_B \mathbf{s}_B + \mathbf{w} \tag{7.1}$$

where $\mathbf{s}_A, \mathbf{s}_B$ are H-MAC CSE pilot signals, $h_A = \alpha_A \, \mathrm{e}^{\mathrm{j}\,\varphi_A}$, $h_B = \alpha_B \, \mathrm{e}^{\mathrm{j}\,\varphi_B}$ are complex channel coefficients, and $\mathbf{w}$ is complex AWGN with σ_w^2 variance per dimension. The pilot signals are assumed to be known to the receiver. The pilot signal space is ideally of the same type as the constellation space of payload NCM. For example, the pilot sequences could use the same constellation alphabet and the same type of basis as the NCM. However, the following treatment is not constrained by this and the pilot signal space can be arbitrary, for example simple sampling space.[1]

The observation model is given by the likelihood

$$p(\mathbf{x}|h_A, h_B) = \frac{1}{\pi^N \sigma_w^{2N}} \exp\left(-\frac{1}{\sigma_w^2}\|\mathbf{x} - (h_A \mathbf{s}_A + h_B \mathbf{s}_B)\|^2\right). \tag{7.2}$$

We also define the observation metric

$$\rho = \|\mathbf{x} - (h_A \mathbf{s}_A + h_B \mathbf{s}_B)\|^2 \tag{7.3}$$

which can be manipulated into

$$\begin{aligned}
\rho = {}&\|\mathbf{x}\|^2 + |h_A|^2 \|\mathbf{s}_A\|^2 + |h_B|^2 \|\mathbf{s}_B\|^2 \\
&- h_A^* \langle \mathbf{x}; \mathbf{s}_A \rangle - h_A \langle \mathbf{x}; \mathbf{s}_A \rangle^* - h_B^* \langle \mathbf{x}; \mathbf{s}_B \rangle - h_B \langle \mathbf{x}; \mathbf{s}_B \rangle^* \\
&+ h_A h_B^* \langle \mathbf{s}_A; \mathbf{s}_B \rangle + h_A^* h_B \langle \mathbf{s}_A; \mathbf{s}_B \rangle^*.
\end{aligned} \tag{7.4}$$

Cramer–Rao Lower Bound – Phase-only CSE in Two-Component H-MAC

The Cramer–Rao lower bound (CRLB) describes the fundamental limits of the CSE MSE performance (Section A.3). CRLB analysis can give a valuable hint for construction of the optimal pilot signals.

We start with a simple case of phase-only CSE, where magnitudes α_A, α_B are assumed to be known. This turns into a classical real-valued CRLB problem. The likelihood is $p(\mathbf{x}|\varphi_A, \varphi_B) = \pi^{-N} \sigma_w^{-2N} \exp(-\rho/\sigma_w^2)$ and the observation metric becomes

$$\begin{aligned}
\rho = {}&\|\mathbf{x}\|^2 + \alpha_A^2 \|\mathbf{s}_A\|^2 + \alpha_B^2 \|\mathbf{s}_B\|^2 \\
&- \alpha_A \, \mathrm{e}^{-\mathrm{j}\,\varphi_A} \langle \mathbf{x}; \mathbf{s}_A \rangle - \alpha_A \, \mathrm{e}^{\mathrm{j}\,\varphi_A} \langle \mathbf{x}; \mathbf{s}_A \rangle^* - \alpha_B \, \mathrm{e}^{-\mathrm{j}\,\varphi_B} \langle \mathbf{x}; \mathbf{s}_B \rangle - \alpha_B \, \mathrm{e}^{\mathrm{j}\,\varphi_B} \langle \mathbf{x}; \mathbf{s}_B \rangle^* \\
&+ \alpha_A \alpha_B \, \mathrm{e}^{\mathrm{j}(\varphi_A - \varphi_B)} \langle \mathbf{s}_A; \mathbf{s}_B \rangle + \alpha_A \alpha_B \, \mathrm{e}^{-\mathrm{j}(\varphi_A - \varphi_B)} \langle \mathbf{s}_A; \mathbf{s}_B \rangle^*.
\end{aligned} \tag{7.5}$$

CRLB requires us to evaluate the following derivatives

---

[1] The constellation space is a special case of a general signal space. It is suitably defined w.r.t. the modulated signals and has a basis allowing a simple form of the expansion coefficients evaluation. Typically, the basis of constellation space is assumed to be an *orthonormal* system of Nyquist pulses $\{\zeta_i(t - nT_S)\}_{i,n}$. In contrast, a general signal space basis does not need to be either orthonormal or Nyquist.

$$\frac{\partial \ln p(\mathbf{x}|\varphi_A, \varphi_B)}{\partial \varphi_A} = -\frac{1}{\sigma_w^2}\Big(j\,\alpha_A\, \mathrm{e}^{-j\,\varphi_A} \langle \mathbf{x}; s_A \rangle - j\,\alpha_A\, \mathrm{e}^{j\,\varphi_A} \langle \mathbf{x}; s_A \rangle^*$$
$$+ j\,\alpha_A\alpha_B\, \mathrm{e}^{j(\varphi_A-\varphi_B)} \langle s_A; s_B \rangle - j\,\alpha_A\alpha_B\, \mathrm{e}^{-j(\varphi_A-\varphi_B)} \langle s_A; s_B \rangle^* \Big), \quad (7.6)$$

$$\frac{\partial^2 \ln p(\mathbf{x}|\varphi_A, \varphi_B)}{\partial \varphi_A^2} = -\frac{1}{\sigma_w^2}\Big(\alpha_A\, \mathrm{e}^{-j\,\varphi_A} \langle \mathbf{x}; s_A \rangle + \alpha_A\, \mathrm{e}^{j\,\varphi_A} \langle \mathbf{x}; s_A \rangle^*$$
$$- \alpha_A\alpha_B\, \mathrm{e}^{j(\varphi_A-\varphi_B)} \langle s_A; s_B \rangle - \alpha_A\alpha_B\, \mathrm{e}^{-j(\varphi_A-\varphi_B)} \langle s_A; s_B \rangle^* \Big)$$
$$= -\frac{2}{\sigma_w^2}\Big(\Re\Big[\alpha_A\, \mathrm{e}^{-j\,\varphi_A} \langle \mathbf{x}; s_A \rangle \Big] - \Re\Big[\alpha_A\alpha_B\, \mathrm{e}^{j(\varphi_A-\varphi_B)} \langle s_A; s_B \rangle \Big]\Big),$$
$$(7.7)$$

$$\frac{\partial \ln p(\mathbf{x}|\varphi_A, \varphi_B)}{\partial \varphi_B} = -\frac{1}{\sigma_w^2}\Big(j\,\alpha_B\, \mathrm{e}^{-j\,\varphi_B} \langle \mathbf{x}; s_B \rangle - j\,\alpha_B\, \mathrm{e}^{j\,\varphi_B} \langle \mathbf{x}; s_B \rangle^*$$
$$- j\,\alpha_A\alpha_B\, \mathrm{e}^{j(\varphi_A-\varphi_B)} \langle s_A; s_B \rangle + j\,\alpha_A\alpha_B\, \mathrm{e}^{-j(\varphi_A-\varphi_B)} \langle s_A; s_B \rangle^* \Big), \quad (7.8)$$

$$\frac{\partial^2 \ln p(\mathbf{x}|\varphi_A, \varphi_B)}{\partial \varphi_B^2} = -\frac{1}{\sigma_w^2}\Big(\alpha_B\, \mathrm{e}^{-j\,\varphi_B} \langle \mathbf{x}; s_B \rangle + \alpha_B\, \mathrm{e}^{j\,\varphi_B} \langle \mathbf{x}; s_B \rangle^*$$
$$- \alpha_A\alpha_B\, \mathrm{e}^{j(\varphi_A-\varphi_B)} \langle s_A; s_B \rangle - \alpha_A\alpha_B\, \mathrm{e}^{-j(\varphi_A-\varphi_B)} \langle s_A; s_B \rangle^* \Big)$$
$$= -\frac{2}{\sigma_w^2}\Big(\Re\Big[\alpha_B\, \mathrm{e}^{-j\,\varphi_B} \langle \mathbf{x}; s_B \rangle \Big] - \Re\Big[\alpha_A\alpha_B\, \mathrm{e}^{j(\varphi_A-\varphi_B)} \langle s_A; s_B \rangle \Big]\Big),$$
$$(7.9)$$

$$\frac{\partial^2 \ln p(\mathbf{x}|\varphi_A, \varphi_B)}{\partial \varphi_A \partial \varphi_B} = -\frac{1}{\sigma_w^2}\Big(\alpha_A\alpha_B\, \mathrm{e}^{j(\varphi_A-\varphi_B)} \langle s_A; s_B \rangle + \alpha_A\alpha_B\, \mathrm{e}^{-j(\varphi_A-\varphi_B)} \langle s_A; s_B \rangle^* \Big)$$
$$= -\frac{2}{\sigma_w^2} \Re\Big[\alpha_A\alpha_B\, \mathrm{e}^{j(\varphi_A-\varphi_B)} \langle s_A; s_B \rangle \Big]. \tag{7.10}$$

The regularity condition is fulfilled which is obtained by evaluating mean values of first derivatives which both become zero

$$\mathrm{E}\left[\frac{\partial \ln p(\mathbf{x}|\varphi_A, \varphi_B)}{\partial \varphi_A} \right] = 0, \quad \mathrm{E}\left[\frac{\partial \ln p(\mathbf{x}|\varphi_A, \varphi_B)}{\partial \varphi_B} \right] = 0. \tag{7.11}$$

Negative means of second and mixed derivatives yield Fisher information matrix

$$\mathbf{J} = \begin{bmatrix} J_{AA} & J_{AB} \\ J_{AB} & J_{BB} \end{bmatrix} \tag{7.12}$$

where

$$J_{AA} = -\mathrm{E}\left[\frac{\partial^2 \ln p(\mathbf{x}|\varphi_A, \varphi_B)}{\partial \varphi_A^2} \right]$$
$$= \frac{2\alpha_A^2 \|s_A\|^2}{\sigma_w^2}, \tag{7.13}$$

$$J_{BB} = -\mathrm{E}\left[\frac{\partial^2 \ln p(\mathbf{x}|\varphi_A, \varphi_B)}{\partial \varphi_B^2}\right]$$

$$= \frac{2\alpha_B^2 \|\mathbf{s}_B\|^2}{\sigma_w^2}, \tag{7.14}$$

$$J_{AB} = -\mathrm{E}\left[\frac{\partial^2 \ln p(\mathbf{x}|\varphi_A, \varphi_B)}{\partial \varphi_A \partial \varphi_B}\right]$$

$$= \frac{2}{\sigma_w^2}\Re\left[\alpha_A \alpha_B\, e^{j(\varphi_A - \varphi_B)}\langle \mathbf{s}_A; \mathbf{s}_B\rangle\right]. \tag{7.15}$$

Finally, the evaluation of main diagonal elements of $\mathbf{J}^{-1}$ gives CRLB on the estimation MSE

$$\mathrm{E}\left[(\hat{\varphi}_A - \varphi_A)^2\right] \geq \left[\mathbf{J}^{-1}\right]_{AA} = \frac{J_{BB}}{\det \mathbf{J}}, \tag{7.16}$$

$$\mathrm{E}\left[(\hat{\varphi}_B - \varphi_B)^2\right] \geq \left[\mathbf{J}^{-1}\right]_{BB} = \frac{J_{AA}}{\det \mathbf{J}}, \tag{7.17}$$

where

$$\det \mathbf{J} = J_{AA}J_{BB} - J_{AB}^2$$

$$= \frac{4\alpha_A^2 \alpha_B^2 \|\mathbf{s}_A\|^2 \|\mathbf{s}_B\|^2}{\sigma_w^4}\left(1 - \left(\Re\left[e^{j(\varphi_A - \varphi_B)}\frac{\langle \mathbf{s}_A; \mathbf{s}_B\rangle}{\|\mathbf{s}_A\|\|\mathbf{s}_B\|}\right]\right)^2\right). \tag{7.18}$$

A substitution gives

$$\mathrm{E}\left[(\hat{\varphi}_A - \varphi_A)^2\right] \geq \frac{\sigma_w^2}{2\alpha_A^2 \|\mathbf{s}_A\|^2 \left(1 - \left(\Re\left[e^{j(\varphi_A - \varphi_B)}\eta\right]\right)^2\right)}, \tag{7.19}$$

$$\mathrm{E}\left[(\hat{\varphi}_B - \varphi_B)^2\right] \geq \frac{\sigma_w^2}{2\alpha_B^2 \|\mathbf{s}_B\|^2 \left(1 - \left(\Re\left[e^{j(\varphi_A - \varphi_B)}\eta\right]\right)^2\right)}, \tag{7.20}$$

where

$$\eta = \frac{\langle \mathbf{s}_A; \mathbf{s}_B\rangle}{\|\mathbf{s}_A\|\|\mathbf{s}_B\|}. \tag{7.21}$$

We now investigate some special cases.

- *Orthogonal pilots* $\langle \mathbf{s}_A; \mathbf{s}_B\rangle = 0$. This case needs to allocate separate subspaces for pilots and turns the problem into two decoupled traditional phase estimators. Its performance is independent of actual phase values.
- *Identical pilots* $\mathbf{s}_A = \mathbf{s}_B$. This is a true H-MAC CSE pilot case that shares the common subspace. However, the performance depends on the relative channel phase $\Delta\varphi = \varphi_A - \varphi_B$. The normalized inner product becomes $\eta = 1$ and

$$\mathrm{E}\left[(\hat{\varphi}_A - \varphi_A)^2\right] \geq \frac{\sigma_w^2}{2\alpha_A^2 \|\mathbf{s}_A\|^2 \sin^2(\varphi_A - \varphi_B)}, \tag{7.22}$$

$$E\left[(\hat{\varphi}_B - \varphi_B)^2\right] \geq \frac{\sigma_w^2}{2\alpha_B^2 \|s_B\|^2 \sin^2(\varphi_A - \varphi_B)}. \tag{7.23}$$

The performance becomes very poor for $\varphi_A - \varphi_B$ close to integer multiples of π, and for $\varphi_A - \varphi_B = k\pi$, $k \in \mathbb{Z}$, the estimation is impossible.

- *Rotated pilots* $s_A = e^{j\psi} s_B$. Using the identical but rotated pilots does not help. The normalized inner product becomes $\eta = e^{j\psi}$ and the problem remains the same. Only the problematic $\varphi_A - \varphi_B$ angles shift by ψ.

Cramer–Rao Lower Bound – Complex Fading Coefficients CSE in Two-Component H-MAC

We now generalize the preceding case to the full estimation of two *complex-valued* fading coefficients $h_A, h_B \in \mathbb{C}$. We need to use CRLB for complex-valued parameters and employ generalized derivative (see Section A.3).

The complex Fisher information matrix

$$\mathbf{J} = \begin{bmatrix} J_{AA} & J_{AB} \\ J_{BA} & J_{BB} \end{bmatrix} \tag{7.24}$$

has elements

$$J_{AA} = -E\left[\frac{\tilde{\partial}^2 \ln p(\mathbf{x}|h_A, h_B)}{\tilde{\partial} h_A^* \tilde{\partial} h_A}\right] = \frac{\|s_A\|^2}{\sigma_w^2}, \tag{7.25}$$

$$J_{AB} = -E\left[\frac{\tilde{\partial}^2 \ln p(\mathbf{x}|h_A, h_B)}{\tilde{\partial} h_A^* \tilde{\partial} h_B}\right] = \frac{\langle s_A; s_B\rangle^*}{\sigma_w^2}, \tag{7.26}$$

$$J_{BA} = -E\left[\frac{\tilde{\partial}^2 \ln p(\mathbf{x}|h_A, h_B)}{\tilde{\partial} h_B^* \tilde{\partial} h_A}\right] = \frac{\langle s_A; s_B\rangle}{\sigma_w^2}, \tag{7.27}$$

$$J_{BB} = -E\left[\frac{\tilde{\partial}^2 \ln p(\mathbf{x}|h_A, h_B)}{\tilde{\partial} h_B^* \tilde{\partial} h_B}\right] = \frac{\|s_B\|^2}{\sigma_w^2}. \tag{7.28}$$

Main diagonal elements of $\mathbf{J}^{-1}$ give CRLB on estimation MSE

$$E\left[|\hat{h}_A - h_A|^2\right] \geq \left[\mathbf{J}^{-1}\right]_{AA} = \frac{J_{BB}}{\det \mathbf{J}}, \tag{7.29}$$

$$E\left[|\hat{h}_B - h_B|^2\right] \geq \left[\mathbf{J}^{-1}\right]_{BB} = \frac{J_{AA}}{\det \mathbf{J}}, \tag{7.30}$$

where

$$\begin{aligned} \det \mathbf{J} &= J_{AA}J_{BB} - J_{AB}J_{BA} \\ &= \frac{\|s_A\|^2\|s_B\|^2 - |\langle s_A; s_B\rangle|^2}{\sigma_w^4}. \end{aligned} \tag{7.31}$$

Finally, we obtain

$$E\left[|\hat{h}_A - h_A|^2\right] \geq \frac{\sigma_w^2}{\|s_A\|^2\left(1 - |\eta|^2\right)}, \tag{7.32}$$

$$\mathrm{E}\left[|\hat{h}_B - h_B|^2\right] \geq \frac{\sigma_w^2}{\|s_B\|^2 \left(1 - |\eta|^2\right)}, \tag{7.33}$$

where

$$\eta = \frac{\langle s_A; s_B \rangle}{\|s_A\| \|s_B\|}. \tag{7.34}$$

Again, we analyze special cases.

- *Orthogonal pilots* $\langle s_A; s_B \rangle = 0$. This case allocates separate subspaces for pilots and turns the problem into two decoupled traditional single-coefficient estimators.
- *Identical pilots* $s_A = s_B$. These H-MAC pilots share the common subspace. The normalized inner product becomes $\eta = 1$ and the estimation is impossible regardless of the actual channel coefficients. Notice the comparison with phase-only CRLB, where this case affected only some particular values of $\Delta\varphi$.
- *Rotated pilots* $s_A = e^{j\psi} s_B$. The normalized inner product becomes $\eta = e^{j\psi}$ and the situation is the same as for $\eta = 1$.

H-MAC CSE Pilot Design

The pilot design needs to balance the pros and cons related to their orthogonality. The orthogonal design requires higher-dimensional dedicated pilot subspace and some form of pilot space resource allocation. This is then rewarded by the ease of estimator implementation and its robust performance. The simplest form of this design can use some form of orthogonal signature sequences with static (and not much efficient) resource allocation assigning a fixed unique sequence to each network node. The required dimension would then be given by the total number of nodes in the network. Pilots in one given H-MAC stage would then need to be much longer than would be actually required by the target estimation fidelity and this would reduce the duty ratio for payload data. On the other hand, the non-orthogonal pilots and some form of dynamic pilot space resource allocation can improve this performance aspect; however, this is at some expense of the estimator complexity and MSE performance penalty.

7.2.3   Channel State Estimators for Linear AWGN H-MAC

Sufficient Statistic

Using the Neyman–Fisher factorization theorem (see Section A.3) and observing the observation metric ρ (7.4), we can easily verify that

$$T(\mathbf{x}) = [\langle \mathbf{x}; s_A \rangle, \langle \mathbf{x}; s_B \rangle]^{\mathrm{T}} \tag{7.35}$$

is the sufficient statistic. It can be neatly interpreted as a two-dimensional projection (matched filtering) of the observation regardless of the actual dimensionality of the pilot signals.

Joint ML Estimator

A joint ML channel coefficient estimator is

$$[\hat{h}_A, \hat{h}_B]^{\mathrm{T}} = \arg\max_{h_A, h_B} p(\mathbf{x}|h_A, h_B) = \arg\min_{h_A, h_B} \rho(h_A, h_B). \tag{7.36}$$

A simple form of the observation metric $\rho(h_A, h_B)$ (7.4) can be used to derive a straightforward joint ML feed-forward solver. However, we need to keep in our mind that the parameters are complex valued and we must use the generalized derivatives in order to find the stationary point.

The solution is obtained by setting the first derivatives to zeros $\tilde{\partial}\rho/\tilde{\partial}h_A \triangleq 0$, $\tilde{\partial}\rho/\tilde{\partial}h_A \triangleq 0$. This gives

$$\frac{\tilde{\partial}\rho(h_A, h_B)}{\tilde{\partial}h_A} = h_A^* \|s_A\|^2 - \langle x; s_A \rangle^* + h_B^* \langle s_A; s_B \rangle \triangleq 0, \tag{7.37}$$

$$\frac{\tilde{\partial}\rho(h_A, h_B)}{\tilde{\partial}h_B} = h_B^* \|s_B\|^2 - \langle x; s_B \rangle^* + h_A^* \langle s_A; s_B \rangle^* \triangleq 0. \tag{7.38}$$

A matrix equation for the estimator solution is

$$\begin{bmatrix} \|s_A\|^2 & \langle s_A; s_B \rangle^* \\ \langle s_A; s_B \rangle & \|s_B\|^2 \end{bmatrix} \begin{bmatrix} \hat{h}_A \\ \hat{h}_B \end{bmatrix} = \begin{bmatrix} \langle x; s_A \rangle \\ \langle x; s_B \rangle \end{bmatrix} \tag{7.39}$$

which easily leads to the closed-form feed-forward joint ML solver of H-MAC coefficients CSE

$$\hat{h}_A = \frac{\langle x; s_A \rangle}{\|s_A\|^2 \left(1 - |\eta|^2\right)} - \frac{\langle x; s_B \rangle \langle s_B; s_A \rangle}{\|s_A\|^2 \|s_B\|^2 \left(1 - |\eta|^2\right)}, \tag{7.40}$$

$$\hat{h}_B = \frac{\langle x; s_B \rangle}{\|s_B\|^2 \left(1 - |\eta|^2\right)} - \frac{\langle x; s_A \rangle \langle s_A; s_B \rangle}{\|s_A\|^2 \|s_B\|^2 \left(1 - |\eta|^2\right)}. \tag{7.41}$$

Notice that, in the special case of *orthogonal* pilots $\langle s_A; s_B \rangle = 0$, we get a trivial decoupled matched filter form

$$\hat{h}_A = \frac{\langle x; s_A \rangle}{\|s_A\|^2}, \tag{7.42}$$

$$\hat{h}_B = \frac{\langle x; s_B \rangle}{\|s_B\|^2}. \tag{7.43}$$

7.3   NCM in Frequency Selective H-MAC Channel

7.3.1   Block-Constant Frequency Selective H-MAC Channel

Many practical WPNC scenarios must assume a frequency selective propagation channel. Generally, the channel is frequently also randomly time-varying. For the purpose of this section, we will, however, assume a *block-constant* fading, where the fading can be considered as time-invariant over the received signal observation window (e.g. NCM encoded packet). The H-MAC channel model is thus a superposition of the components with AWGN where each component is a linear time-invariant frequency selective channel described by the convolution operation. This model can also nicely describe, as a special case, a standard flat fading where we can easily incorporate *arbitrary delays* on each component signal. Therefore it generalizes the treatment of Section 7.2 in allowing *uneven delays* for each component.

The input–output channel model for K component H-MAC in temporal domain for the observation window $t \in [0, T)$ is

$$u(t) = \sum_{k=1}^{K} s_k(t) * h_k(t) \tag{7.44}$$

and

$$x(t) = u(t) + w(t) \tag{7.45}$$

where $w(t)$ is AWGN. The transmitted signals of H-MAC component nodes are $s_k(t)$, and $h_k(t)$ are impulse responses of component channels. A *special* case of flat fading with uneven delays is modeled by impulse responses

$$h_k(t) = h_k \delta(t - \tau_k) \tag{7.46}$$

where $h_k = \alpha_k\, e^{j\varphi_k}$ is the channel fading coefficient and τ_k is the delay.

Under an assumption of *finite* bandwidth signals $s_k(t)$, we can equivalently use a *tapped delay line* discrete time model. Assume that T_p is a sampling period compliant with the bandwidth of the component signals $s_k(t)$. Then we use sampling theorem expansion for signals $s_k(t)$ inside the convolution integral

$$\begin{aligned}
u_n &= \sum_{k=1}^{K} \int s_k(nT_p - \tau) h_k(\tau)\, d\tau \\
&= \sum_{k=1}^{K} \int \left(\sum_m s_k(mT_p) \operatorname{sinc}\left(\frac{nT_p - \tau - mT_p}{T_p} \right) \right) h_k(\tau)\, d\tau \\
&\stackrel{(a)}{=} \sum_{k=1}^{K} \sum_{\ell} s_{k,n-\ell} \int \operatorname{sinc}\left(\frac{\tau - \ell T_p}{T_p} \right) h_k(\tau)\, d\tau
\end{aligned} \tag{7.47}$$

where (a) uses a substitution $n - m = \ell$, $u_n = u(nT_p)$, $s_{k,n} = s_k(nT_p)$, and we also used an even symmetry of $\operatorname{sinc}(.)$ function. Finally, the discrete time model is

$$u_n = \sum_{k=1}^{K} \sum_{\ell} s_{k,n-\ell} h_{k,\ell} \tag{7.48}$$

where discrete time channel impulse coefficient

$$h_{k,\ell} = \int \operatorname{sinc}\left(\frac{\tau - \ell T_p}{T_p} \right) h_k(\tau)\, d\tau \tag{7.49}$$

is a sampling *projection* (up to a scaling) of the channel impulse response (see also [54] for extensions to randomly time-variant channel).

Notice that the limited bandwidth assumption was needed *only* for the transmitted signals $s_k(t)$. The impulse response can be *arbitrary*. This will be needed when using this model for the special case with flat fading and arbitrary delay. If the impulse response also complies with sampling conditions, then it simply holds $h_{k,\ell} = T_p h_k(\ell T_p)$. Otherwise, it is just a projection.

The *special* case of $h_k(t) = h_k \delta(t - \tau_k)$ then gives

$$h_{k,\ell} = h_k \operatorname{sinc}\left(\frac{\tau_k - \ell T_p}{T_p}\right) \tag{7.50}$$

and we see that for non-integer τ_k the discrete time channel impulse becomes a *series* of discrete pulses. In this case, the series is theoretically infinite and non-causal and any practical usage will inevitably require some form of approximation, e.g. shortening.

7.3.2 NCM with OFDM Waveform

OFDM and Zero-Forcing Frequency Domain Equalizer

The OFDM is a popular waveform with many useful properties. One of the most widely used properties is the capability of OFDM to easily handle the signal propagation through the frequency selective channel. The method is based on transmitting the signal with a cyclic prefix and using a zero-forcing frequency domain type of equalizer on the receiver side. The zero-forcing equalization is not generally optimal but its simplicity makes it a popular choice. Despite the general belief, this method is *not* bound to the OFDM waveform at all. However, its usage with the OFDM waveform comes "for free" since the DFT operations needed for the zero-forcing equalizer are performed by the OFDM receiver anyway.

The OFDM waveform in discrete time[2] is

$$\mathbf{s} = N \operatorname{DFT}^{-1}[\mathbf{q}] \tag{7.51}$$

where $\mathbf{q} = [q_0, \dots, q_{N-1}]^{\mathrm{T}}$ is the vector of constellation points in sub-carriers and N is the total number of sub-carriers. The receiver sub-carrier matched filtering on the received signal in AWGN $x(t) = s(t) + w(t)$ is then performed by the sampling $\mathbf{x}$ and the DFT operation

$$\mathbf{z} = \frac{1}{N} \operatorname{DFT}[\mathbf{x}]. \tag{7.52}$$

We now briefly introduce the cyclic prefix and zero-forcing frequency domain equalizer. All treatment now assumes a discrete time sample space processing. Assume a payload signal $\mathbf{s} = [s_0, \dots, s_{N-1}]^{\mathrm{T}}$ that is extended by the cyclic prefix of the length M into $\mathbf{s}' = [s_{N-M}, \dots, s_{N-1}, s_0, \dots, s_{N-1}]^{\mathrm{T}}$ and the channel impulse response $\mathbf{h} = [h_0, \dots, h_{L-1}]^{\mathrm{T}}$ of the length L. The output of the linear channel will be a discrete time convolution $\mathbf{u}' = \mathbf{s}' * \mathbf{h}$ where

$$u_n = \sum_{\ell=0}^{L-1} s'_{n-\ell} h_\ell. \tag{7.53}$$

When we select only the components with indices $n \in [0 : N-1]$, i.e. $\mathbf{u} = [u_0, \dots, u_{N-1}]^{\mathrm{T}}$, the result clearly becomes a *cyclic* convolution

$$\mathbf{u} = \mathbf{s} \circledast \mathbf{h} \tag{7.54}$$

---

[2] The discrete time representation ignores any artifacts associated with sampling theorem formal violation, particularly the effects on the sub-carrier signal space orthogonality. These effects become negligible for large $N \gg 1$, which will be assumed here.

if the length of the impulse response fits into the prefix length, more precisely if $L - 1 \leq M$. The fact that it turns a standard discrete channel convolution into the cyclic one, which would not happen without the cyclic prefix, allows us to use DFT for the frequency domain description

$$\mathbf{U} = \mathbf{S} \boxdot \mathbf{H} \tag{7.55}$$

where $\boxdot$ is element-wise Hadamard product, and $\mathbf{U} = \text{DFT}[\mathbf{u}]$, $\mathbf{S} = \text{DFT}[\mathbf{s}]$, $\mathbf{H} = \text{DFT}[\mathbf{h}]$.

The zero-forcing frequency domain equalizer is then

$$\mathbf{S} = \mathbf{U} \boxdot^{-1} \mathbf{H}, \tag{7.56}$$

i.e. element-wise division. Notice that the zero-forcing cyclic prefix-based equalizer is not constrained only to OFDM. However, the DFT-based OFDM matched filter demodulation provides the frequency domain representation for free.

In a linear channel with the impulse response $\mathbf{h}$ and AWGN, the input–output model is

$$\mathbf{x} = \mathbf{s} \circledast \mathbf{h} + \mathbf{w} \tag{7.57}$$

and the matched filter observation becomes

$$\mathbf{z} = \mathbf{q} \boxdot \mathbf{H} + \frac{1}{N}\mathbf{W} \tag{7.58}$$

where $\mathbf{W} = \text{DFT}[\mathbf{w}]$. In fact, it turns the system, as seen at the matched filter output, into a multiplexed set of parallel independent sub-channels each with the flat fading coefficient H_n.

NCM with OFDM Waveform in H-MAC

The principles shown above for a classical single-user OFDM with the cyclic prefix and the zero-forcing frequency equalization can be directly used also for the NCM. It works the same way either for the true frequency selective channels or for the flat fading with uneven component delays modeled by the multipath channel.

Assume that $\mathbf{s}_k = N \text{DFT}^{-1}[\mathbf{q}_k]$ are pure payload transmitted signals obtained from sub-carrier constellation points $\mathbf{q}_k$ for the kth component signal in H-MAC. They are extended by cyclic prefixes into $\mathbf{s}'_k$ and transmitted. The discrete time sample space model of the received signal with stripped-down cyclic prefix is

$$\mathbf{x} = \sum_{k=1}^{K} \mathbf{s}_k \circledast \mathbf{h}_k + \mathbf{w} \tag{7.59}$$

where K is the number of component nodes in H-MAC, and $\mathbf{h}_k$ are impulse responses of component channels. Viewed from the perspective of sub-carrier symbols, the system matched filter output model becomes

$$\mathbf{z} = \frac{1}{N} \text{DFT}[\mathbf{x}] = \sum_{k=1}^{K} \mathbf{q}_k \boxdot \mathbf{H}_k + \frac{1}{N}\mathbf{W} \tag{7.60}$$

where $\mathbf{H}_k = \text{DFT}[\mathbf{h}_k]$ and $\mathbf{W} = \text{DFT}[\mathbf{w}]$. Each sub-carrier is thus

$$z_n = \sum_{k=1}^{K} q_{k,n} H_{k,n} + \frac{1}{N} w_n \tag{7.61}$$

and from the perspective of the sub-carrier, we can design the NCM as for an ordinary flat fading channel.

An important consequence of this result is its application to the special case of flat fading H-MAC channels with *uneven* delays. This is a problem specific to WPNC that cannot be ignored. A precise symbol-timing alignment of the signals received from multiple H-MAC component nodes is extremely difficult to guarantee even if we have allowed some type of a closed-loop (feed-back) timing adjustment of the transmitted signals. However, the NCM with OFDM waveform easily accepts arbitrary timing misalignment among the signals received from H-MAC components up to approximately[3] the length of the cyclic prefix.

7.4 NCM Design for Parametric Channels

7.4.1 Parameter Invariant and Uniformly Most Powerful Design

The performance of NCM is strongly affected by the H-MAC channel parametrization, namely by the dependence of the H-constellation on the relative fading (see Section 3.5.2). It would be highly desirable if we could eliminate, or at least minimize, these effects by some smart NCM design, simply to make the NCM parametrization resistant. Before we try to do this, however, we need to clarify what is understood under various forms of the resistance against the H-MAC channel parametrization. We focus purely on the *relative* fading of the *linear* channel, since the common fading only rotates the overall received combined signal and does not make any problems (see Example 3.4). The common fading represents a common scaling coefficient that can be factorized out of the channel-combined signal regardless of the number of the H-MAC components. As such, it does not change the *shape* of the H-constellation and all general concepts from classical point-to-point communications apply. In contrast with this, the relative fading changes nonlinearly the shape of the constellation and we must apply WPNC specific solutions.

Parameter Invariant NCM

Under the *parameter invariant NCM* term, we understand a design which guarantees that some processing algorithms or performance indicators do *not* depend on the *relative* fading. The invariance of the processing algorithms means that the receiver *processing* will not have to adopt to the relative fading; however, the resulting *performance* can still of course depend on the relative fading.[4] The considered processing algorithms or performance indicators might have many forms.

---

[3] The projection of Dirac delta function into the sampling functions has theoretically infinite length and we must adopt some approximation by its truncation.

[4] For example, similarly to the classical PSK being influenced by the attenuation in the channel – the demodulator structure is unaffected; however, the bit error rate performance is affected.

- *H-constellation invariance* This form essentially means that the H-constellation shape remains the same for any relative fading. However, it can globally rotate or linearly scale with the common fading. This is the strictest form of the invariance effectively saying that the relative fading does not have any effect.
- *H-constellation decision regions invariance* This form slightly relaxes the previous one by allowing the shift of some H-constellation points; however, it is only those points, and it happens in such a way that *cannot* affect the decision boundaries between different HNC map values. *Sub-variants* can consider various forms (or approximations) of the *metric* used to define the decision regions, e.g. the true H-metric or H-distance approximation, etc. (see Section 4.4).
- *H-performance metric invariance* This requires that some *performance* metric is strictly (or approximately) independent of the relative fading. There is a variety of performance metrics that could be used with various degrees of their connection to the overall end-to-end WPNC network performance. Some of the popular metrics are
  - H-symbol observation likelihood,
  - H-distance,
  - H-mutual information.

Uniformly Most Powerful Design

The uniformly most powerful (UMP) NCM design allows either the structure or performance indicators to depend on the relative fading. However, we require that this particular NCM will perform *better* than any other possible NCM for *all possible relative fading values*. Possibly, we may restrict the validity of the statement to some *subspace* of the relative fading, e.g. to become uniformly most powerful against phase rotation.

Critical State Performance Optimization

Frequently, the parameter invariant or the uniformly most powerful design is difficult to achieve. In this case we can try at least to design the NCM that avoids or minimizes the effect of some *critical* situation. The most typical example is the NCM design minimizing the number of occurrences and the impact of *unresolved singular fading states*.

7.4.2   H-Distance Criterion Parametric Design

Now we derive conditions for *channel symbol-wise H-distance* performance criterion NCM design. The choice of H-distance and the symbol-wise metric is of course not the optimal one (see Section 4.4); however, its simplicity allows to obtain some easily interpretable results.

Two-Component Linear H-MAC AWGN System Model

We will assume the two-component linear H-MAC AWGN system model. The constellation space symbol-wise model of the channel combined payload signal is

$$\mathbf{u}(c_A, c_B) = \mathbf{s}_A(c_A) + h\mathbf{s}_B(c_B) \tag{7.62}$$

where we assumed only the *relative* fading $h \in \mathbb{C}$ (see Section 3.5.2 and Example 3.4). Common fading is easy to cope with: it only rotates the whole constellation, and we can drop it for a simplicity. Component transmitted constellation points are $s_A, s_B \in \mathbb{C}^{N_s \times 1}$, where N_s is the constellation dimensionality per symbol, and the corresponding component channel code symbols are c_A, c_B. For a simplicity, the channel symbol sequential number is dropped.

Our choice of having the relative channel fading associated with s_B was purely ad hoc. We could have also done that vice versa. This implies that all statements in the following should be understood as symmetrically applicable to both s_A and s_B.

The observation model of the channel-combined signal is AWGN $\mathbf{x} = \mathbf{u} + \mathbf{w}$. In this channel, the true decoding metric for an isomorphic layered NCM is the marginalized likelihood (4.12). We can use the H-distance approximation (4.20) under some specific conditions (see details in Section 4.4). Although the H-distance is an approximation with limited validity, it provides easy-to-understand results. We must, however, be careful in their application. Under this approximation (see Section 4.5 for more exact treatment involving also *self-distance* spectrum), the error rate performance will be given by H-constellation H-distances.

H-Distance Spectrum

The H-distance of the H-constellation observed at the channel-combined signals is defined as

$$\rho^2(c_A, c_B, c_A', c_B') = \|\mathbf{u} - \mathbf{u}'\|^2 \tag{7.63}$$

where $\mathbf{u} = \mathbf{u}(c_A, c_B)$, $\mathbf{u}' = \mathbf{u}(c_A', c_B')$, and c_A, c_B, c_A', c_B' are such that the channel-combined H-alphabet points belong to two *different* HNC map values

$$c = \chi_c(c_A, c_B) \neq \chi_c(c_A', c_B') = c'. \tag{7.64}$$

In other words, values $\rho^2(c_A, c_B, c_A', c_B')$ belong to the H-distance spectrum (Section 4.5.4).

We define $\Delta s_A = s_A(c_A) - s_A(c_A')$ and $\Delta s_B = s_B(c_B) - s_B(c_B')$. A simple manipulation with H-distance reveals that

$$\begin{aligned}
\rho^2(c_A, c_B, c_A', c_B') &= \|\Delta s_A + h \Delta s_B\|^2 \\
&= \|\Delta s_A\|^2 + |h|^2 \|\Delta s_B\|^2 + 2\Re\left[h^*\langle \Delta s_A; \Delta s_B\rangle\right].
\end{aligned} \tag{7.65}$$

Complex Relative Channel Gain Invariant H-Distance Spectrum

The NCM with H-distances invariant to full complex-valued relative channel gain h would require $\rho^2(c_A, c_B, c_A', c_B')$ not being a function of h, i.e.

$$\frac{\tilde{\partial} \rho^2}{\tilde{\partial} h} = h^* \|\Delta s_B\|^2 + \langle \Delta s_A; \Delta s_B\rangle^* \triangleq 0 \tag{7.66}$$

where we used a generalized derivative. This means

$$h\|\Delta s_B\|^2 = -\langle \Delta s_A; \Delta s_B\rangle, \tag{7.67}$$

$$h\|\Delta s_A\|^2 = -\langle \Delta s_B; \Delta s_A\rangle, \tag{7.68}$$

where the second equation comes from the symmetry of the problem.

The complex relative gain invariant NCM design w.r.t. H-distance spectrum is *impossible*. Should the above equations hold for all $h \neq 0$ and some $\|\Delta s_A\|^2 \neq 0$ and $\|\Delta s_B\|^2 \neq 0$, the value of $\langle \Delta s_A; \Delta s_B \rangle$ would have to adjust to the actual channel state. This is impossible without the channel state information being available on the transmitter side. This would not be needed if $\|\Delta s_A\|^2 = 0$, $\|\Delta s_B\|^2 = 0$, and $\langle \Delta s_A; \Delta s_B \rangle = 0$ for *all* c_A, c_B, c_A', c_B' such that the corresponding H-symbols differ $c \neq c'$. But this would mean that all H-alphabet points belonging to different HNC map values would have to have the transmitted points identical on *both* components. This is again a contradiction.

Relative Channel Phase Invariant H-Distance Spectrum

Clearly, some dependence on the relative fading $h = |h| \, e^{j\psi}$ is unavoidable. We can naturally try to narrow the condition to the phase-only one.

LEMMA 7.1 (NCM with Phase Invariant H-Distance Spectrum)   *The NCM has relative channel phase invariant H-distance spectrum if and only if*

$$\langle \Delta s_A; \Delta s_B \rangle = 0 \tag{7.69}$$

for all $\Delta s_A = s_A(c_A) - s_A(c_A')$, $\Delta s_B = s_B(c_B) - s_B(c_B')$ *such that* $c = \chi_c(c_A, c_B) \neq \chi_c(c_A', c_B') = c'$.

Proof   For the forward implication, a substitution of the condition $\langle \Delta s_A; \Delta s_B \rangle = 0$ into (7.65) reveals

$$\rho^2(c_A, c_B, c_A', c_B') = \|\Delta s_A\|^2 + |h|^2 \|\Delta s_B\|^2 \tag{7.70}$$

which clearly does not depend on the relative channel phase ψ.

For the converse, we manipulate the cross-term in (7.65) into

$$\Re\left[h^* \langle \Delta s_A; \Delta s_B \rangle \right] = |h \langle \Delta s_A; \Delta s_B \rangle| \cos\left(\sphericalangle \langle \Delta s_A; \Delta s_B \rangle - \psi \right). \tag{7.71}$$

If the ρ^2 does not depend on ψ then the above expression does not depend on it either. This implies that either $\sphericalangle \langle \Delta s_A; \Delta s_B \rangle - \psi = \text{const}$ for arbitrary ψ, which would require the channel state information on the transmitter side, or that the magnitude $|\langle \Delta s_A; \Delta s_B \rangle| = 0$, i.e. $\langle \Delta s_A; \Delta s_B \rangle = 0$. $\square$

The lemma allows us to easily identify a *sufficient* condition for compliant HNC maps, as will be shown next.

LEMMA 7.2 (Phase Invariant H-Distance Compliant HNC Maps)   *Two-component NCM with HNC map $c = \chi_c(c_A, c_B)$ and one-to-one symbol-wise constellation maps $s_A = s_A(c_A)$, $s_B = s_B(c_B)$ has a relative phase invariant H-distance spectrum if*

$$\chi_c(c_A, c_B) = \chi_c(c_A', c_B') \ \forall c_A, c_B, c_A', c_B' : c_A \neq c_A' \wedge c_B \neq c_B'. \tag{7.72}$$

Proof   Under one-to-one constellation mapping conditions, it holds that $\Delta s_A \neq 0 \iff c_A \neq c_A'$ and $\Delta s_B \neq 0 \iff c_B \neq c_B'$. The phase invariance (Lemma 7.1) requires $\langle \Delta s_A; \Delta s_B \rangle = 0$ for any $c \neq c'$. In order to have $c \neq c'$, at least one pair of component symbols must differ. If one of the component symbol pairs is equal, and the other one is not, i.e. $c_A = c'_A$ and $c_B \neq c'_B$ or $c_A \neq c'_A$ and $c_B = c'_B$, then $\Delta s_A = \mathbf{0}$ or $\Delta s_B = \mathbf{0}$, and as a consequence also $\langle \Delta s_A; \Delta s_B \rangle = 0$. If both components differ $c_A \neq c'_A \wedge c_B \neq c'_B$ we cannot simplistically rely on zero values of Δs_A or Δs_B. If the case of $c_A \neq c'_A \wedge c_B \neq c'_B$, which is equivalent to $\Delta s_A \neq \mathbf{0} \wedge \Delta s_B \neq \mathbf{0}$, belongs to a common HNC map value $c = c'$ then the condition of $\langle \Delta s_A; \Delta s_B \rangle = 0$ for any $c \neq c'$ will be fulfilled. $\qquad \square$

REMARK 7.1 (Sufficient Condition for HNC vs. General Solution) Notice that the lemma forms only a *sufficient* (but *not* necessary) condition. On the other hand it holds regardless of the constellation alphabet. A general solution for the zero inner product $\langle \Delta s_A; \Delta s_B \rangle = 0$ can of course be achieved for non-zero components but only for some specific particular constellations.

REMARK 7.2 (Binary Alphabets) Component constellations of NCM with *binary* alphabets are particular easy examples fulfilling the sufficient condition lemma. Assume, without the loss of generality, that the channel symbols alphabet is $c_A, c_B \in \mathbb{F}_2$. All cases where $c_A \neq c'_A \wedge c_B \neq c'_B$ are these

$$
\begin{array}{cccccc}
c_A & c_B & c'_A & c'_B & c & c' \\
0 & 0 & 1 & 1 & 0 & 0 \\
0 & 1 & 1 & 0 & 1 & 1 \\
1 & 0 & 0 & 1 & 1 & 1 \\
1 & 1 & 0 & 0 & 0 & 0
\end{array}
\qquad (7.73)
$$

As we easily see, the XOR HNC map $c = \chi_c(c_A, c_B)$ *complies* with the relative phase invariant design.

REMARK 7.3 (Higher Cardinality Alphabets) A fulfillment of the sufficient condition lemma is *impossible* for higher cardinality alphabets $M_s > 2$. This can be shown by a contradiction. Assume, without the loss of generality, integer valued symbols $c_A, c_B \in [0 : M_s - 1], c \in [0 : M - 1]$. Let $\chi_c(0, 0) = 0$. Then it must hold for all $c_A \neq 0 \wedge c_B \neq 0$ that $\chi_c(c_A, c_B) = 0$. Let $\chi_c(0, 1) = 1$. Then for all $c_A \neq 0 \wedge c_B \neq 1$ it must hold $\chi_c(c_A, c_B) = 1$.

If we had a binary alphabet, there would be clearly no conflict of having $\chi_c(1, 1) = 0$, implied by first condition, and $\chi_c(1, 0) = 1$. However, for $M_s > 2$, the first condition implies $\chi_c(1, 2) = 0$, and the second condition implies $\chi_c(1, 2) = 1$, which makes a contradiction.

REMARK 7.4 (H-Distance Criterion) A high *caution* should be paid in interpreting the invariance w.r.t. H-distance criterion. For example, a binary BPSK NCM with XOR HNC map is perfectly H-distance invariant to the relative phase. However, when we confront that against a true performance measure by the hierarchical mutual information (see Figure 6.5b), we see that the H-distance criterion can be used only for a high-SNR regime. This is in full coherence with the discussion in Section 4.4.

Uniformly Most Powerful Design for Relative Phase

The UMP design for a relative channel phase with H-distance performance criterion means finding such particular NCM $(\bar{s}_A, \bar{s}_B, \bar{\chi}_c)$ that the H-distance (7.65) for this NCM would be greater than or equal to that for any other NCM (s_A, s_B, χ_c). There is only one part of the H-distance expression that depends on the phase ψ and it practically turns into discovering whether we can find $\bar{s}_A, \bar{s}_B, \bar{\chi}_c$ such that

$$\Re\left[|h|\,e^{-j\psi}\langle\Delta\bar{s}_A;\Delta\bar{s}_B\rangle\right] \geq \Re\left[|h|\,e^{-j\psi}\langle\Delta s_A;\Delta s_B\rangle\right] \tag{7.74}$$

or

$$\cos\left(\sphericalangle\langle\Delta\bar{s}_A;\Delta\bar{s}_B\rangle - \psi\right) \geq \cos\left(\sphericalangle\langle\Delta s_A;\Delta s_B\rangle - \psi\right) \tag{7.75}$$

for any $\psi \in [0, 2\pi)$.

Evidently, for *general* $\langle\Delta s_A;\Delta s_B\rangle$ and *general* alphabet size M_s, no such NCM exists. In the case of $\langle\Delta s_A;\Delta s_B\rangle \neq 0$, whatever choice of $\bar{s}_A, \bar{s}_B, \bar{\chi}_c$ we make, it fixes the value on angle $\sphericalangle\langle\Delta\bar{s}_A;\Delta\bar{s}_B\rangle$ and there will always be some ψ value violating the UMP condition. The relative phase invariant design with $\langle\Delta s_A;\Delta s_B\rangle = 0$ is also *not* generally the UMP design, since there will be always some NCM with $\langle\Delta s_A;\Delta s_B\rangle \neq 0$ where $\cos\left(\sphericalangle\langle\Delta s_A;\Delta s_B\rangle - \psi\right) > 0$ for some ψ.

The *exception* is the *binary* alphabet NCM with XOR HNC map. It is phase invariant and UMP w.r.t. H-distance. The binary alphabet size and XOR map implies that $\langle\Delta s_A;\Delta s_B\rangle = 0$ for any situation with $c \neq c'$ regardless of the constellation. So it *cannot* happen that the cross-term $2\Re\left[h^*\langle\Delta s_A;\Delta s_B\rangle\right]$ in (7.65) would be positive.

7.4.3   Tx-Based Adaptation and Diversity-Based Solutions

A Tx-based adaptation technique for the parametric H-MAC is based on a conceptually simple principle. The transmitter *pre-compensates* the transmitted signal in such a way that the channel-combined signal viewed at the receiver appears as if there were no (or some favorable) channel parametrization. For example, if the channel introduces the phase rotations, resulting in some relative phase, the transmitters pre-rotate the signals that they combine at the channel as if there were no relative phase. Clearly, this technique requires a feedback channel providing the channel state information and the technique is prone to all the problems known from such feedback adaptations in classical single-user systems. These are typically the precision and the causality of provided estimates w.r.t. the channel dynamics, amount of the feedback information, a proper dynamic model of the channel, and many others.

Apart from the problems inherited from classical point-to-point adaptive systems with feedback, there is an additional, WPNC specific, phenomenon. The adaptation problem does *not* need to be *solvable* even under perfect supporting conditions. For example, assume two source nodes SA, SB that are both received by two relays R1, R2 in one stage, and for simplicity assume only phase rotation in the channels. We denote adjustable transmitter phases for sources φ_A, φ_B, phase shift of SA-R1 channel $\varphi_{A,1}$ and similarly of other links $\varphi_{A,2}, \varphi_{B,1}, \varphi_{B,2}$, and the desired relative phase at relays ψ_1, ψ_2. Our target is then to find φ_A, φ_B such that

$$(\varphi_A + \varphi_{A,1}) - (\varphi_B + \varphi_{B,1}) = \psi_1, \tag{7.76}$$

$$(\varphi_A + \varphi_{A,2}) - (\varphi_B + \varphi_{B,2}) = \psi_2. \tag{7.77}$$

This clearly has a solution only if $\psi_1 - (\varphi_{A,1} - \varphi_{B,1}) = \psi_2 - (\varphi_{A,2} - \varphi_{B,2})$ at least in $\mathrm{mod}2\pi$ sense.

Diversity-based solutions, in the temporal, frequency, or spatial domain, can diversify the effects of fading parametrization in individual eigenspaces of the H-MAC. This is, however, achieved at the expense of utilizing multiple eigendimensions for the diversity gain instead of dedicating them to the payload data flow (as for the multiplexing–diversity tradeoff in space-time communications).

8 NCM Design for Partial HSI and Asymmetric H-MAC

8.1 Introduction

As shown in Chapter 5 in general, and in Chapter 6 in more concrete forms, the design of NCM is relatively straightforward and its performance provides an advantage only under specific conditions. It particularly applies to the form of HNC maps and some requirements on the SNR balance of component channels in H-MAC. For example, the isomorphic layered NCM for HDF strategy gives an H-rate advantage if the HNC map is a GF-based linear minimal cardinality map and the component channels have roughly the same SNRs. When applied for example to a butterfly network, it implies symmetric distances on the source-to-relay channels and a perfect HSI with the same rate as the H-rate at the relay. All relays that use a minimal cardinality linear GF HNC map must be on the same GF vector space as the source messages. The resulting equations and therefore also the HI processed by the relay have the same codebook size as the component messages. Similarly, CF strategy requires a common nested lattice code on sources and the HNC map also implies that all flows of hierarchical information from relays to the destination will have the same codebook size.

Clearly, many practical WPNC networks will need to operate with asymmetric SNR conditions or asymmetric parametrization on the H-MAC, and also the destination might have available only a limited HI or HSI (w.r.t. source codebook sizes) to compute the desired data. The "brute force" solution is to use minimal common source rates, which are bottlenecked by the weakest channel, either on the H-MAC or on the HSI side. This chapter, however, offers a solution allowing us to design NCM for partial HSI and asymmetric H-MAC. It is based on splitting the source data streams into several lower-rate independent components. The split sources can be viewed as virtual component nodes. The split allows us to use different strategies for selected virtual components. We can also group a selected subset of virtual nodes and apply a given strategy that matches to the selected group. For example, we can use layered NCM for the group of virtual nodes having the same rate and having a similar SNR. We can form multiple groups in this way and combine them by another coding strategy. For example, if the groups have very asymmetric rates or SNRs, we can use a classical multi-user strategy. Generally it leads to solutions using various hybrid combined classical multi-user and hierarchical coding designs.

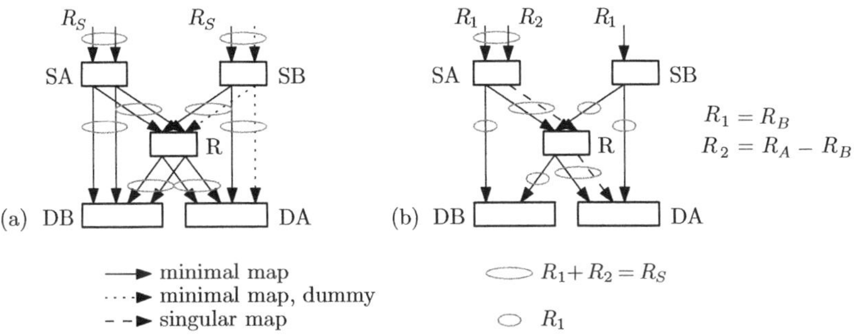

Figure 8.1 Example scenario for partial HSI and asymmetric H-MAC.

Although the combination of the classical multi-user and the H-coding might seem to be straightforward at first sight, we will see that some aspects of the H-coding, namely a self-dispersion of the equivalent H-channel, create some unexpected challenges and results.

An example scenario is shown in Figure 8.1. We assume source SA with the total rate R_A and source SB with the rate $R_B < R_A$. A solution using a single minimum cardinality HNC map would need to use a dummy message with the rate R'_B (e.g. zero-padding) to align source rates to $R_S = R_A = R_B + R'_B$ (Figure 8.1a). All links must support the rate R_S. Notice that the zero-padding of the message fed into the $R_B + R'_B$ codebook does *not* generally make a good R_B codebook. The solution using NCM for multi-map H-MAC (Figure 8.1b) uses a combination of the minimal map with the rate R_1 and a singular (single-user) map with the rate R_2, where $R_S = R_1 + R_2$. As we see, we can substantially reduce the rates.

8.2 NCM for Multi-Map H-MAC

8.2.1 Design Goals

The NCM design for a partial HSI (or partial HI in other destination observations) scenario and an asymmetric H-MAC must solve two aspects.

Extended Cardinality Relay HNC Map

A final destination uses the set of HNC maps available in some (typically full message decisions) information measure to solve them for the desired message. If some of these HSIs or HIs have limited cardinality, the total cardinality of the remaining ones must compensate for that, either by having extended cardinality or by having redundant sets of HIs. Consider, for example, a butterfly network with two sources with codebooks of equal sizes $M_A = M_B = M$. If the source–destination HSI is, e.g. owing to a long propagation path, providing only a partial HSI represented by the codebook size $M_{A'} < M$, then the relay must provide an *extended* cardinality message HNC map with $M_R > M$.

Asymmetric Component Codebooks

The case of asymmetric H-MAC channels, e.g. having substantially different SNRs, calls for a solution with asymmetric component codebook rates. These can be then properly tuned to the channel SNRs.

8.2.2   Structured NCM for Multi-Map H-MAC

The solution fulfilling the above-stated goals is a *structured* NCM using simultaneously *multiple* HNC maps as a hybrid composition. The building-block maps used for this composition are such that we can easily design the NCM. The maps are combined in such a way that the resulting composite map has *extended* cardinality. The maps can be arbitrarily asymmetric w.r.t. the component messages and thus allow asymmetric total rates of component codebooks. The particular structure of the composition can be any form of *classical* multi-user design, e.g. time-division, or superposition coding, etc.

Building-block NCMs are *isomorphic layered NCMs*. If the isomorphic layered NCM is *regular*, then we know (Section 5.7.4) that the hierarchical mutual information rates are *achievable* by *layered H-decoding* and the system is conveniently modeled by the *equivalent hierarchical channel* (Figure 5.12). We also showed that the *linear isomorphic NCM* (Section 4.7.2) can be practically used.

Structured NCM Codebook Construction

DEFINITION 8.1 (Structured NCM Codebook)   Assume an H-MAC channel with L component nodes and true component messages $\tilde{b} = \{b_1, \ldots, b_L\}$. Each message b_ℓ is split into several virtual sub-messages $b_{\ell,m}$ creating *one-to-one* mapping to the original message $b_\ell \mapsto \{b_{\ell,1}, b_{\ell,2}, \ldots\}$. The virtual sub-messages are grouped into the K sets $\tilde{d}_k$, $k \in [1 : K]$, where each $b_{\ell,m}$ belongs to exactly one $\tilde{d}_k$, and the sets are mutually *independent* $\tilde{d}_k \perp \tilde{d}_i$, $k \neq i$, $k, i \in [1 : K]$. The set of K NCM codebooks $\tilde{c}_k^N = \tilde{C}_k(\tilde{d}_k)$, $k \in [1 : K]$, $\tilde{d}_k = \{d_{k,1}, d_{k,2}, \ldots\}$, $\tilde{c}_k^N = \{c_{k,1}^N, c_{k,2}^N, \ldots\}$, with symbol-wise constellations space mappers $s_{k,j,n} = \mathcal{A}_{s,k}(c_{k,j,n})$, $n \in [1 : N]$, and the associated HNC maps $d_k = \chi_k(\tilde{d}_k)$, are then combined in the second-layer classical MAC coding strategy $\overset{\circ}{\mathcal{C}}$.

The set of $\{\tilde{C}_k\}_k$ with their associated HNC maps $\{\chi_k\}_k$ and the second-layer coding strategy $\overset{\circ}{\mathcal{C}}$ is called a *structured NCM codebook*.

The structured NCM has two core structuring features. The first layer of structuring is a set of NCMs where the input messages for each set are *independent*. The input messages are *virtual* messages obtained by splitting the real ones, e.g. by multiplexing the bit stream of the real source node into the sub-streams, which can be seen as new virtual nodes. The corresponding messages are then also called "virtual" to emphasize this fact. The second layer of structuring is done by classical MAC coding strategy $\overset{\circ}{\mathcal{C}}$ (e.g. time-sharing, or superposition coding, see Section 8.3) defined w.r.t. the outputs of the first layer. Figure 8.2 shows the structuring example.

Equivalent Multi-Map H-MAC Channel

Assume a structured NCM where *all* component NCMs are *isomorphic layered* NCMs. This will be referred to as a structured isomorphic layered NCM. Each isomorphic NCM

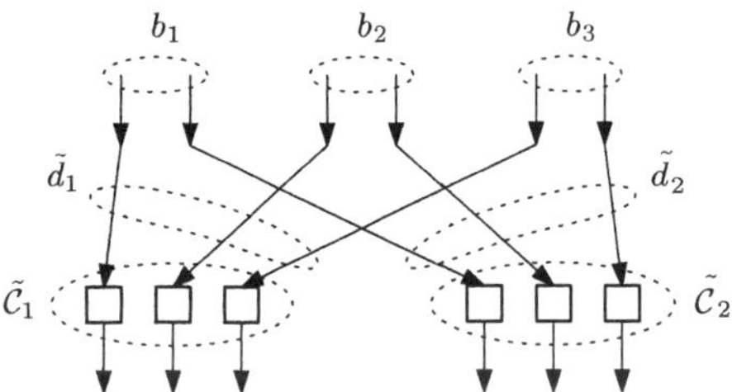

Figure 8.2 Structured NCM. There are three physically existing nodes with messages b_1, b_2, b_3. Each of these messages is split into two independent ones effectively creating six virtual node messages. Each is individually encoded. This creates a six-component channel where we, however, group the components into component groups $\tilde{d}_1, \tilde{d}_2$. The coding strategy is applied in two *structured* subsequent levels. In the first level, we apply some NCM on each component group $\tilde{d}_1$ and $\tilde{d}_2$ separately. Then, in the second level, we combine the resulting NCM encoded groups using a classical multi-user technique. Because of the structured application of a two-level strategy first within the sub-groups and in-between the sub-groups, we call it a *structured NCM*.

$\tilde{C}_k$ is associated with its own H-codebook C_k, HNC maps $d_k = \chi_k(\tilde{d}_k)$, symbol-wise code map $c_k = \chi_{c,k}(\tilde{c}_k)$, and H-constellation $\mathcal{U}_k$.[1] The structured isomorphic layered NCM in H-MAC can be modeled as a set of equivalent hierarchical channels (similar to Figure 5.12) where the second-layer *observation channel* is a *classical multi-user MAC* (Figure 8.3). This will be called a *multi-map H-MAC*. It is, however, important to stress that the model has two layers. The first is the H-MAC for each component NCM and the resulting H-constellations are observed as a classical multi-user MAC.

For simplicity of the treatment, we will consider only two-component $K = 2$ structured NCM in the following text. The generalization is straightforward.

8.2.3   Achievable H-rate Region for Multi-Map H-MAC

We now find the achievable H-rate region for multi-map H-MAC with two components $K = 2$. The derivation relies, similarly as in Section 5.7.4, on isomorphic and regularity properties.

THEOREM 8.2 (Achievable H-Rate Region for Multi-Map H-MAC) *Assume isomorphic layered regular structured NCM $\tilde{c}_k^N = \tilde{C}_k(\tilde{d}_k)$ with symbol-wise one-to-one constellation mapper $s_{k,j,n} = \mathcal{A}_{s,k}(c_{k,j,n})$ for the two-map ($k \in \{1,2\}$, $K = 2$) H-MAC memoryless channel described by PDF $p(x^N|\tilde{s}_1^N, \tilde{s}_2^N) = \prod_n p(x_n|\tilde{s}_{1,n}, \tilde{s}_{2,n})$. H-codebooks are $(2^{NR_k}, N)$ codes $c_k^N = C_k(d_k)$, $k \in \{1,2\}$. Corresponding HNC symbol-wise maps are $d_k = \chi_k(\tilde{d}_k)$, $c_k = \chi_{c,k}(\tilde{c}_k)$ and H-constellations are $\mathcal{U}_k(c_k)$. The achievable H-rate*

---

[1] Notice that we use slightly "overloaded" notation w.r.t. a standard single map NCM.

(a) In a *single map HNC*, $c_n = \chi_c(\tilde{c}_n)$, $\tilde{c}_n = (\ldots, c_{k,n}, \ldots)$ where c_n means the nth symbol in the sequence of HNC map c, and $c_{k,n}$ is the nth symbol of the kth component.

(b) In a *structured* NCM with explicit notation, $c_{k,n} = \chi_{c,k}(\tilde{c}_{k,n})$, $\tilde{c}_{k,n} = (\ldots, c_{k,m,n}, \ldots)$, where $c_{k,n}$ is the nth symbol of the kth map, and $c_{k,m,n}$ is the nth symbol of the mth component in the kth node group. The complete N-dimensional vector for the kth map is c_k^N.

(c) If it is clear from the context, we may *drop* the explicit sequence number index n. For example, the symbol-wise kth group map is $c_k = \chi_{c,k}(\tilde{c}_k)$.

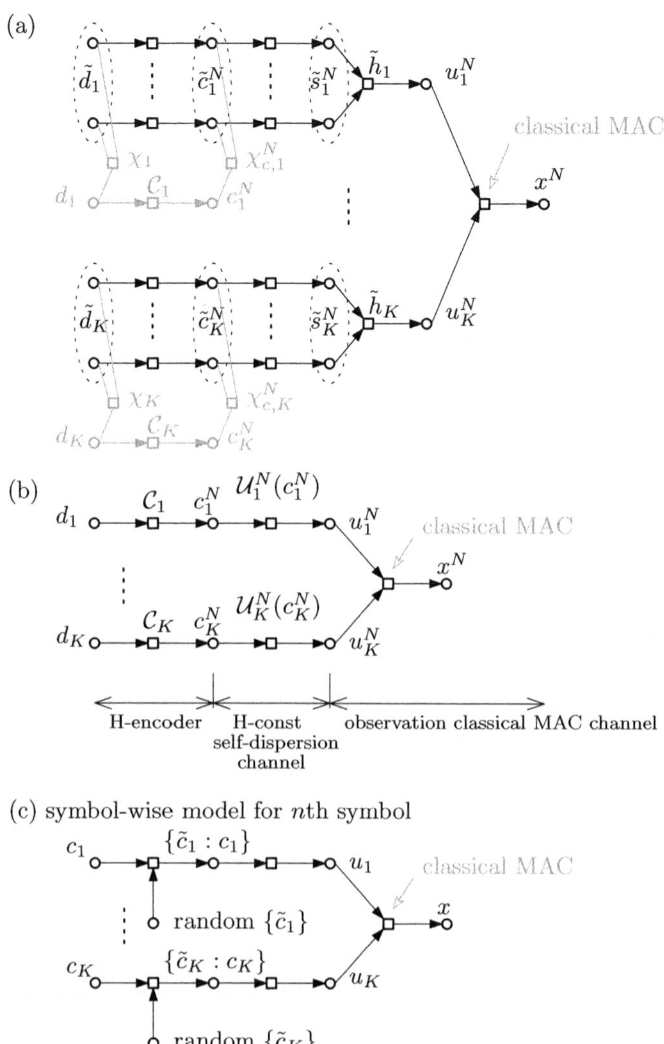

Figure 8.3 Equivalent multi-map H-MAC channel. Part (a) depicts the way virtual message groups $\tilde{d}_1, \ldots, \tilde{d}_K$ form the real observation components $u_1^N, \ldots, u_K^N$ and the overall receiver observation x^N. *Within* each group, we interpret the channel-combined signal as *isomorphic NCM* with virtual H-message d_k and H-codeword c_k^N. *Among* the groups, we do *not* define any relationship. We simply consider the contributions from groups as a classical MAC observation. Each isomorphic NCM can be modeled using an equivalent H-channel. This is shown in part (b). Each equivalent H-channel is formed by the H-encoder followed by the H-constellation self-dispersion channel (Section 5.7.4, Figure 5.12). The outputs of equivalent H-channels are then combined by classical MAC observation. Part (c) shows the symbol-wise model from the perspective of H-code symbols $c_1, \ldots, c_K$. The self-dispersion part is modeled using random group components $\tilde{c}_k$ consistent with a given H-code symbol c_k.

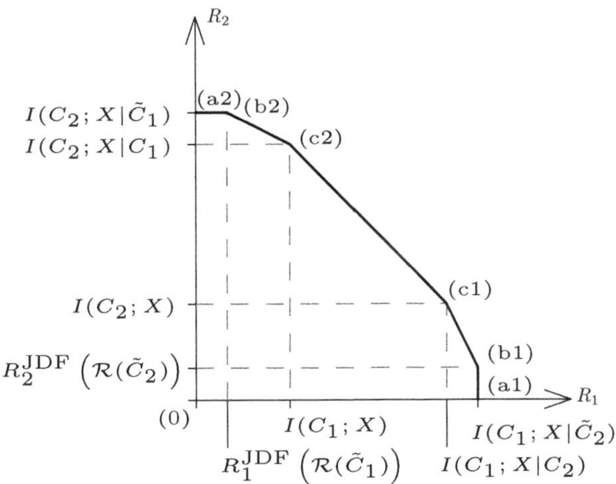

Figure 8.4 Achievable H-rate region for multi-map H-MAC.

region $\mathcal{R}(C_1, C_2)$ is a convex hull or H-rate vectors (R_1, R_2) defined by the following corner points (Figure 8.4):

(a1) $R_1 = I(C_1; X|\tilde{C}_2)$, $R_2 = 0$,
(b1) $R_1 = I(C_1; X|\tilde{C}_2)$, $R_2 = R_2^{\mathrm{JDF}}\left(\mathcal{R}(\tilde{C}_2)\right)$,
(c1) $R_1 = I(C_1; X|C_2)$, $R_2 = I(C_2; X)$,
(a2,b2,c2) similar to (a1,b1,c1) with swapped indices 1 and 2,
(0) $R_1 = R_2 = 0$.

At points (b1) and (b2), $R_k^{\mathrm{JDF}}\left(\mathcal{R}(\tilde{C}_k)\right)$ is the hierarchical rate for JDF strategy where the classical multi-user rate region is $\mathcal{R}(\tilde{C}_k)$.

Proof The proof follows similar lines and reuses some statements and interim results of the proof of Theorem 5.10, so we will show only the outline here. We will show the decoding strategy for all corner points of the region. In all cases, the decoder is a joint typicality decoder. All codebooks are assumed to be symbol-wise IID and drawn according to a proper symbol constellation PDF $p(s_{k,j}^N) = \prod_{n=1}^{N} p(s_{k,j,n})$. Because the constellation mappers $\mathcal{A}_{s,k}$ are symbol-wise and one-to-one maps, we can equivalently use the $\tilde{C}_k$ codesymbols in the following treatment, e.g. in the channel model $p(x^N|\tilde{c}_1^N, \tilde{c}_2^N)$.

The proof is based on the equivalent multi-map H-MAC channel model shown in Figure 8.3c. We recall that the individual branches for $c_1, \ldots, c_K$ form H-constellation dispersion channels as in a standard H-MAC channel (Figure 5.12). Therefore, all conditions required there, i.e. all NCMs being *isomorphic* and *layered*, must hold here as well. In the following treatment, we will employ both the hierarchical and the joint symbols and related entities as the rates and the mutual information. It is important to properly distinguish them. A core principle that is used throughout the whole proof is the *self-dispersion* of the equivalent channel (Section 5.7.4, Figure 5.12). It means that the conditioning by H-symbol still leaves the *randomness* in the H-constellation. This disappears only for the self-folded NCM.

(a1) This point corresponds to a standard single-branch (single-map) H-MAC. The mutual information $I(C_1; X|\tilde{C}_2)$ conditioned by a complete set of components $\tilde{C}_2$ corresponds to a perfect cancellation of any influence from the NCM branch number $k = 2$. As is shown in Theorem 5.10, this rate is achievable by any *regular* NCM.

(b1) The JDF strategy is used first to perform a classical multi-user decoding for components $\tilde{C}_2$ involved in HNC maps $\chi_{c,2}$. The classical multi-user decoding for $\tilde{C}_2$ has achievable rate region $\mathcal{R}(\tilde{C}_2)$ for the component coderates of branch $k = 2$ NCM. Notice that the components of interest are *only* $\tilde{C}_2$. All components $\tilde{C}_1$ associated with the map $\chi_{c,1}$ are marginalized as an unwanted interference in $\mathcal{R}(\tilde{C}_2)$ evaluation. By the definition of the structured NCM, this interference is independent of $\tilde{C}_1 \perp\!\!\!\perp \tilde{C}_2$. The reliably individually decoded components $\tilde{D}_2$ are then used for $D_2 = \chi_2(\tilde{D}_2)$. A corresponding achievable H-rate for D_2 message is $R_2 = R_2^{\mathrm{JDF}}\left(\mathcal{R}(\tilde{C}_2)\right)$. Notice, that the rates of $\mathcal{R}(\tilde{C}_2)$ themselves do not directly appear in the H-region (Figure 8.4) but only indirectly through $R_2^{\mathrm{JDF}}\left(\mathcal{R}(\tilde{C}_2)\right)$. The reliably decoded set of components $\tilde{D}_2$, and thus also $\tilde{C}_2$, is then used for the interference cancellation. The residual $\tilde{C}_2$ interference-free observation is thus a single-branch equivalent H-channel with the achievable H-rate $R_1 = I(C_1; X|\tilde{C}_2)$.

(c1) We first decode the H-message D_2 using layered single-branch layered H-decoding where the independent interference $\tilde{C}_1$, $\tilde{C}_1 \perp\!\!\!\perp D_2$, is marginalized as a part of the channel model,

$$p(x|c_2) = \frac{1}{p(c_2)} \sum_{\tilde{c}_1,(\tilde{c}_2:c_2)} p(x|\tilde{c}_1, \tilde{c}_2)p(\tilde{c}_1)p(\tilde{c}_2). \tag{8.1}$$

Then (Theorem 5.10), the H-rate $R_2 = I(C_2; X)$ is achievable for a regular NCM in branch 2.

Having available the H-message D_2, and thus also the H-codeword C_2^N, does *not* mean that the observation is completely interference free from branch 2 (see Figure 8.3c). There is still a randomness in the H-constellation dispersion part of the channel, i.e. $\{\tilde{C}_2 : C_2\}$. For notational brevity, we denote this as $\bar{C}_k = \{\tilde{C}_k : C_k\}$. Although not explicitly kept in the notation, C_k and $\tilde{C}_k$ can generally be mutually dependent. From the perspective of branch 1, we can, however, consider $\bar{C}_2$ as an interference and marginalize it in the channel PDF. The interference is independent, $\bar{C}_2 \perp\!\!\!\perp \tilde{C}_1$. The resulting observation is again a single-branch equivalent H-channel with observation

$$p(x|c_1, c_2) = \frac{1}{p(c_1)p(c_2)} \sum_{(\tilde{c}_1:c_1),(\tilde{c}_2:c_2)} p(x|\tilde{c}_1, \tilde{c}_2)p(\tilde{c}_1)p(\tilde{c}_2) \tag{8.2}$$

where c_2 is known. The achievable H-rate is $R_1 = I(C_1; X|C_2)$.

The H-rate for point (b1) is better than or equal to the H-rate of point (c1), i.e.

$$I(C_1; X|C_2) \leq I(C_1; X|\tilde{C}_2). \tag{8.3}$$

This easily comes from realizing that $I(C_1; X|\tilde{C}_2) = I(C_1; X|C_2, \bar{C}_2)$, i.e. that H-symbol C_2 and the H-dispersion $\bar{C}_2$ are together equivalent to all components $\tilde{C}_2$. Then we have

$$I(C_1; X|C_2, \bar{C}_2) = \mathcal{H}[C_1|C_2, \bar{C}_2] - \mathcal{H}[C_1|X, C_2, \bar{C}_2], \tag{8.4}$$

and

$$I(C_1; X|C_2) = \mathcal{H}[C_1|C_2] - \mathcal{H}[C_1|X, C_2]. \tag{8.5}$$

Components in both branches are independent of $\tilde{C}_1 \perp\!\!\!\perp \tilde{C}_2$ and thus also the H-symbol of one branch is independent with the other one $C_1 \perp\!\!\!\perp \tilde{C}_2$, or equivalently $C_1 \perp\!\!\!\perp (C_2, \bar{C}_2)$, $p(c_1|c_2, \bar{c}_2) = p(c_1)$ and also $C_1 \perp\!\!\!\perp C_2$, $p(c_1|c_2) = p(c_1)$. As a consequence, $\mathcal{H}[C_1|C_2, \bar{C}_2] = \mathcal{H}[C_1]$ and $\mathcal{H}[C_1|C_2] = \mathcal{H}[C_1]$, and thus

$$I(C_1; X|C_2, \bar{C}_2) = \mathcal{H}[C_1] - \mathcal{H}[C_1|X, C_2, \bar{C}_2], \tag{8.6}$$

$$I(C_1; X|C_2) = \mathcal{H}[C_1] - \mathcal{H}[C_1|X, C_2]. \tag{8.7}$$

Since the conditioning can only reduce the entropy, we have

$$\mathcal{H}[C_1|X, C_2, \bar{C}_2] \leq \mathcal{H}[C_1|X, C_2], \tag{8.8}$$

which proves the statement.

(a2), (b2), (c2), (0), and convex hull The proof for points (a2), (b2), (c2) is the same as for (a1), (b1), (c1), and only swaps indices of branches 1 and 2. The zero rate for (0) is trivial. All other points on the convex hull are achievable by time-sharing. $\qquad\square$

REMARK 8.1 (Self-Folded NCM) In a specific case of self-folded NCMs, the hierarchical self-dispersion is not present $\bar{C}_{\bar{k}} = \emptyset$ and $I(C_k; X|\tilde{C}_{\bar{k}}) = I(C_k; X|C_{\bar{k}})$, where $k \in \{1, 2\}$ and $\bar{k} = 3 - k$, see Figure 8.5a. The values $I(C_k; X|\tilde{C}_{\bar{k}})$ and $I(C_k; X|C_{\bar{k}})$ differ only under the presence of *non-zero* self-dispersion. The higher the self-dispersion, the higher the difference.

REMARK 8.2 (Hierarchical Interference Cancellation) The decoding strategy behind the rates $R_k = I(C_k; X|C_{\bar{k}})$ that appear in points (c1), (c2) can be named a *hierarchical interference cancellation*. The knowledge of H-code symbols $C_{\bar{k}}$ surely helps to remove some of the interference but does *not* remove it completely (as in the classical case). There still remains a randomness caused by the H-constellation dispersion. This means that the interference canceler only removes the degrees of freedom constrained by the H-codebook, thus we call it the *hierarchical* interference canceler.

The performance of the hierarchical interference canceler cannot be better than the performance of the classical interference canceler, which has full $\tilde{C}_{\bar{k}}$ available. It always holds (8.3)

$$I(C_k; X|C_{\bar{k}}) \leq I(C_k; X|\tilde{C}_{\bar{k}}) \tag{8.9}$$

with the equality achieved for the self-folded NCM.

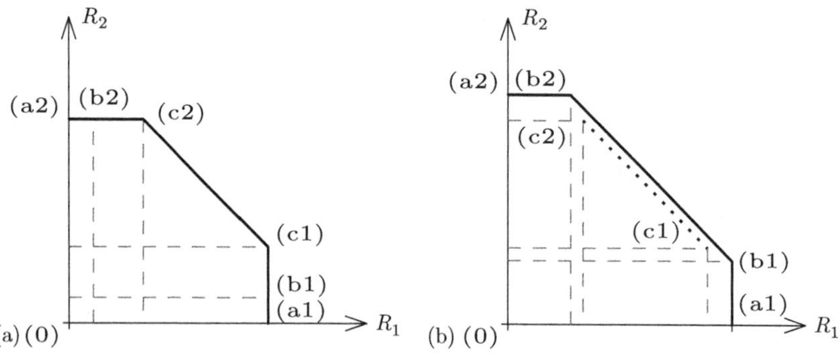

Figure 8.5 Achievable H-rate region for multi-map H-MAC under specific cases. Part (a) shows the case when the self-dispersion is not present. Part (b) shows the case of $R_{ck} < R_{bk}$, i.e. the combination of classical multi-user and single-branch H-decoding performs better than two-branch H-decoding.

REMARK 8.3 (JDF vs. HDF)   Consider the sum-H-rate of points (bk) and (ck), i.e.

$$R_{bk} = I(C_k; X | \tilde{C}_{\bar{k}}) + R_{\bar{k}}^{\text{JDF}} \left(\mathcal{R}(\tilde{C}_{\bar{k}}) \right), \tag{8.10}$$

$$R_{ck} = I(C_k; X | C_{\bar{k}}) + I(C_{\bar{k}}; X). \tag{8.11}$$

We can interpret the sum-H-rate R_{bk} as the rate using a mixture of classical and single-branch H-decoding. The sum-H-rate R_{ck} uses two-branch H-decoding. Depending on the values R_{bk} and R_{ck}, the (c1)–(c2) line might, but does not have to, form the border of the convex hull (Figure 8.5b). The line (c1)–(c2) must have the slope -1 (see Remark 8.4). Also notice that, in order for (c1)–(c2) line to be in the convex hull interior, it does not necessarily mean that the HDF single branch H-rate ($I(C_{\bar{k}}; X)$) is worse than the JDF H-rate ($R_{\bar{k}}^{\text{JDF}}(\mathcal{R}(\tilde{C}_{\bar{k}}))$). This reflects the trade-off between the more efficient H-decoding in branch $\bar{k}$ (R_2 coordinate of (c1) vs. (b1)) which, however, might leave not fully resolved interference (the self-dispersion interference remains even if we know the interfering H-message) for the branch k (R_1 coordinate of (c1) vs. (b1)).

Under some specific system conditions (e.g. low SNR, highly asymmetric channels, singular HNC maps), it may happen that $R_k^{\text{JDF}}(\mathcal{R}(\tilde{C}_k)) > I(C_k; X | \tilde{C}_{\bar{k}})$ (R_1 coordinate of (b2) vs. (b1)).

REMARK 8.4 (Cut-Set Bound for H-Rates Does Not Apply)   Clearly, the following expression holds regardless of the interpretation of the involved variables

$$\begin{aligned} R_{ck} &= I(C_k; X | C_{\bar{k}}) + I(C_{\bar{k}}; X) \\ &= \mathcal{H}[C_k | C_{\bar{k}}] - \mathcal{H}[C_k | X, C_{\bar{k}}] + \mathcal{H}[C_{\bar{k}}] - \mathcal{H}[C_{\bar{k}} | X] \\ &= \mathcal{H}[C_k, C_{\bar{k}}] - \mathcal{H}[C_k, C_{\bar{k}} | X] \\ &= I(C_k, C_{\bar{k}}; X). \end{aligned} \tag{8.12}$$

This is the achievable sum-H-rate expressed in terms of joint mutual information $I(C_k, C_{\bar{k}}; X)$. It has the same form as it would appear in the cut-set bound expressed w.r.t. H-rates R_1, R_2. However, we *cannot* interpret it this way, because the sum-H-rate R_{bk} can exceed this rate. The underlying reason for this is given by the fact that the equivalent channel, apart from H-messages, also contains the interfering self-dispersion part.

REMARK 8.5 (Linear Minimal HNC Map) The theorem required the regularity of the involved NCMs. Linear GF-based HNC maps (Theorem 5.11) fulfill that. But, in turn, it implies the minimum cardinality HNC map and equal component codebook rates. Then the JDF H-rate is given by the bottleneck component rate

$$R_k^{\text{JDF}}\left(\mathcal{R}(\tilde{C}_k)\right) = \min\left(I(C_{k,1}; X|C_{k,2}), I(C_{k,2}; X|C_{k,1}), \frac{1}{2}I(C_{k,1}, C_{k,2}; X)\right). \quad (8.13)$$

REMARK 8.6 (Special Case: Singular HNC Map) A singular (single component only) HNC map is a practically important case. It can be used in the asymmetric H-MAC case. One branch has participating sub-messages from both nodes and forms a standard HNC, and the second branch is formed only by a singular map containing an additional sub-message from only one node (e.g. having a higher SNR). The singular map is the map that has only a *single* component, i.e. $\tilde{d}_k = \{d_{k,1}\}$ and $\tilde{c}_k = \{c_{k,1}\}$. It fits under a linear GF-based HNC map model, where the only scaling coefficient is unity. From the point of view of Theorem 8.2, it means that $C_k = \tilde{C}_k = C_{k,1}$ and $\mathcal{R}(\tilde{C}_k)$ is a single-user rate region and then

$$R_k^{\text{JDF}}\left(\mathcal{R}(\tilde{C}_k)\right) = I(C_{k,1}; X). \quad (8.14)$$

As a consequence, points (b$\bar{k}$) and (c$\bar{k}$) coincide

$$I(C_{\bar{k}}; X|\tilde{C}_k) = I(C_{\bar{k}}; X|C_k) = I(C_{\bar{k}}; X|C_{k,1}), \quad (8.15)$$

$$R_k^{\text{JDF}}\left(\mathcal{R}(\tilde{C}_k)\right) = I(C_k; X) = I(C_{k,1}; X). \quad (8.16)$$

If the singularity of the HNC map applies to both branches, we get a classical two-user MAC.

REMARK 8.7 (H-Codebook with One-Dimensional H-Message) Theorem 8.2 (similar to Theorem 5.10) *implicitly* assumes the H-codebook with one-dimensional H-message. It means that the H-codebook is not a product codebook and has only one degree of freedom. This is needed in order to describe the achievable rate using a *scalar* variable and one value of the mutual information. Otherwise, the codebook rate would be described by a set of conditions over various subsets of the involved degrees of freedom. This case is discussed in the next remark.

REMARK 8.8 (Special Case: Full and Minimal HNC Map) A combination of the full and minimal HNC maps allows us to have only a partial HSI (or other HI) at the destination in the case of symmetric H-MAC. Assume the full map in the branch k and the minimal one in the branch $\bar{k}$. The H-codebook for the full map is a J_k-fold product codebook $\tilde{C}_k = C_{k,1} \times C_{k,2} \cdots \times C_{k,J_k}$. It is an isomorphic and regular one, and has a zero self-dispersion. The HNC map symbol is equal to the component set $C_k = \tilde{C}_k$ and our target performance is determined by a "sum-rate" $R_k^{\text{JDF}}(\mathcal{R}(\tilde{C}_k)) =$

$\sum_{i=1}^{J_k} \mathcal{R}_i(\tilde{C}_k)$, where $\mathcal{R}_i(\tilde{C}_k)$ is the achievable classical multi-user rate for the codebook $C_{k,i}$.

But because the codebook is a J_k-fold product codebook used on independent sub-messages, we can no longer use a single-user type of the condition for the achievable rate, even if our interest is only a "sum-rate" associated with $\tilde{C}_k$. We must properly use all, from the first-order up to the J_kth order, conditions on the achievable rate. This applies to all points on the R_k axis. R_k rates for points (b$\bar{k}$) and (c$\bar{k}$) then coincide to the value $R_k^{\mathrm{JDF}}(\mathcal{R}(\tilde{C}_k))$. The R_k rate for point (ck) is then $R_k^{\mathrm{JDF}}(\mathcal{R}(\tilde{C}_k|C_{\bar{k}}))$ and for point (bk) it is $R_k^{\mathrm{JDF}}(\mathcal{R}(\tilde{C}_k|\tilde{C}_{\bar{k}}))$.

When the full map is used for conditioning the $R_{\bar{k}}$ rates, then obviously

$$I(C_{\bar{k}}; X|\tilde{C}_k) = I(C_{\bar{k}}; X|C_k) \tag{8.17}$$

because $C_k = \tilde{C}_k$. The $R_{\bar{k}}$ rates for points (b$\bar{k}$) and (c$\bar{k}$) then coincide too. Thus, similarly as for the singular map, the points (b$\bar{k}$) and (c$\bar{k}$) coincide.

REMARK 8.9 (Real Rates of Sources)   Our prime interest is the real rates of the source messages b_ℓ. The H-rate region for structured NCM identifies achievable rates for d_k messages. These virtual structured rates must then be related to the real message rates in a way depending on the particular form of the rate splitting. Because the construction of the structured NCM assumed independent d_k messages, this interpretation has a simple form of a summation of proper R_k rates. See Definition 8.1 defining the structuring of real messages into the virtual ones.

REMARK 8.10 (Union of H-Rate Regions Over Transmission Parameters)   Achievable H-rates depend on the transmission parameters, namely the *type* and the *symbol energy* of the constellation alphabets $\mathcal{A}_{s,k}$. By a proper setting of $\mathcal{A}_{s,k}$ properties, we can adjust the relative strength of the individual components. The achievable H-rate regions are a union of all regions for all $\mathcal{A}_{s,k}$ fulfilling some given constraint. It is typically the total power of transmitted signal from one real node.

8.3   Structured NCM Design

8.3.1   Layered Block-Structured NCM

A layered structured NCM where the observation classical MAC channel part of the equivalent model (Figure 8.3c,b) is the *time-sharing*, is called a *block-structured* NCM. The scheme has a simplistic implementation. Its achievable rate region is clearly a subset of the convex hull of Theorem 8.2 given by a triangle of points (0), (a1), (a2); see Figure 8.4. The only degree of freedom is the proportion of the time allocated to each phase. The H-rate points (a1) and (a2) correspond to a single-branch single-map NCM (see Chapter 6). They also form a subset of the H-rate points of the more advanced superposition-based strategy.

8.3.2 Layered Superposition-Structured NCM

A layered structured NCM where the observation classical MAC channel part of the equivalent model (Figure 8.3c,b) is the *superposition coding*, is called a *superposition-structured* NCM. It is the strategy that fully corresponds to Theorem 8.2 and its achievable rate region is given by Figure 8.4.

H-rates Region for Singular and Minimal Map in AWGN Channel

As in Chapter 6, a numerical evaluation of the H-rates region requires us to specify concrete scenarios and system models. The most important assumption is a particular *finite channel alphabet* and a particular relay reception *channel model*. We evaluate the H-rate region for a special case of an asymmetric scenario with finite channel alphabets where one HNC map is a minimal GF linear function and the second map is a singular single-user map. We will also assume a linear memoryless channel with AWGN. The system model assumes virtual two-component node SA, with total code rate R_A, and SB, with total code rate R_B, where the channel between SB and the relay R is weaker, i.e. $R_A \geq R_B$. The rate splitting is done only at SA, $R_A = R_1 + R_2$ where $R_B = R_1$. The power splitting at SA makes the minimum map components (i.e. s_{A1}, s_B) with equal powers.

Let the true H-MAC observation be

$$x_n = s_{A,n} + h s_{B,n} + w_n \tag{8.18}$$

where we have assumed a unity common fading, and w_n is a complex-valued AWGN with σ_w^2 variance per dimension. The relative fading between $s_{A1,n}$ and $s_{B,n}$ is allowed to have a general phase $h = |h| \exp(j\,\psi)$ and a limited magnitude $|h| \leq 1$ that reflects the weaker channel SB-R. The magnitude $|h|$ needs to be known at SA in order to make a proper power split. Node SA transmits

$$s_{A,n} = |h| s_{A1,n}(c_{A1,n}) + e^{j\phi}\sqrt{1 - |h|^2} s_{A2,n}(c_{A2,n}). \tag{8.19}$$

The source SA uses the superposition coding strategy for the sharing of the minimal and the singular map streams. For finite alphabet codebooks (unlike for Gaussian codebooks), the properties of the superposition coding depend on the relative phase and we must model that by the phase ϕ. The power scaling is set to make the complete SA signal $s_{A,n}$ power to be the same as the equally powered components $s_{A1,n}, s_{A2,n}$ and, at the same time, the scaled component $s_{A1,n}$ (participating in the minimal map with $s_{B,n}$) to have the same power as $s_{B,n}$. The received signal model for the nth symbol at the relay is

$$x_n = u_n + w_n \tag{8.20}$$

where the overall channel combined payload signal $u_n = s_{A,n} + h s_{B,n}$ is

$$u_n = |h| s_{A1,n}(c_{A1,n}) + e^{j\phi}\sqrt{1 - |h|^2} s_{A2,n}(c_{A2,n}) + h s_{B,n}(c_{B,n}). \tag{8.21}$$

The transmission strategy allocating an equal symbol energy to the signals s_{A1} and s_B that participate in the minimal HNC map is clearly not the only option. We can optimize the setup of the transmission parameters for some particular target performance goal. This, however, is not considered in this section.

The component virtual finite alphabet codebooks are $c_{A1}^N = C_1(d_{A1})$, $c_B^N = C_1(d_B)$, $c_{A2}^N = C_2(d_{A2})$ where the message b_A is split into d_{A1}, d_{A2} and the message $b_B = d_B$ is left as it is. The codebook C_k has the rate R_k. All codesymbols $c_{A1,n}, c_{B,n}, c_{A2,n}$ are symbol-wise mapped on channel symbols $s_{A1,n}, s_{B,n}, s_{A2,n}$ by maps $\mathcal{A}_{A1}, \mathcal{A}_B, \mathcal{A}_{A2}$ respectively, each having a cardinality M. The HNC map $c_{1,n} = \chi_c(c_{A1,n}, c_{B,n})$ is assumed to be GF linear minimal. The second HNC map is singular $c_{2,n} = c_{A2,n}$.

The evaluation of the H-rate region requires us to find

$$I(C_1; X | \tilde{C}_2) = I(C_1; X | C_{A2}), \tag{8.22}$$

$$I(C_1; X | C_2) = I(C_1; X | C_{A2}), \tag{8.23}$$

$$R_2^{\mathrm{JDF}}(\mathcal{R}(\tilde{C}_2)) = I(C_{A2}; X), \tag{8.24}$$

$$I(C_2; X) = I(C_{A2}; X), \tag{8.25}$$

for region corners (a1, b1, c1). Notice that, owing to the singularity of HNC map in branch 2, $I(C_1; X | \tilde{C}_2) = I(C_1; X | C_2)$ and also $R_2^{\mathrm{JDF}}(\mathcal{R}(\tilde{C}_2)) = I(C_2; X)$ thus the points (b1) and (c1) are *identical*. Next, we need to find

$$I(C_2; X | \tilde{C}_1) = I(C_{A2}; X | C_{A1}, C_B), \tag{8.26}$$

$$I(C_2; X | C_1) = I(C_{A2}; X | C_1), \tag{8.27}$$

$$R_1^{\mathrm{JDF}}(\mathcal{R}(\tilde{C}_1)) = \min\left(I(C_{A1}; X | C_B), I(C_B; X | C_{A1}), \frac{1}{2} I(C_{A1}, C_B; X) \right), \tag{8.28}$$

$$I(C_1; X) = I(C_1; X), \tag{8.29}$$

for corners (a2, b2, c2). It involves the set of mutual information expressions

$$I(C_1; X | C_{A2}) = \mathcal{H}[X | C_{A2}] - \mathcal{H}[X | C_1, C_{A2}], \tag{8.30}$$

$$I(C_{A2}; X) = \mathcal{H}[X] - \mathcal{H}[X | C_{A2}], \tag{8.31}$$

$$I(C_{A2}; X | C_{A1}, C_B) = \mathcal{H}[X | C_{A1}, C_B] - \mathcal{H}[X | C_{A1}, C_{A2}, C_B], \tag{8.32}$$

$$I(C_{A2}; X | C_1) = \mathcal{H}[X | C_1] - \mathcal{H}[X | C_1, C_{A2}], \tag{8.33}$$

$$I(C_{A1}; X | C_B) = \mathcal{H}[X | C_B] - \mathcal{H}[X | C_{A1}, C_B], \tag{8.34}$$

$$I(C_B; X | C_{A1}) = \mathcal{H}[X | C_{A1}] - \mathcal{H}[X | C_{A1}, C_B], \tag{8.35}$$

$$I(C_{A1}, C_B; X) = \mathcal{H}[X] - \mathcal{H}[X | C_{A1}, C_B], \tag{8.36}$$

$$I(C_1; X) = \mathcal{H}[X] - \mathcal{H}[X | C_1]. \tag{8.37}$$

The numerical evaluation of the entropies can be done, e.g., by Monte-Carlo evaluation of integrals, Section 5.7.7. All involved PDFs

$$p(x_n), \tag{8.38}$$

$$p(x_n | c_{A1,n}), \ p(x_n | c_{A2,n}), \ p(x_n | c_{B,n}), \ p(x_n | c_{1,n}), \tag{8.39}$$

$$p(x_n | c_{1,n}, c_{A2,n}), \ p(x_n | c_{A1,n}, c_{B,n}) \tag{8.40}$$

can be obtained by a proper marginalization from $p(x_n|c_{A1,n}, c_{A2,n}, c_{B,n}) = p_W(x_n - u_n(c_{A1,n}, c_{A2,n}, c_{B,n}))$ where we assume uniform and IID a priori PDFs

$$p(c_{A1,n}) = p(c_{A2,n}) = p(c_{B,n}) = \frac{1}{M}, \tag{8.41}$$

and where M is a common size of the channel alphabet. The complex Gaussian PDF for N_u-dimensional constellation symbols is

$$p_W(w_n) = \frac{1}{\pi^{N_u} \sigma_w^{2N_u}} \exp(-\frac{1}{\sigma_w^2} \|w_n\|^2). \tag{8.42}$$

Common constellations, such as MPSK, QAM, have one complex dimension, $N_u = 1$. Complex envelope constellation space Gaussian noise has $\sigma_w^2 = 2N_0$ variance per dimension, where N_0 is a single-sided power spectrum density of the real-valued noise. The SNR is defined (as in Chapter 6) w.r.t. SA only

$$\begin{aligned}
\gamma_x &= \frac{\mathcal{E}_{s,A}}{N_0} \\
&= \frac{\mathrm{E}\left[\left| |h|s_{A1,n} + e^{j\phi}\sqrt{1 - |h|^2} s_{A2,n}\right|^2\right]}{2N_0} \\
&= \frac{|h|^2 \mathrm{E}\left[|s_{A1,n}|^2\right] + (1 - |h|^2)\mathrm{E}\left[|s_{A2,n}|^2\right]}{2N_0}
\end{aligned} \tag{8.43}$$

where we properly respected the $1/2$ scaling between the complex envelope and the signal on the carrier symbol energies.

In order to evaluate whether the point (c2) or (b2) dominates the H-rate region, we must evaluate the sum-rates

$$\begin{aligned}
R_{b2} &= I(C_2; X|\tilde{C}_1) + R_1^{\mathrm{JDF}}\left(\mathcal{R}(\tilde{C}_1)\right) \\
&= I(C_{A2}; X|C_{A1}, C_B) \\
&\quad + \min\left(I(C_{A1}; X|C_B), I(C_B; X|C_{A1}), \frac{1}{2}I(C_{A1}, C_B; X)\right), \tag{8.44} \\
R_{c2} &= I(C_2; X|C_1) + I(C_1; X) \\
&= I(C_{A2}; X|C_1) + I(C_1; X). \tag{8.45}
\end{aligned}$$

If $R_{c2} > R_{b2}$ then the point (c2) dominates.

Total Rates Region for Both Sources

The preceding section derived the rate region for the H-rates R_1, R_2. However, the ultimate performance metric is given by the total source node rates $R_A = R_1 + R_2$ and $R_B = R_1$. Rate region (R_A, R_B) can be obtained by a linear transformation

$$\begin{bmatrix} R_A \\ R_B \end{bmatrix} = \begin{bmatrix} 1 & 1 \\ 1 & 0 \end{bmatrix} \begin{bmatrix} R_1 \\ R_2 \end{bmatrix}. \tag{8.46}$$

The H-rates and directly transformed total rates regions are shown in Figure 8.6a and Figure 8.6b. The H-rates corner points (a1, b1, c1, c2, b2, a2) correspond to total rate points (a1′, b1′, c1′, c2′, b2′, a2′).

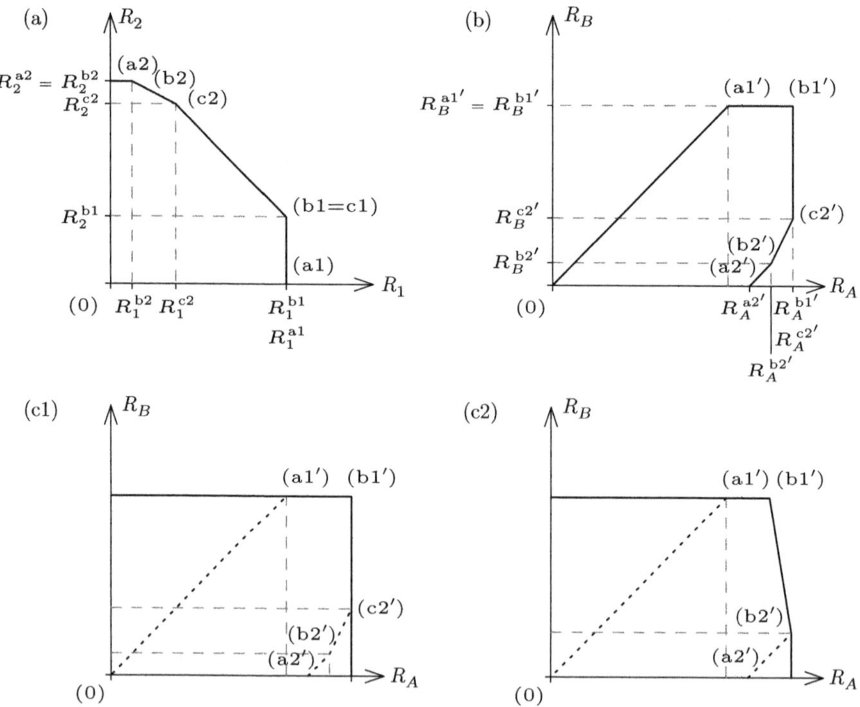

Figure 8.6 Superposition-structured NCM – virtual H-rates region (a); strict-sense transformed total rates region (b); and amended total rates region for $R_A^{b2'} < R_A^{c2'}$ (c1) and for $R_A^{b2'} > R_A^{c2'}$ (c2).

However, the strict-sense linear transformation of H-rates produces a total rates region, which can be further trivially extended. If any given point (r_1, r_2) in the rate region is achievable, then also all points $(\alpha_1 r_1, \alpha_2 r_2)$, where $\alpha_1, \alpha_2 \in [0, 1]$, are achievable. The achievable strategy is simply not to use a portion of the achievable rate, e.g. by zero-filling data. This amended region is shown in Figure 8.6c1,c2. It is worth noting that from the properties of H-rate region, it must hold that $R_B^{a1'} = R_B^{b1'}$ and $R_A^{b1'} = R_A^{c1'} = R_A^{c2'}$; however, $R_A^{b2'} \lessgtr R_A^{c2'}$ could be either less than or greater than. This is shown in Figure 8.6c1, c2.

Reference Scenarios

In order to show performance advantages of a superposition-structured NCM in asymmetric scenarios, we will use two reference scenarios. Both of them directly operate on full (non-split) source messages described by rates R_A, R_B. The first reference scenario is a classical multi-user MAC. It allows us to use JDF strategy, i.e. a classical MAC followed by a discrete NC, with arbitrary HNC maps. Its performance is simply given by the classical MAC achievable channel rates. The second reference scenario is a common NCM using a single HNC map directly on source SA and SB data. The form of NCM and the type of channel alphabet is kept the same as for the superposition-structured

NCM. The performance of this scenario is analyzed in Chapter 6. We show the split rates R_1, R_2 that demonstrate how the individual map parts contribute to the total rates. We also show the total rates R_A, R_B, which are the only ones that show the practical performance and also can be compared with the reference cases.

It is also important to distinguish two cases for the performance comparisons.

Partial HSI limited The partial HSI limited case is the situation where we use the superposition-structured NCM because the final destination does not have the full HSI (or other full independent HI). Clearly a common single HNC *minimal* cardinality map NCM cannot be used in this situation at all. Because the finite alphabet scenarios require a linear minimal GF-based map in order to be isomorphic, the only reference usable scenario is a classical MAC with optional JDF with a suitable HNC map.

H-MAC asymmetry limited The case limited by the channel asymmetry in H-MAC assumes that the bottleneck is not in destination HSI/HI but in our capability to effectively utilize all H-MAC capabilities under the channel quality asymmetry. Both reference scenarios are generally applicable here.

BPSK with Singular and XOR HNC Maps

This numerical example uses BPSK channel alphabets $\mathcal{A}_{A1} = \mathcal{A}_B = \mathcal{A}_{A2} = \{\pm 1\}$, binary codes $c_{A1,n}, c_{B,n}, c_{A2,n} \in \mathbb{F}_2$, and the only possible GF HNC map, the XOR one. Figure 8.7 shows the numerical results for a variety of channel parameters.

4PSK with Singular and Bit-Wise XOR HNC Maps

This case is defined by $M = 4$, $s_{A,n}, s_{B,n} \in \mathcal{A}_s = \{1, j, -1, -j\}$, $\mathcal{A}_s(0) = 1$, $\mathcal{A}_s(1) = j$, $\mathcal{A}_s(2) = -1$, $\mathcal{A}_s(3) = -j$, where we use "natural" constellation indexing. The codes are assumed to be binary $c_{A1,n}, c_{B,n}, c_{A2,n} \in \mathbb{F}_2$ with symbols grouped into pairs and used as inputs to the constellation mappers (Section 6.3.4). The HNC map is a bit-wise XOR. Figure 8.8 shows the numerical results for a variety of channel parameters.

Performance

The numerical results show several common characteristics. Clearly, the more asymmetric the H-MAC the larger the performance advantage that is obtained from the superposition-structured NCM. For a more symmetric case, the bottleneck appears in the superposed singular map. This behavior can be seen on R_2 or on $[R_A, 0]$ coordinates. The superposition-structured NCM R_2 rate weakens for the more symmetric cases.

We can also see that both reference schemes have the rate R_A limited by the other, weaker, channel. The superposition-structured NCM takes the advantage in high R_A region in more asymmetric cases.

The weaker-channel rate R_B provides worse performance for the superposition-structured NCM than for the reference schemes. This can be explained by the fact that the weaker channel must participate in the HNC map and, on top of it, it suffers from the singular map interference. The higher the asymmetry, the higher the energy that is allocated to the singular map.

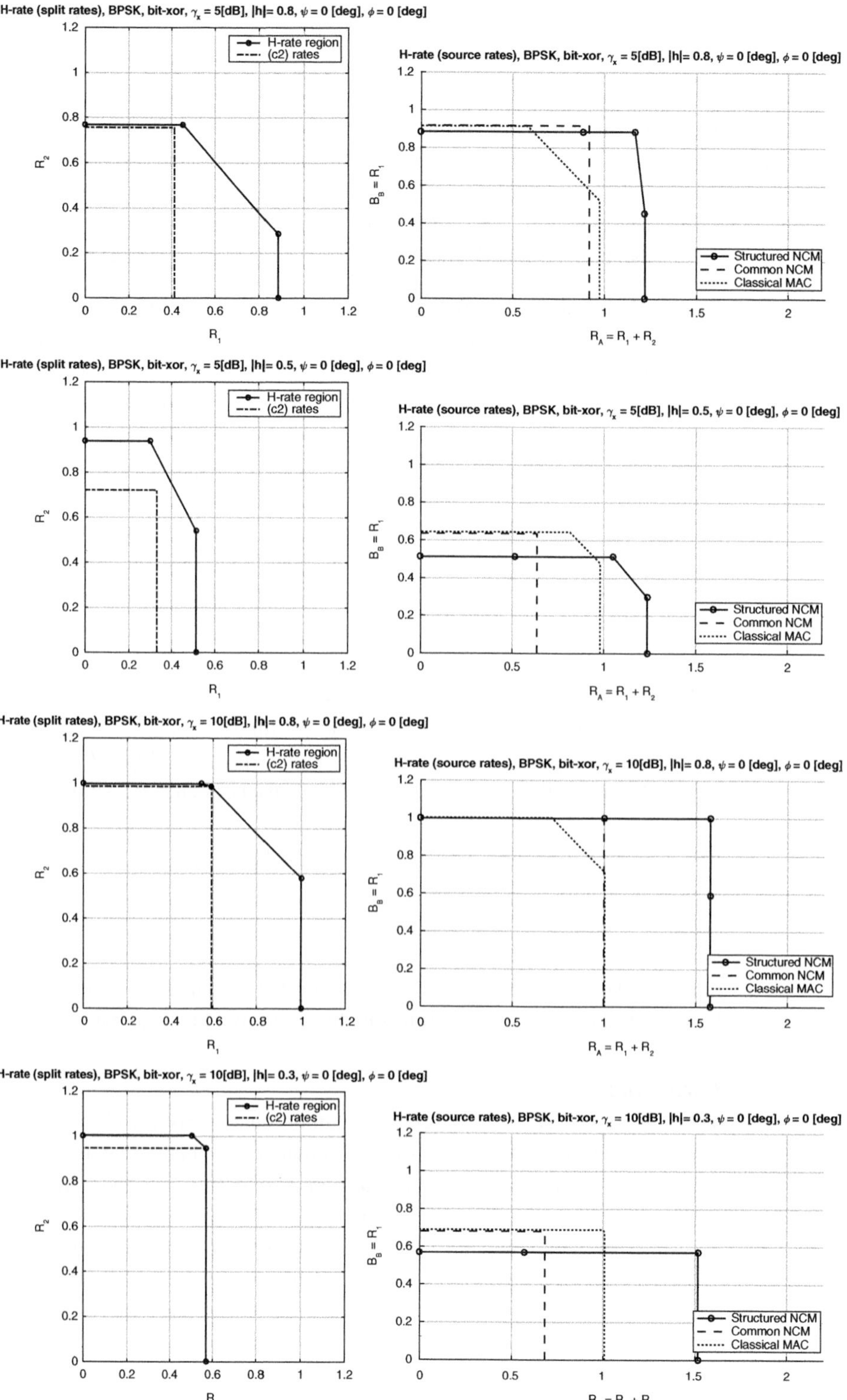

Figure 8.7 H-rate regions for superposition-structured NCM – BPSK with singular and XOR HNC maps. The left-hand column shows H-rates R_1, R_2 region of the virtual sources (the dash-dot line separately shows the coordinates of the (c2) point), the right-hand column shows amended total source rates R_A, R_B region. The channel parameters $h = |h| \exp(\mathrm{j}\, \psi)$, ϕ, and SNR $\gamma_x = \mathcal{E}_{s,A}/N_0$ are related to the system model (8.21) and (8.43).

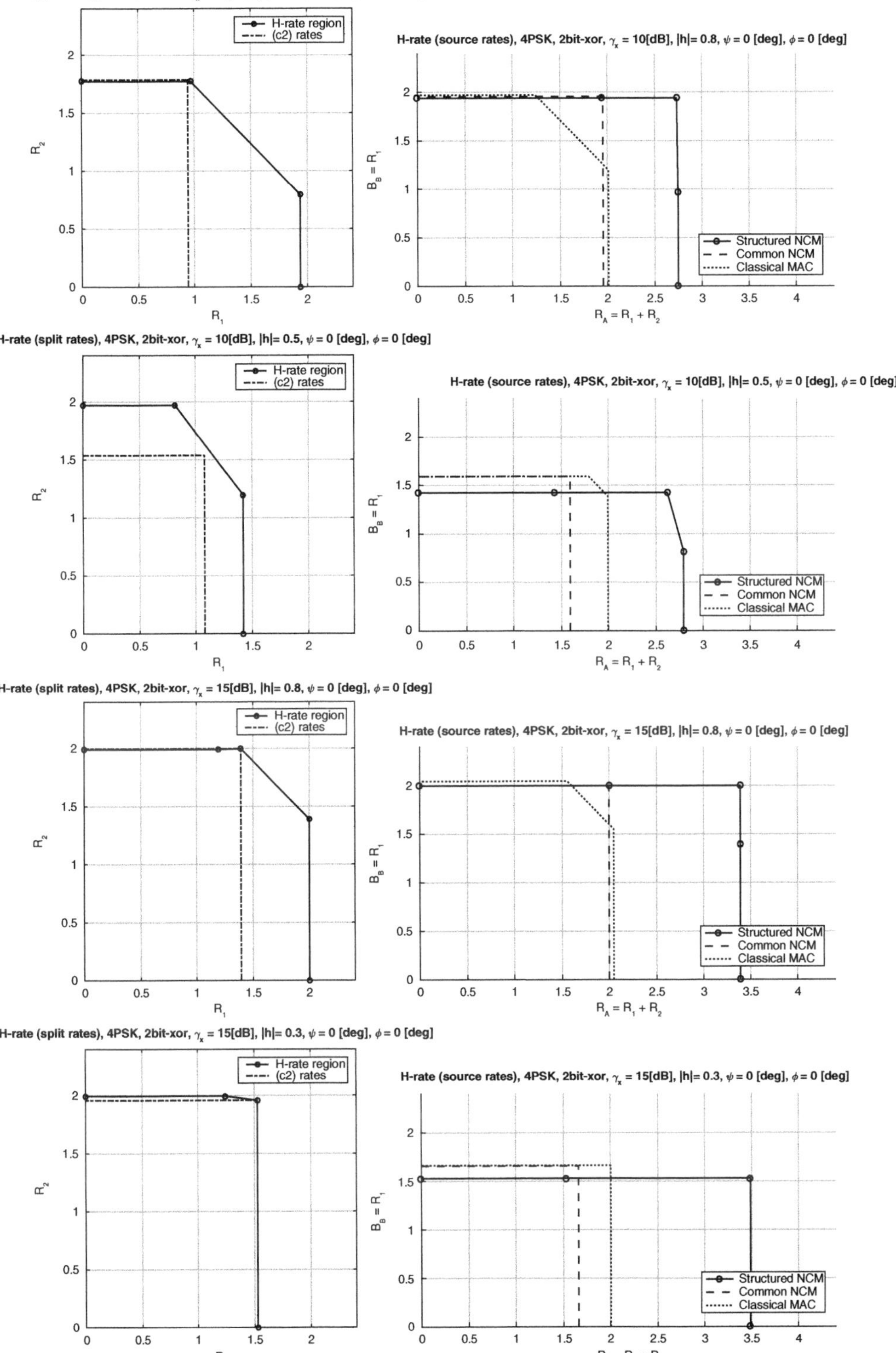

Figure 8.8 H-rate regions for superposition-structured NCM – 4PSK with singular and bit-wise XOR HNC maps. The left-hand column shows H-rates R_1, R_2 region of the virtual sources (the dash-dot line separately shows the coordinates of the (c2) point), the right-hand column shows amended total source rates R_A, R_B region. The channel parameters $h = |h|\exp(\mathrm{j}\,\psi)$, ϕ, and SNR $\gamma_x = \mathcal{E}_{s,A}/N_0$ are related to the system model (8.21) and (8.43).

8.3.3   CF-Based Superposition-Structured NCM

The CF is an isomorphic NCM and we can use it as a building component of the structured NCM. A usage of CF has both advantages and disadvantages. On one hand it is not constrained by the low and finite cardinality of the channel alphabet as is the layered superposition-structured NCM (Section 8.3.2) and its highly dimensional codewords have a potential to approach the capacity. On the other hand the usage of a *common* nested lattice imposes some constraints on the codebook rates w.r.t. to the computation rates. A part of the structuring strategy is to make use of the classical superposition coding in the H-MAC. As discussed in [47], the CF technique with multiple maps used to solve for one component at the relay does not generally achieve the classical MAC capacity. This affects additionally the performance when using the superposition structuring.

In asymmetric scenarios with split virtual sub-messages, one of the HNC maps is singular and the second map is the one with minimal cardinality. The singular map means that the single sub-message data are fully decoded at the relay. In the context of [47], it forms a *private* message, which is fully decoded at the relay and cannot participate (e.g. since the other nodes do not have available its codebook) in the processing at other nodes. The remaining sub-messages participating in minimal HNC map are then *public*.

CF with Common Lattice
System Model

The first, and the simplest, possibility is to apply CF in its canonic common-lattice form (Section 5.6) on the structured virtual users as in (8.21) and simply treat each virtual user as an independent entity. The channel model and the transmission power settings are identical with Section 8.3.2. The observation whole-codeword vector model is

$$\mathbf{x} = h_{A1}\mathbf{s}_{A1} + h_{A2}\mathbf{s}_{A2} + h_B\mathbf{s}_B + \mathbf{w} \tag{8.47}$$

where the virtual structured channel is

$$\mathbf{h} = [h_{A1}, h_{A2}, h_B]^{\mathrm{T}} = [|h|, e^{j\phi}\sqrt{1 - |h|^2}, |h|e^{j\psi}]^{\mathrm{T}}. \tag{8.48}$$

The lattice codewords with dithers are defined as in Section 5.6, particularly (5.91). The common nested lattice $\Lambda_s \subset \Lambda_c$ codebook $\mathcal{C}_0$ has the rate R_0. The virtual component rates are $R_1 = R_2 = R_0$ and the real node rates are $R_A = R_1 + R_2 = 2R_0$ and $R_B = R_1 = R_0$. The model thus describes a situation with two users with unbalanced rates, one R_0 and the other one $2R_0$. The latter data stream is demultiplexed into two virtual users with rates $R_1 = R_2 = R_0$. One of these streams is encoded as a nested lattice codeword with this rate, which is then superimposed on a second joint CF codeword formed from the other two codewords.

Computation Rates

Assume that the desired CF coefficient vectors, corresponding to the desired HNC data message maps, are $\mathbf{a}_1 = [a_{A1}, 0, a_B]^{\mathrm{T}}$, $\mathbf{a}_2 = [0, a_{A2}, 0]^{\mathrm{T}}$, $a_{A1}, a_{A2}, a_B \in \mathbb{Z}_j$, where the first is the minimal map for two components and the second is a singular single component map. The maps have some zero components. A zero component essentially means that the component corresponding to the zero coefficient behaves as the unresolved interference with the effective power $P_{\mathrm{IFC},i} = P_s |\alpha h_i|^2$ where P_s is the transmitted power, α is CF preprocessor scaling, and h_i is the coefficient of the corresponding channel. This situation directly reflects the superposition nature of the structured NCM. We can directly use the desired maps and the computation rate with MMSE optimized α is ($k \in \{1, 2\}$)

$$R_c(\mathbf{a}_k) = \lg^+ \frac{P_s \|\mathbf{h}\|^2 + P_w}{P_s \left(\|\mathbf{h}\|^2 \|\mathbf{a}_k\|^2 - \left|\mathbf{h}^{\mathrm{H}}\mathbf{a}_k\right|^2 \right) + P_w \|\mathbf{a}_k\|^2}. \tag{8.49}$$

The non-zero components are still open to an optimization. The codebook rate must comply with $R_0 \leq R_c(\mathbf{a}_k)$ for both $k \in \{1, 2\}$.

Another option, which widens the optimization space for coefficients, is to use auxiliary maps $\tilde{\mathbf{a}}_k$ such that the intended maps can be obtained by a linear transformation

$$\underbrace{\begin{bmatrix} \mathbf{a}_1^{\mathrm{T}} \\ \mathbf{a}_2^{\mathrm{T}} \end{bmatrix}}_{\mathbf{A}} = \underbrace{\begin{bmatrix} t_{11} & t_{12} \\ t_{21} & t_{22} \end{bmatrix}}_{\mathbf{T}} \underbrace{\begin{bmatrix} \tilde{\mathbf{a}}_1^{\mathrm{T}} \\ \tilde{\mathbf{a}}_2^{\mathrm{T}} \end{bmatrix}}_{\tilde{\mathbf{A}}} \tag{8.50}$$

where $\mathbf{T}$ is a full-rank $\mathbb{Z}^{2\times 2}$ transformation matrix. The auxiliary coefficients do not have to have zero components and can provide higher computation rates

$$R_c(\tilde{\mathbf{a}}_k) = \lg^+ \frac{P_s \|\mathbf{h}\|^2 + P_w}{P_s \left(\|\mathbf{h}\|^2 \|\tilde{\mathbf{a}}_k\|^2 - \left|\mathbf{h}^{\mathrm{H}}\tilde{\mathbf{a}}_k\right|^2 \right) + P_w \|\tilde{\mathbf{a}}_k\|^2}. \tag{8.51}$$

In the CF style of notation, P_s and P_w are defined as second moments per dimension in the *complex envelope* constellation space. In order to relate this with notation used by layered superposition-structured NCM, we get $P_w = 2N_0$ and $P_s = 2\mathcal{E}_{s,A}$, which gives

$$\gamma_x = \frac{\mathcal{E}_{s,A}}{N_0} = \frac{P_s}{P_w} \tag{8.52}$$

and

$$R_c(\tilde{\mathbf{a}}_k) = \lg^+ \frac{\gamma_x \|\mathbf{h}\|^2 + 1}{\gamma_x \left(\|\mathbf{h}\|^2 \|\tilde{\mathbf{a}}_k\|^2 - \left|\mathbf{h}^{\mathrm{H}}\tilde{\mathbf{a}}_k\right|^2 \right) + \|\tilde{\mathbf{a}}_k\|^2}. \tag{8.53}$$

We also have $\|\mathbf{h}\|^2 = 1 + |h|^2$ for $|h| \leq 1$.

Rate Region

This canonic CF-based solution has a rectangular R_A, R_B rate region with

$$R_A \leq 2R_{c,0}, \quad R_B \leq R_{c,0}. \tag{8.54}$$

The optimized computation rate over the auxiliary coefficients consistent with the desired ones is given by the smaller from the two computation rates for two different auxiliary maps $\tilde{\mathbf{a}}_1 \neq \tilde{\mathbf{a}}_2$

$$R_{c,0} = \min_{k \in \{1,2\}} R_{c,k}. \tag{8.55}$$

Both rates $R_{c,k}$ must use maps $\tilde{\mathbf{a}}_k$ consistent (see the next paragraph) with $\mathbf{a}_1, \mathbf{a}_2$

$$R_{c,k} = \max_{\tilde{\mathbf{a}}_k:(\mathbf{a}_1,\mathbf{a}_2)} R_c(\tilde{\mathbf{a}}_k). \tag{8.56}$$

The virtual structured codebooks must be identical.

Consistent Auxiliary Maps

Auxiliary maps $\tilde{\mathbf{A}} = [\tilde{\mathbf{a}}_1, \tilde{\mathbf{a}}_2]^{\mathrm{T}}$ must be consistent with the desired maps $\mathbf{A} = [\mathbf{a}_1, \mathbf{a}_2]^{\mathrm{T}}$. It means that the linear transform by the matrix $\mathbf{T}$ transforms them into the specific form containing only zeros and *non-zero* complex integers

$$\begin{bmatrix} a_{A1} & 0 & a_B \\ 0 & a_{A2} & 0 \end{bmatrix} = \begin{bmatrix} t_{11} & t_{12} \\ t_{21} & t_{22} \end{bmatrix} \begin{bmatrix} \tilde{a}_{1,A1} & \tilde{a}_{1,A2} & \tilde{a}_{1,B} \\ \tilde{a}_{2,A1} & \tilde{a}_{2,A2} & \tilde{a}_{2,B} \end{bmatrix}, \tag{8.57}$$

$a_{A1}, a_{A2}, a_B \in \mathbb{Z}_j \setminus \{0\}$. Because the map $\mathbf{a}_2$ contains only one non-zero coefficient, we can, without the loss of generality, assume $a_{A2} = 1$. In order to verify the consistency of $\tilde{\mathbf{A}}$, we need to verify the existence of the transform $\mathbf{T}$ such that

$$t_{11}\tilde{a}_{1,A1} + t_{12}\tilde{a}_{2,A1} = a_{A1} \in \mathbb{Z}_j \setminus \{0\}, \tag{8.58}$$

$$t_{11}\tilde{a}_{1,A2} + t_{12}\tilde{a}_{2,A2} = 0, \tag{8.59}$$

$$t_{11}\tilde{a}_{1,B} + t_{12}\tilde{a}_{2,B} = a_B \in \mathbb{Z}_j \setminus \{0\} \tag{8.60}$$

for the first map, and

$$t_{21}\tilde{a}_{1,A1} + t_{22}\tilde{a}_{2,A1} = 0, \tag{8.61}$$

$$t_{21}\tilde{a}_{1,A2} + t_{22}\tilde{a}_{2,A2} = 1, \tag{8.62}$$

$$t_{21}\tilde{a}_{1,B} + t_{22}\tilde{a}_{2,B} = 0 \tag{8.63}$$

for the second map. Coefficients a_{A1}, a_B are some non-zero complex integers.

The solution consistent with the first map can be found for *arbitrary* $\tilde{\mathbf{A}}$ with

$$|\tilde{a}_{1,A1}| + |\tilde{a}_{2,A1}| \neq 0 \ \wedge \ |\tilde{a}_{1,B}| + |\tilde{a}_{2,B}| \neq 0. \tag{8.64}$$

This allows non-zero a_{A1} and a_B. There are four degrees of freedom ($t_{11}, t_{12}, a_{A1}, a_B$) and three equations, and thus there is an infinite number of solutions. The one with smallest modulus complex integer transform coefficients is

$$t_{11} = \tilde{a}_{2,A2}, \ \ t_{12} = -\tilde{a}_{1,A2}, \ \text{for} \ \tilde{a}_{1,A2} \neq 0, \ \tilde{a}_{2,A2} \neq 0, \tag{8.65}$$

$$t_{11} = 0, \ \ t_{12} = 1, \ \text{for} \ \tilde{a}_{1,A2} \neq 0, \ \tilde{a}_{2,A2} = 0, \tag{8.66}$$

$$t_{11} = 1, \ \ t_{12} = 0, \ \text{for} \ \tilde{a}_{1,A2} = 0, \ \tilde{a}_{2,A2} \neq 0. \tag{8.67}$$

The case of $\tilde{a}_{1,A2} = 0$, $\tilde{a}_{2,A2} = 0$ is not treated since it is clearly inconsistent with the second map.

The second map has two degrees of freedom (t_{21}, t_{22}) and three equations, thus the solution exists only under some conditions. There are two conditions with a homogenous RHS. The involved coefficients must be linearly dependent and thus

$$\det \begin{bmatrix} \tilde{a}_{1,A1} & \tilde{a}_{2,A1} \\ \tilde{a}_{1,B} & \tilde{a}_{2,B} \end{bmatrix} = 0. \tag{8.68}$$

If this holds, then the transform coefficients can be found from the first two equations

$$\begin{bmatrix} \tilde{a}_{1,A1} & \tilde{a}_{2,A1} \\ \tilde{a}_{1,A2} & \tilde{a}_{2,A2} \end{bmatrix} \begin{bmatrix} t_{21} \\ t_{22} \end{bmatrix} = \begin{bmatrix} 0 \\ 1 \end{bmatrix}. \tag{8.69}$$

It has a solution if

$$\det \begin{bmatrix} \tilde{a}_{1,A1} & \tilde{a}_{2,A1} \\ \tilde{a}_{1,A2} & \tilde{a}_{2,A2} \end{bmatrix} \neq 0. \tag{8.70}$$

This also inherently covers the case of $\tilde{a}_{1,A2} = 0$, $\tilde{a}_{2,A2} = 0$ mentioned above.

Summarizing the results, we see that if conditions (8.64), (8.68), and (8.70) are fulfilled, then we can always find a linear transform $\mathbf{T}$ producing desired complex integer valued coefficients $\mathbf{A}$. If the determinant conditions are not fulfilled, we must use other pairs of auxiliary coefficients with potentially lower minimal computation rates.

Performance

The common lattice CF-based structured NCM performance is numerically evaluated (Figure 8.9) under comparable conditions as in Section 8.3.2. The coefficients were optimized by an exhaustive search in the space limited by the condition (5.115) while taking care of (8.64), (8.68), and (8.70) to use only those auxiliary $\tilde{\mathbf{a}}_1, \tilde{\mathbf{a}}_2$, which can be transformed into a proper form $\mathbf{a}_1 = [a_{A1}, 0, a_B]^T$ and $\mathbf{a}_2 = [0, a_{A2}, 0]^T$. In order to constrain the set of combinations of $\tilde{\mathbf{a}}_1, \tilde{\mathbf{a}}_2$ to a numerically manageable number, we restricted the search for consistent coefficients only to those having the four highest computation rates. If the consistent pair cannot be found in the highest four rates, we set the resulting rate to zero (it would not be better than the fifth rate anyway). The coefficient pairs providing the two highest rates are surprisingly frequently *not* consistent with target desired maps $\mathbf{A}$. A comparison of a CF-based structured NCM solution with layered finite alphabet solutions (classical MAC, common NCM, superposition-structured NCM, Figures 8.7 and 8.8) is in Figure 8.10 and Figure 8.11. Although the superposition-structured NCM has only simple finite low-cardinality alphabets, it is superior to CF in performance in many cases.

CF with Successive Decoding and Interference Cancellation

The constraint of identical codebooks can be removed by using principles of CF with successive decoding and interference cancellation (see Section 9.3) inspired by [47]. The usage of (1) the codebook rates matched to a particular channel state, and (2) capability of performing successive interference cancellations, can increase the achievable rate region.

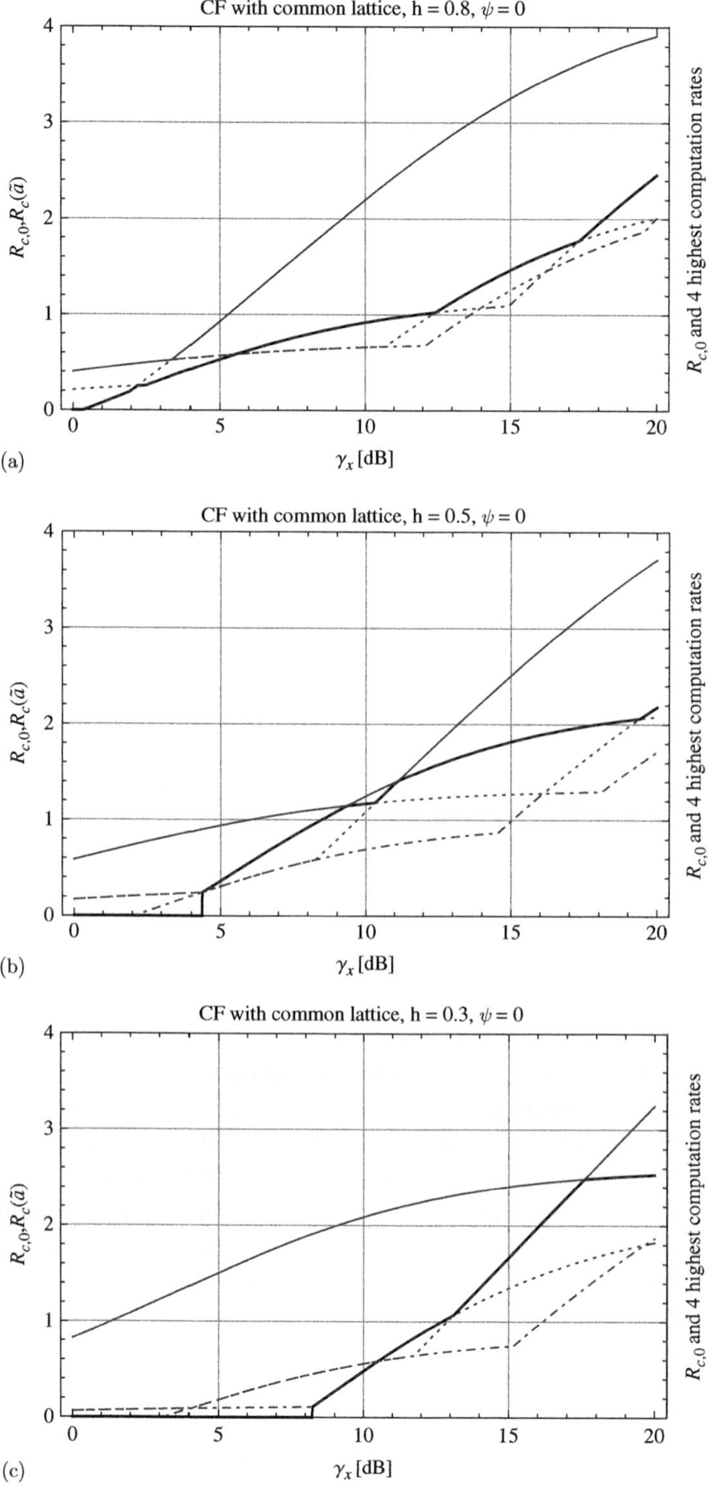

Figure 8.9 Performance of common lattice CF-based structured NCM. The resulting rate $R_{c,0}$ is shown as a solid black line. In order to see the participation of the individual maps, we also plot in gray the four highest computation rates among all $\tilde{\mathbf{a}}_k$ regardless of the consistency conditions. The channel parameters are defined by (8.48).

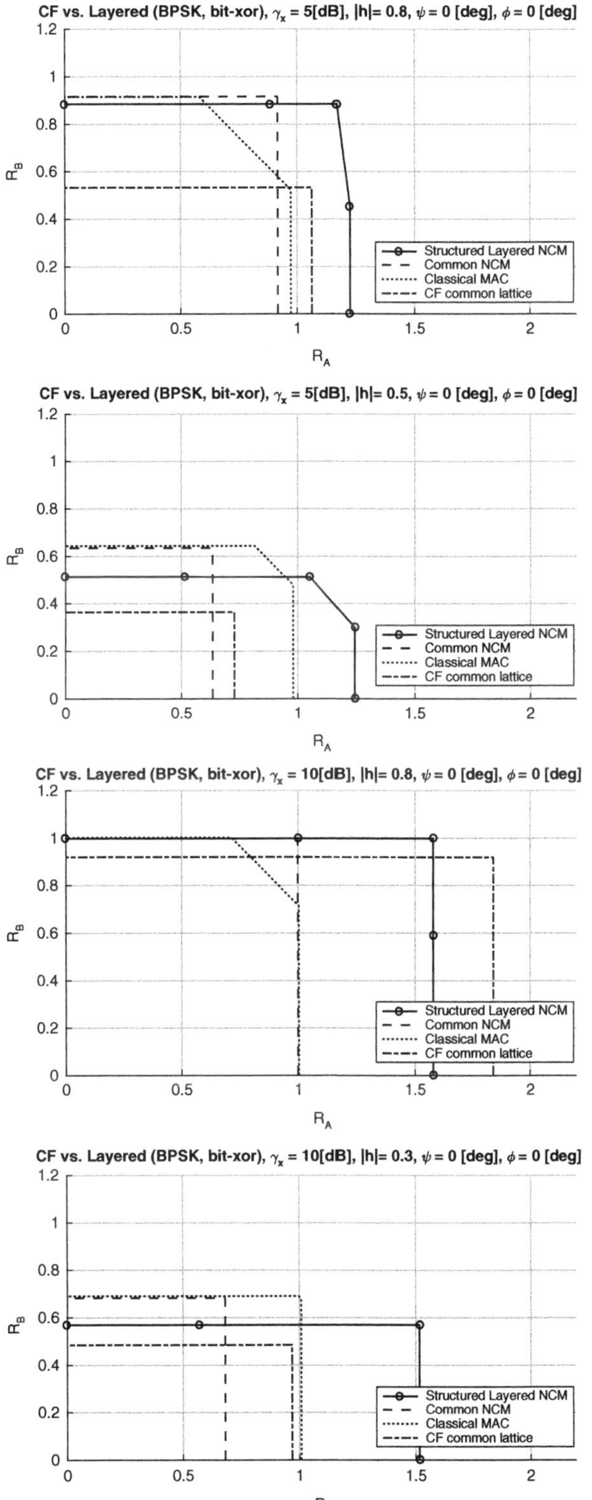

Figure 8.10 A comparison of CF-based structured NCM solution with finite BPSK alphabet solutions (classical MAC, common NCM, layered superposition-structured NCM).

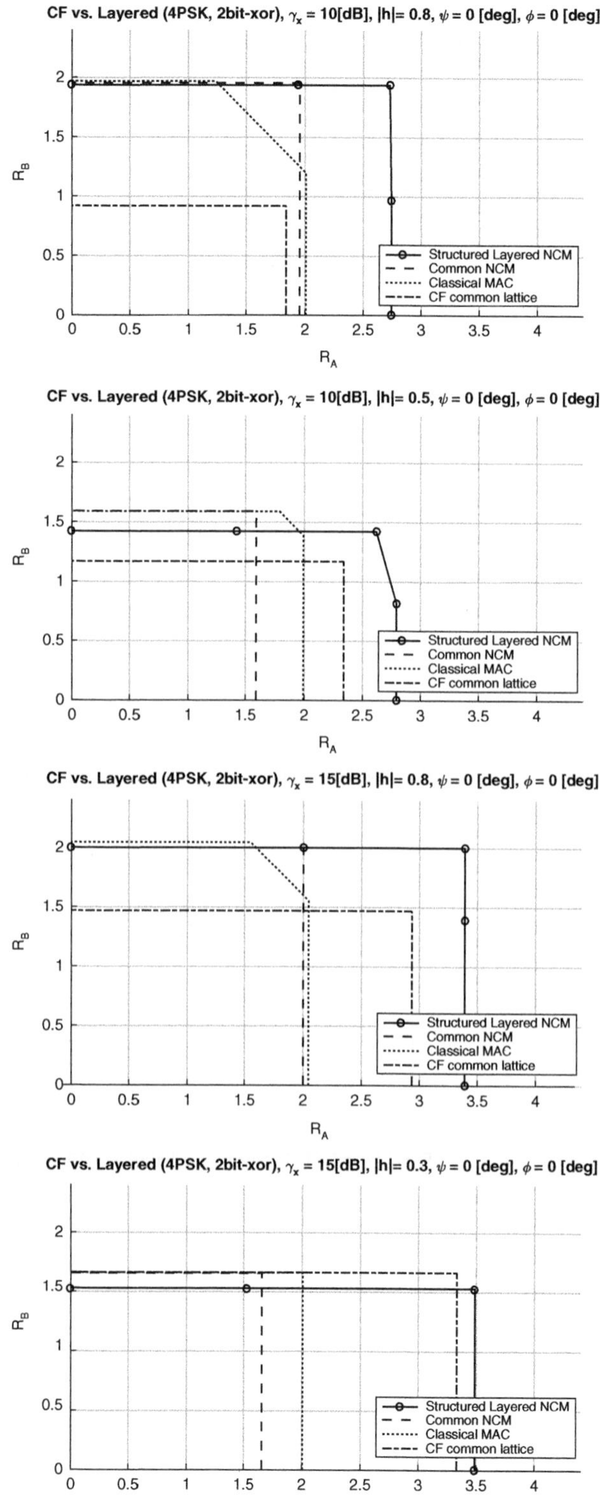

Figure 8.11 A comparison of CF-based structured NCM solution with finite 4PSK alphabet solutions (classical MAC, common NCM, layered superposition-structured NCM).

9 Joint Hierarchical Interference Processing

9.1 Introduction

This chapter deals with a very WPNC specific problem. It can be explained with reference to the classical interference mitigation technique, where we utilize the knowledge of the codebook and the signal space structure of the interfering components in order to minimize their impact on the desired data. But in the WPNC solution context provided in this chapter (cf. Chapter 8), it has two fundamental *distinguishing* aspects. First, the structure of the interfering and/or the desired signals/codebooks is not known uniquely. Both of them are allowed to be *hierarchical* codebooks which have multiple signal space representations of H-data. Second, the interfering and desired signals are *not* independent. They share some common, potentially hierarchical-only information, and we expect that they can *support* each other – it creates a *friendly* hierarchical interference. These aspects, the hierarchical self-dispersion and supporting aspects of the interaction, require us to adopt a fundamentally different approach including a revisiting of some coding theorems. We will show that friendly hierarchical interference can help in the desired data decoding. We will show the solutions for both Compute and Forward and Layered NCMs.

9.2 Joint Hierarchical Interference Processing

A relay in a WPNC network can receive signals from many component nodes. Some system scenarios allow us to define and utilize (particularly in the HDF class of scenarios) *multiple HNC maps* at *one* receiving relay. For example, the CF-based NCM fulfills the isomorphic property for multiple HNC map coefficient sets. Another example is a specific case corresponding to the structured scenarios of Chapter 8, where several subsets of real or virtual component nodes form multiple NCMs. However, the situation is specific in the sense that the component messages for these multiple HNC maps are *independent* and we observe signals internally coupled by HNC maps as a *superposition*. The system model treated in *this chapter* assumes that the multiple maps are generally defined over a *common* set of component messages. In all these cases, we can utilize the fact that the signals have defined *multiple* internal mutual structures (formed by NCMs) and can be used to help the relay processing. In order to clearly distinguish this case, we will call it *Joint Hierarchical Interference Processing*.

The mutual internal structure shared by the signals, however, has a specific form. It does *not* reveal the structure (the codebooks) of *individual* received components. It only reveals their joint mutual structure defined by the HNC maps. There are many degrees of freedom and only some of them are jointly shared. Therefore we call the interacting (interfering) signal *Hierarchical Interference* (H-Ifc). A major distinguishing aspect of the hierarchical interference is the fact that there is still an ambiguity in the signal for given knowledge of the isomorphic hierarchical codebook structure. This is a simple consequence of the H-codebook being a *many-to-one* function of the component codebooks.

The form of the help obtained from the availability of the multiple hierarchical signal codebook structures can be broadly classified into two cases. In the first case, we try to *remove* the influence of the hierarchical interference from the received signal. This case will be called *Hierarchical Interference Cancellation* (H-Ifc cancellation). The cancellation cannot be perfect, even with perfect knowledge of the H-codeword, because there is still an ambiguity given by the many-to-one HNC mapping. However, we can try to minimize its impact. In the second case, our H-codebook internal structure knowledge is used to improve the performance of some other H-codebook decoding, or it is used to provide some target HNC map that would otherwise be inaccessible. This case will be called *Hierarchical Interference Processing* (H-Ifc processing).

9.3   Joint Hierarchical Interference Processing in CF-Based NCM

The CF-based NCM (Section 5.6) using nested lattices is very suitable for a utilization of multiple HNC maps. The standard CF assumes essentially that the number of involved source nodes and relays is high and we have plenty of choices to optimize HNC map coefficients that maximize the computation rate.

However, in practical situations with a small number of nodes, we are rather limited and frequently only a few HNC maps (e.g. constrained by (5.115)) are allowed in order to guarantee the solvability at the final destination. In order to respect all subsequent stages of the network (not just the H-MAC of the first stage), it frequently dictates further constraints on the map that is aimed to be processed by the relay. A *particular* HNC map is typically *desired* to be processed by the relay. The modulo lattice processing and the corresponding linear GF-based HNC maps make it quite easy to be used with different allowed combination coefficients that, however, might not be optimal on their own. There are several possible ways to utilize these multiple maps. They differ by the targeted optimized utility and also by the constraints imposed by this step.

The system model (particularly the transmitted signals and channel model) and notation used in the next treatment follows closely the one used in Section 5.6. For simplicity, we will treat the situation on a single relay. The generalization for multiple relays is straightforward. The HNC coefficient map $\mathbf{a}$ and its other corresponding entities (e.g. H-codeword $\mathbf{c}$, lattice scaling α, etc.) defined in Section 5.6 are now termed as a *desired* map (H-codeword, etc.). We also define auxiliary maps $\mathbf{a}'_\ell$, $\ell \in [1 : L]$, where L is a number of available/used auxiliary maps. The auxiliary H-codewords are

$$\mathbf{c}'_\ell = \left(\sum_{k=1}^{K} a'_{\ell,k} \mathbf{c}_k \right) \bmod \Lambda_s \qquad (9.1)$$

and similarly (using the "prime" notation) for other entities. If only one auxiliary map is used, we simply drop the index and use $\mathbf{a}'$. We also define a column-based matrix of auxiliary coefficients $\mathbf{A}' = [\mathbf{a}'_1, \ldots, \mathbf{a}'_L]$.

We also slightly generalize the source component lattice codebooks. We will assume that there is a set of nested lattices $\Lambda_s \subseteq \Lambda_{c,K} \subseteq \cdots \subseteq \Lambda_{c,2} \subseteq \Lambda_{c,1}$ where the pair $(\Lambda_s, \Lambda_{c,k})$ forms a nested lattice R_k rate code on the kth source node. The source nodes are sorted such that $R_1 \geq R_2 \geq \cdots \geq R_k$. We also denote $R_0 = \max_k R_k = R_1$ and $\Lambda_{c,0} = \Lambda_{c,1}$ for a convenient relationship to the common rate and lattice used in the standard CF as treated in Section 5.6.

9.3.1   Integer-Constrained H-Ifc Cancellation

The simplest variant of using auxiliary maps is to reduce, at least partially, given the *integer* coefficient constraint, the interference represented by auxiliary maps and make the signal more "matched" to the desired map [45]. If the channel coefficients h_k were also integers then this could lead to "perfect" cancellation of those integer combinations that are not required by the desired map, and therefore the computation rate of the desired map would be improved. However, the channel coefficients are not integers and this procedure only partially cancels this impact.

Auxiliary Maps

Assume that we first decode each of the auxiliary H-codewords $\mathbf{c}'_\ell$ separately by a standard CF technique. All source node rates must comply with the MMSE optimized computation rate (5.112) of the particular auxiliary map

$$R_0 \leq R_c(\tilde{\alpha}'_\ell, \mathbf{a}'_\ell), \quad \ell \in [1:L]. \tag{9.2}$$

The codewords with added dither vectors allow us to obtain a reliable decision on the modulo lattice weighted sum of transmitted signals

$$\mathbf{s}'_\ell = \left(\sum_{k=1}^{K} a'_{\ell,k}(\mathbf{c}_k + \mathbf{u}_k) \right) \bmod \Lambda_s. \tag{9.3}$$

Notice an important fact that, having decoded the auxiliary H-codewords we have access only to the *modulo lattice reduced* H-codewords and subsequently only to the *modulo lattice reduced* signal $\mathbf{s}'_\ell$. So we do *not* have available the actual superposed value produced by the channel. We will show that this reduction does not matter if the H-Ifc cancellation coefficients are *integers*.

Decoding with H-Ifc Reduction

The actual receiver performs the estimate

$$\hat{\mathbf{c}} = \left(Q_{\Lambda_{c,0}}(\mathbf{y}) \right) \bmod \Lambda_s \tag{9.4}$$

where an integer linear combination of auxiliary $\mathbf{s}'_\ell$ is used to reduce the interference

$$\mathbf{y} = \alpha \mathbf{x} - \mathbf{u} - \sum_{\ell=1}^{L} g_\ell \mathbf{s}'_\ell. \tag{9.5}$$

Equivalent Channel

The receiver *equivalent* processing (compare that with (5.102) used in standard CF) is

$$\mathbf{y}' = \left(\alpha\mathbf{x} - \mathbf{u} - \sum_{\ell=1}^{L} g_\ell \mathbf{s}'_\ell \right) \bmod \Lambda_s \tag{9.6}$$

where $g_\ell \in \mathbb{Z}$. Now we proceed to the derivation of the equivalent channel (compare it with the one derived for a standard CF). The receiver processing output can be manipulated using the properties of $\bmod \Lambda_s$

$$
\begin{aligned}
\mathbf{y}' &= \left(\alpha \left(\sum_{k=1}^{K} h_k(\mathbf{c}_k + \mathbf{u}_k) \bmod \Lambda_s + \mathbf{w} \right) - \mathbf{u} - \sum_{\ell=1}^{L} g_\ell \mathbf{s}'_\ell \right) \bmod \Lambda_s \\
&= \left(\mathbf{c} - \mathbf{c} - \mathbf{u} + \alpha \left(\sum_{k=1}^{K} h_k(\mathbf{c}_k + \mathbf{u}_k) \bmod \Lambda_s + \mathbf{w} \right) \right. \\
&\qquad\qquad \left. - \sum_{\ell=1}^{L} g_\ell \left(\sum_{k=1}^{K} a'_{\ell,k}(\mathbf{c}_k + \mathbf{u}_k) \right) \bmod \Lambda_s \right) \bmod \Lambda_s \\
&\overset{(a)}{=} \left(\mathbf{c} - \sum_{k=1}^{K} a_k(\mathbf{c}_k + \mathbf{u}_k) \bmod \Lambda_s + \alpha \sum_{k=1}^{K} h_k(\mathbf{c}_k + \mathbf{u}_k) \bmod \Lambda_s \right. \\
&\qquad\qquad \left. - \sum_{k=1}^{K} \sum_{\ell=1}^{L} g_\ell a'_{\ell,k}(\mathbf{c}_k + \mathbf{u}_k) \bmod \Lambda_s + \alpha\mathbf{w} \right) \bmod \Lambda_s \\
&= \left(\mathbf{c} + \sum_{k=1}^{K} \left(\alpha h_k - a_k - \sum_{\ell=1}^{L} g_\ell a'_{\ell,k} \right) (\mathbf{c}_k + \mathbf{u}_k) \bmod \Lambda_s + \alpha\mathbf{w} \right) \bmod \Lambda_s. \tag{9.7}
\end{aligned}
$$

Notice that step (a) required $g_\ell \in \mathbb{Z}$ in order to utilize $\bmod \Lambda_s$ property (A.130). We use an equivalent dither to form an equivalent modulo lattice channel (as for standard CF). Since $\mathbf{u}_k \sim \mathcal{U}(\mathcal{V}(\Lambda_s))$ then also $(\mathbf{c}_k + \mathbf{u}_k) \bmod \Lambda_s \sim \mathcal{U}(\mathcal{V}(\Lambda_s))$ for arbitrary $\mathbf{c}_k$, and we substitute the actual dither by the equivalent one $\mathbf{u}_{k,\mathrm{eq}} = (\mathbf{c}_k + \mathbf{u}_k) \bmod \Lambda_s$ that has the same stochastic properties $\mathbf{u}_{k,\mathrm{eq}} \sim \mathcal{U}(\mathcal{V}(\Lambda_s))$ and is *independent* of $\mathbf{c}_k$. Equivalent dither is zero mean and has the same power as the transmitted signal, $\frac{1}{N}\mathrm{E}\left[\|\mathbf{u}_{k,\mathrm{eq}}\|^2\right] = P(\Lambda_s) = P_s$.

The equivalent hierarchical modulo lattice channel is then

$$\mathbf{y}_{\mathrm{eq}} = \left(\mathbf{c} + \mathbf{w}_{\mathrm{eq}} \right) \bmod \Lambda_s \tag{9.8}$$

where the equivalent noise is

$$\mathbf{w}_{\mathrm{eq}} = \sum_{k=1}^{K} \left(\alpha h_k - a_k - \sum_{\ell=1}^{L} g_\ell a'_{\ell,k} \right) \mathbf{u}_{k,\mathrm{eq}} + \alpha\mathbf{w}. \tag{9.9}$$

The equivalent noise has variance per dimension

$$P_{w_{eq}} = \frac{1}{N} \mathrm{E}\left[\|\mathbf{w}_{eq}\|^2\right]$$

$$= \frac{1}{N} \sum_{k=1}^{K} \left|\alpha h_k - a_k - \sum_{\ell=1}^{L} g_\ell a'_{\ell,k}\right|^2 \mathrm{E}\left[\|\mathbf{u}_{k,eq}\|^2\right] + \frac{|\alpha|^2}{N} \mathrm{E}\left[\|\mathbf{w}\|^2\right]$$

$$= P_s \left\|\alpha\mathbf{h} - \mathbf{a} - \sum_{\ell=1}^{L} g_\ell \mathbf{a}'_\ell\right\|^2 + |\alpha|^2 P_w. \tag{9.10}$$

Computation Rate

The equivalent channel has the same form as in the standard CF with the change that the coefficient vector $\mathbf{a}$ is replaced by a new *effective* combination coefficient vector

$$\bar{\mathbf{a}} = \mathbf{a} + \sum_{\ell=1}^{L} g_\ell \mathbf{a}'_\ell. \tag{9.11}$$

All results obtained for the standard CF thus hold here if the effective $\bar{\mathbf{a}}$ is substituted instead of $\mathbf{a}$. This holds for the computation rate with MMSE optimized scaling coefficient $\hat{\alpha}$ (see (5.112))

$$R_c(\hat{\alpha}, \bar{\mathbf{a}}) = \lg^+ \frac{P_s\|\mathbf{h}\|^2 + P_w}{P_s\left(\|\mathbf{h}\|^2\|\bar{\mathbf{a}}\|^2 - \left|\mathbf{h}^H\bar{\mathbf{a}}\right|^2\right) + P_w\|\bar{\mathbf{a}}\|^2}. \tag{9.12}$$

Special Case

Notice that in a *special case* of $L = 1$ and $\mathbf{a}'_1$ being the vector with *only one* non-zero coefficient on the ith position, $a'_{1,i} = 1$, $a'_{1,i'} = 0$, $i' \neq i$, the modulo lattice operation does not change the value of the H-Ifc removed signal

$$\mathbf{s}'_1 = \left(\sum_{k=1}^{K} a'_{1,k}(\mathbf{c}_k + \mathbf{u}_k)\right) \bmod \Lambda_s$$

$$= \left(a'_{1,i}(\mathbf{c}_i + \mathbf{u}_i)\right) \bmod \Lambda_s$$

$$= (\mathbf{c}_i + \mathbf{u}_i) \bmod \Lambda_s$$

$$= \mathbf{s}_i. \tag{9.13}$$

The auxiliary map thus corresponds to a single-user decoding of the ith user in a standard MAC channel where other users are considered as random interferers. For the second step of the decoding, the H-Ifc cancellation can thus perfectly remove the contribution from the ith node

$$\mathbf{y}' = (\alpha\mathbf{x} - \mathbf{u} - \alpha h_i \mathbf{s}_i) \bmod \Lambda_s. \tag{9.14}$$

The computation rate is then the one for the new effective channel $\mathbf{h}'$ with $h_i = 0$ and $h'_{i'} = h_i$ for $i' \neq i$.

Performance

The two-step decoding strategy using effective combination coefficients $\bar{\mathbf{a}}$ allows us more degrees of freedom in choosing the coefficients. It can finally provide a higher computation rate for desired $\mathbf{a}$ or even can provide a positive rate in the situation where the standard CF for the desired map would have a zero rate. The advantage is thus *greater freedom* in the combination coefficients but, on the other hand, *all source rates must still comply* (as in the standard CF) with the computation rates in *both* stages

$$R_k \leq R_c(\hat{\alpha}'_\ell, \mathbf{a}'_\ell), \quad \ell \in [1:L], \tag{9.15}$$

$$R_k \leq R_c(\hat{\alpha}, \bar{\mathbf{a}}). \tag{9.16}$$

9.3.2   Successive Nulling of HNC Map Coefficients

Another option improving the CF performance targets on the usage of multi-level nested lattice codes with different rates. We attempt to relax the necessity of all source rates being limited by a *common* computation rate. The idea [47] stands on the ordering of the source rates $R_1 \geq R_2 \geq \cdots \geq R_k$, where the higher rate corresponds to the finer lattice $\Lambda_s \subseteq \Lambda_{c,K} \subseteq \cdots \subseteq \Lambda_{c,2} \subseteq \Lambda_{c,1}$, and the multi-step receiver processing. It uses CF decoding modulo lattice decoding in multiple steps outlined here.

(1) At each step $\ell \in [1:L]$, $L = K$, we first set auxiliary coefficients $\mathbf{a}'_\ell$ that have high computation rates $R_{c,\ell}(\hat{\alpha}_\ell, \mathbf{a}'_\ell)$. The auxiliary coefficients $\mathbf{a}'_\ell$ must be such that the *desired* H-codeword $\mathbf{c}$ with the desired map $\mathbf{a}$ can be obtained by a *linear combination* from the auxiliary H-codewords $\mathbf{c}'_\ell = \left(\sum_{k=1}^{K} a'_{\ell,k} \mathbf{c}_k \right) \bmod \Lambda_s$, $\ell \in [1:L]$.

(2) At step ℓ, we subtract a specific combination of H-codewords obtained in previous steps, in such a way that all finer lattices $\Lambda_{c,\ell-1}, \Lambda_{c,\ell-2}, \ldots, \Lambda_{c,1}$ do *not* participate in the decoding and $\Lambda_{c,\ell}$ participates in decoding. This means that the resulting effective combination coefficient map $\mathbf{a}''_\ell$ for the ℓth step has *zeros* on all positions $a''_{\ell,k} = 0$ with indices $k \in \{1, \ldots, (\ell-1)\}$ and $a''_{\ell,\ell} \neq 0$.

(3) Since the finest lattice that remains after the previous step is $\Lambda_{c,\ell}$, we can use lattice quantizer $Q_{\Lambda_{c,\ell}}$ and the rate of the ℓth source node can be R_ℓ.

(4) The previously removed finer lattice points are, after quantization, added back in order to obtain the original $\mathbf{c}'_\ell$.

(5) The trick is that we can show that the *equivalent channel appears as if there was a map* $\mathbf{a}''_\ell$ while the *effective noise remains the same as it was for* $\mathbf{a}'_\ell$. Thus the computation rate $R_{c,\ell}(\hat{\alpha}_\ell, \mathbf{a}'_\ell))$ remains *unaffected* by the finer lattice points removal and we can decode as if there were *no sources with finer lattices* $\Lambda_{c,\ell-1}, \Lambda_{c,\ell-2}, \ldots, \Lambda_{c,1}$. This means that the coderate can be $R_\ell \leq R_{c,\ell}(\hat{\alpha}_\ell, \mathbf{a}'_\ell)$ and it is *not* dictated by the common smallest computation rate as is the case of standard CF ($R_0 \leq R_{c,\ell}(\hat{\alpha}_\ell, \mathbf{a}'_\ell)$, for all ℓ).

Now we provide more details.

Receiver Successive CF Decoding with Map Coefficients Nulling

At the ℓth step, $\ell = 1, \ldots, L$, $L = K$, evaluate

$$\mathbf{y}_\ell = \alpha_\ell \mathbf{x} - \mathbf{u}'_\ell - \sum_{j=1}^{\ell-1} g_{\ell j} \mathbf{c}'_j \tag{9.17}$$

where $\mathbf{c}'_j = \left(\sum_{k=1}^K a'_{j,k} \mathbf{c}_k \right) \bmod \Lambda_s$ and $\mathbf{u}'_\ell = \left(\sum_{k=1}^K a'_{\ell,k} \mathbf{u}_k \right) \bmod \Lambda_s$. Notice that, unlike the standard integer-constrained H-Ifc cancellation (Section 9.3.1), we subtract H-codewords *without the dithers*. The $g_{\ell j} \in \mathbb{Z}$ coefficients are such that

$$a'_{\ell,k} - \sum_{j=1}^{\ell-1} g_{\ell j} a'_{j,k} = 0, \quad \text{for } k \in \{1, \ldots, \ell-1\}, \tag{9.18}$$

$$a'_{\ell,k} - \sum_{j=1}^{\ell-1} g_{\ell j} a'_{j,k} \neq 0, \quad \text{for } k = \ell. \tag{9.19}$$

This defines an effective auxiliary map

$$\mathbf{a}''_\ell = \mathbf{a}'_\ell - \sum_{j=1}^{\ell-1} g_{\ell j} \mathbf{a}'_j. \tag{9.20}$$

Coefficients $g_{\ell j}$ are obtained by the triangularization procedure (see details in [47]).

The auxiliary H-codeword estimate is

$$\hat{\mathbf{c}}'_\ell = \left(Q_{\Lambda_{c,\ell}}(\mathbf{y}_\ell) \right) \bmod \Lambda_s \tag{9.21}$$

and the desired H-codeword $\hat{\mathbf{c}}$, corresponding to the map $\mathbf{a}$, is obtained by a linear combination from $\hat{\mathbf{c}}'_\ell$ auxiliary H-codewords.

Equivalent Channel

The receiver *equivalent* processing (as in (5.102)) is

$$\mathbf{y}'_\ell = \mathbf{y}_\ell \bmod \Lambda_s = \left(\alpha_\ell \mathbf{x} - \mathbf{u}'_\ell - \sum_{j=1}^{\ell-1} g_{\ell j} \mathbf{c}'_j \right) \bmod \Lambda_s. \tag{9.22}$$

The equivalent channel is obtained by using properties of $\bmod \Lambda_s$

$$\mathbf{y}'_\ell = \left(\mathbf{c}'_\ell - \left(\sum_{k=1}^K a'_{\ell,k} \mathbf{c}_k \right) \bmod \Lambda_s + \alpha_\ell \left(\sum_{k=1}^K h_k (\mathbf{c}_k + \mathbf{u}_k) \bmod \Lambda_s + \mathbf{w} \right) \right.$$

$$\left. - \left(\sum_{k=1}^K a'_{\ell,k} \mathbf{u}_k \right) \bmod \Lambda_s - \sum_{j=1}^{\ell-1} g_{\ell j} \mathbf{c}'_j \right) \bmod \Lambda_s$$

$$= \left(\mathbf{c}'_\ell - \sum_{j=1}^{\ell-1} g_{\ell j} \mathbf{c}'_j + \sum_{k=1}^K \left(\alpha_\ell h_k - a'_{\ell,k} \right) (\mathbf{c}_k + \mathbf{u}_k) \bmod \Lambda_s + \alpha_\ell \mathbf{w} \right) \bmod \Lambda_s$$

$$= \left(\left(\sum_{k=1}^{K} \mathbf{c}_k \underbrace{\left(a'_{\ell,k} - \sum_{j=1}^{\ell-1} g_{\ell j} a'_{j,k} \right)}_{a''_{\ell,k}} \right) \bmod \Lambda_s \right.$$

$$\left. + \sum_{k=1}^{K} \left(\alpha_\ell h_k - a'_{\ell,k} \right) \left(\mathbf{c}_k + \mathbf{u}_k \right) \bmod \Lambda_s + \alpha_\ell \mathbf{w} \right) \bmod \Lambda_s. \tag{9.23}$$

We see that the H-codeword corresponds to the map $\mathbf{a}''_\ell$ whereas the equivalent noise part is exactly the same as it was for the map $\mathbf{a}'_\ell$. The equivalent channel is thus

$$\mathbf{y}_{\mathrm{eq},\ell} = \left(\mathbf{c}''_\ell + \mathbf{w}_{\mathrm{eq},\ell} \right) \bmod \Lambda_s \tag{9.24}$$

where the effective auxiliary H-codeword

$$\mathbf{c}''_\ell = \left(\sum_{k=1}^{K} a''_{\ell,k} \mathbf{c}_k \right) \bmod \Lambda_s \tag{9.25}$$

does not contain any lattice points from $\Lambda_{c,\ell-1}, \Lambda_{c,\ell-2}, \dots, \Lambda_{c,1}$. Therefore we can use quantizer $Q_{\Lambda_{c,\ell}}$ and the effective processed coderate is R_ℓ. The equivalent noise is

$$\mathbf{w}_{\mathrm{eq},\ell} = \sum_{k=1}^{K} \left(\alpha_\ell h_k - a'_{\ell,k} \right) \mathbf{u}_{k,\mathrm{eq}} + \alpha_\ell \mathbf{w} \tag{9.26}$$

and thus the associated computation rate is $R_{c,\ell}(\hat{\alpha}_\ell, \mathbf{a}'_\ell)$.

Performance

The condition for reliable decoding is then $R_\ell \leq R_{c,\ell}(\hat{\alpha}_\ell, \mathbf{a}'_\ell)$ where the computation rate in each decoding step ℓ restricts only the rate R_ℓ. Thus the finer lattices with higher rates have a chance of being used if there exist such auxiliary coefficients $\mathbf{a}'_\ell$ that allow the higher computation rate $R_{c,\ell}(\hat{\alpha}_\ell, \mathbf{a}'_\ell)$. This is in contrast with common flat rate $R_0 \leq R_{c,\ell}(\hat{\alpha}_\ell, \mathbf{a}'_\ell)$, for all ℓ, required by the standard CF. The reverse calculation of the maps $\mathbf{c}'_\ell$ from the effective maps $\mathbf{c}''_\ell$ is not necessary. Since both relationships, the $\mathbf{c}'_\ell$ to $\mathbf{c}''_\ell$ and also $\mathbf{c}'_\ell$ to $\mathbf{c}$, are linear, we can directly compute the desired $\mathbf{c}$ from $\mathbf{c}''_\ell$.

9.3.3   Joint Hierarchical Successive CF Decoding

The highest level of utilizing the auxiliary decoded HNC maps is to use them for *full* H-Ifc cancellation [42], [58]. The form of the algorithm performing the desired map $\mathbf{a}$ decoding together with H-Ifc cancellation using the auxiliary maps $\{\mathbf{a}'_\ell\}_{\ell=1}^{L}$ will be called the *Joint Hierarchical Successive CF Decoding (Joint H-SCFD)*.

The full H-Ifc cancellation requires the receiver to recover a full modulo-lattice *unconstrained* combined received signal but, on the other hand, the combination coefficients of the subtracted H-Ifc cancellation do *not* need to be integers. Compare that with integer-constrained H-Ifc cancellation, where the H-Ifc cancellation signal was modulo lattice combination (9.3, 9.5) and the derivation of the equivalent channel required that combination coefficients were integers (9.7).

Recovery of Unconstrained Signal Combination

The recovery of a full combined signal for some given auxiliary map $\mathbf{a}'_\ell$, however, requires some additional steps and constraints. The CF decoding of the auxiliary H-codeword $\mathbf{c}'_\ell$ corresponding to the auxiliary map $\mathbf{a}'_\ell$ recovers only a modulo lattice combination of the transmitted signals (including proper dithers)

$$
\mathbf{s}'_\ell = \left(\sum_{k=1}^{K} a'_{\ell,k}(\mathbf{c}_k + \mathbf{u}_k) \right) \bmod \Lambda_s
$$

$$
= \left(\sum_{k=1}^{K} a'_{\ell,k}(\mathbf{c}_k + \mathbf{u}_k) \bmod \Lambda_s \right) \bmod \Lambda_s. \tag{9.27}
$$

However, we need a full *unconstrained* combination

$$
\bar{\mathbf{s}}'_\ell = \sum_{k=1}^{K} a'_{\ell,k}(\mathbf{c}_k + \mathbf{u}_k) \bmod \Lambda_s \tag{9.28}
$$

where $\mathbf{s}_k = (\mathbf{c}_k + \mathbf{u}_k) \bmod \Lambda_s$ are true transmitted signals. Using modulo lattice operation properties, we get

$$
\mathbf{s}'_\ell = \bar{\mathbf{s}}'_\ell - Q_{\Lambda_s}(\bar{\mathbf{s}}'_\ell). \tag{9.29}
$$

We now show how and under what conditions we can recover $\bar{\mathbf{s}}'_\ell$ from $\mathbf{s}'_\ell$. We subtract $\mathbf{s}'_\ell$ from α'_ℓ scaled received signal

$$
\mathbf{z}_\ell = \alpha'_\ell \mathbf{x} - \mathbf{s}'_\ell
$$

$$
= \sum_{k=1}^{K} \alpha'_\ell h_k(\mathbf{c}_k + \mathbf{u}_k) \bmod \Lambda_s + \alpha'_\ell \mathbf{w} - \bar{\mathbf{s}}'_\ell + Q_{\Lambda_s}\left(\sum_{k=1}^{K} a'_{\ell,k}\mathbf{s}_k \right)
$$

$$
= \sum_{k=1}^{K} (\alpha'_\ell h_k - a'_{\ell,k})(\mathbf{c}_k + \mathbf{u}_k) \bmod \Lambda_s + \alpha'_\ell \mathbf{w} + Q_{\Lambda_s}\left(\sum_{k=1}^{K} a'_{\ell,k}\mathbf{s}_k \right). \tag{9.30}
$$

The first two parts of the expression clearly form an equivalent noise (see also (5.105))

$$
\mathbf{w}_{\mathrm{eq},\ell} = \sum_{k=1}^{K} \left(\alpha'_\ell h_k - a'_{\ell,k} \right) \mathbf{u}_{k,\mathrm{eq}} + \alpha'_\ell \mathbf{w} \tag{9.31}
$$

with variance per dimension (5.106)

$$
P_{w_{\mathrm{eq},\ell}} = P_s \| \alpha'_\ell \mathbf{h} - \mathbf{a}'_\ell \|^2 + |\alpha'_\ell|^2 P_w. \tag{9.32}
$$

In order for the achievable computation rate (5.111) to be non-zero, for whatever α'_ℓ is used by the auxiliary CF decoder (e.g. the MMSE optimized or any other), it must hold that $P_s > P_{w_{\mathrm{eq},\ell}}$. If this holds, then

$$
Q_{\Lambda_s}(\mathbf{z}_\ell) = Q_{\Lambda_s}\left(\sum_{k=1}^{K} a'_{\ell,k}\mathbf{s}_k \right) \tag{9.33}
$$

with high probability. Finally, we reconstruct the desired unconstrained signal combination

$$
\bar{\mathbf{s}}'_\ell = \mathbf{s}'_\ell + Q_{\Lambda_s}(\mathbf{z}_\ell). \tag{9.34}
$$

Decoding with H-Ifc Cancellation

The actual receiver obtains the estimate by

$$\hat{\mathbf{c}} = \left(Q_{\Lambda_{c,0}}(\mathbf{y})\right) \bmod \Lambda_s \tag{9.35}$$

where an unconstrained linear combination of all available auxiliary H-coded signals $\{\bar{\mathbf{s}}'_\ell\}_{\ell=1}^L$ is used to cancel the interference associated with auxiliary maps $\{\mathbf{a}'_\ell\}_{\ell=1}^L$

$$\mathbf{y} = \alpha\mathbf{x} - \mathbf{u} - \sum_{\ell=1}^L \beta_\ell \bar{\mathbf{s}}'_\ell \tag{9.36}$$

where $\beta_\ell \in \mathbb{C}$. We also define $\boldsymbol{\beta} = [\beta_1, \ldots, \beta_L]^{\mathrm{T}}$.[1]

Multi-Tap Hierarchical Equalizer

Notice that we generally allowed complex weighting of the H-signals. This can be viewed as a multi-tap equalizer, which removes the influence of other decodable structures (H-codes) from the received signal. Each of those structures is allowed to have its own β_ℓ scaling. This relaxes one of the problems of standard CF (discussed at the end of Section 5.6), i.e. *common* scalar scaling of all superposed received mismatched lattices. For the H-Ifc cancellation, we now have the chance to scale each of the known structures by its own coefficient.

Equivalent Channel

The receiver *equivalent* processing (similarly as in (5.102)) is

$$\mathbf{y}' = \mathbf{y} \bmod \Lambda_s = \left(\alpha\mathbf{x} - \mathbf{u} - \sum_{\ell=1}^L \beta_\ell \bar{\mathbf{s}}'_\ell\right) \bmod \Lambda_s. \tag{9.37}$$

The equivalent channel for the target H-codeword $\mathbf{c}$ corresponding to the map $\mathbf{a}$ is obtained by using the properties of $\bmod\,\Lambda_s$

$$
\begin{aligned}
\mathbf{y}' &= \left(\mathbf{c} - \left(\sum_{k=1}^K a_k\mathbf{c}_k\right) \bmod \Lambda_s + \alpha\left(\sum_{k=1}^K h_k(\mathbf{c}_k + \mathbf{u}_k) \bmod \Lambda_s + \mathbf{w}\right)\right. \\
&\quad \left. - \left(\sum_{k=1}^K a_k\mathbf{u}_k\right) \bmod \Lambda_s - \sum_{\ell=1}^L \beta_\ell\left(\sum_{k=1}^K a'_{\ell,k}(\mathbf{c}_k + \mathbf{u}_k) \bmod \Lambda_s\right)\right) \bmod \Lambda_s \\
&= \left(\mathbf{c} + \sum_{k=1}^K \left(\alpha h_k - \sum_{\ell=1}^L \beta_\ell a'_{\ell,k} - a_k\right)(\mathbf{c}_k + \mathbf{u}_k) \bmod \Lambda_s + \alpha\mathbf{w}\right) \bmod \Lambda_s.
\end{aligned}
\tag{9.38}
$$

Notice that the lack of modulo lattice operation around $\sum_{k=1}^K a'_{\ell,k}(\mathbf{c}_k + \mathbf{u}_k) \bmod \Lambda_s$ allowed us to use *non-integer* β_ℓ, which would otherwise be impossible.

---

[1] The form of the processing in [42] uses only one auxiliary map $L = 1$ and does not include the equivalent equalizing transformation. Also it first removes the projection of the auxiliary H-signal to the received signal and then it forms an effective signal consisting of the complex weighted sum of the first step result and again the auxiliary H-signal. So the projection coefficient and second step coefficient simply multiply in the resulting expression. The coefficients are then optimized. This treatment is identical to the simple direct subtraction of the β_ℓ scaled signal as in this book and in [58].

The equivalent channel, from the perspective of H-codeword $\mathbf{c}$ associated with target map $\mathbf{a}$, is then

$$\mathbf{y}_{\text{eq}} = \left(\mathbf{c} + \mathbf{w}_{\text{eq}}\right) \bmod \Lambda_s \qquad (9.39)$$

where the equivalent noise is

$$\mathbf{w}_{\text{eq}} = \sum_{k=1}^{K} \left(\alpha h_k - \sum_{\ell=1}^{L} \beta_\ell a'_{\ell,k} - a_k\right) \mathbf{u}_{k,\text{eq}} + \alpha \mathbf{w}. \qquad (9.40)$$

The variance per dimension of the equivalent noise is

$$\begin{aligned}
P_{w_{\text{eq}}}(\alpha, \boldsymbol{\beta}) &= P_s \sum_{k=1}^{K} \left|\alpha h_k - \sum_{\ell=1}^{L} \beta_\ell a'_{\ell,k} - a_k\right|^2 + |\alpha|^2 P_w \\
&= P_s \left\|\alpha \mathbf{h} - \sum_{\ell=1}^{L} \beta_\ell \mathbf{a}'_\ell - \mathbf{a}\right\|^2 + |\alpha|^2 P_w \\
&= P_s \left\|\alpha \mathbf{h} - \mathbf{A}'\boldsymbol{\beta} - \mathbf{a}\right\|^2 + |\alpha|^2 P_w.
\end{aligned} \qquad (9.41)$$

Optimization of Equivalent Noise Power

For a given desired combination map $\mathbf{a}$ and the auxiliary map matrix $\mathbf{A}'$, we can now find $\hat{\alpha}, \hat{\boldsymbol{\beta}}$ that minimize the equivalent noise power

$$\hat{\alpha}, \hat{\boldsymbol{\beta}} = \min_{\alpha, \boldsymbol{\beta}} P_{w_{\text{eq}}}(\alpha, \boldsymbol{\beta}). \qquad (9.42)$$

The utility $P_{w_{\text{eq}}}(\alpha, \boldsymbol{\beta})$ is a real-valued function of complex coefficients and we *must* use a generalized derivative, over both the scalar and the vector (Section A.3.5), to find a stationary point. We get

$$\frac{\tilde{\partial} P_{w_{\text{eq}}}(\alpha, \boldsymbol{\beta})}{\tilde{\partial}\alpha} = \left(\alpha^* \|\mathbf{h}\|^2 - \boldsymbol{\beta}^{\mathrm{H}} \mathbf{A}'^{\mathrm{H}} \mathbf{h} - \mathbf{a}^{\mathrm{H}} \mathbf{h}\right) P_s + \alpha^* P_w, \qquad (9.43)$$

$$\frac{\tilde{\partial} P_{w_{\text{eq}}}(\alpha, \boldsymbol{\beta})}{\tilde{\partial}\boldsymbol{\beta}} = \left(\mathbf{A}'^{\mathrm{T}} \mathbf{A}'^* \boldsymbol{\beta}^* + \alpha^* \mathbf{A}'^{\mathrm{T}} \mathbf{h}^* + \mathbf{A}'^{\mathrm{T}} \mathbf{a}^*\right) P_s. \qquad (9.44)$$

Setting these derivatives to zeros (zero scalar and zero vector) and solving for $\hat{\alpha}, \hat{\boldsymbol{\beta}}$ gives

$$\left(P_s \|\mathbf{h}\|^2 + P_w\right) \hat{\alpha} - P_s \mathbf{h}^{\mathrm{H}} \mathbf{A}' \hat{\boldsymbol{\beta}} = P_s \mathbf{h}^{\mathrm{H}} \mathbf{a}, \qquad (9.45)$$

$$\mathbf{A}'^{\mathrm{H}} \mathbf{h} \hat{\alpha} - \mathbf{A}'^{\mathrm{H}} \mathbf{A}' \hat{\boldsymbol{\beta}} = \mathbf{A}'^{\mathrm{H}} \mathbf{a}. \qquad (9.46)$$

We can now solve these two equations keeping in mind that α is scalar and $\boldsymbol{\beta}$ is a vector. Assuming that the auxiliary map matrix is *full-rank* $\text{rank}(\mathbf{A}') = L$ then we can utilize matrix inversion $(\mathbf{A}'^{\mathrm{H}} \mathbf{A}')^{-1}$. As an interim result we express

$$\hat{\boldsymbol{\beta}} = (\mathbf{A}'^{\mathrm{H}} \mathbf{A}')^{-1} \mathbf{A}'^{\mathrm{H}} \left(\hat{\alpha} \mathbf{h} - \mathbf{a}\right). \qquad (9.47)$$

Then the MMSE optimized coefficients are

$$\hat{\alpha} = \frac{P_s \mathbf{h}^{\mathrm{H}} (\mathbf{I} - \mathbf{P}_{\mathbf{A}'}) \mathbf{a}}{P_s \mathbf{h}^{\mathrm{H}} (\mathbf{I} - \mathbf{P}_{\mathbf{A}'}) \mathbf{h} + P_w}, \qquad (9.48)$$

$$\hat{\beta} = \mathbf{A}'^\dagger \left(\frac{P_s \mathbf{h}\mathbf{h}^H(\mathbf{I} - \mathbf{P}_{\mathbf{A}'})}{P_s \mathbf{h}^H(\mathbf{I} - \mathbf{P}_{\mathbf{A}'})\mathbf{h} + P_w} - \mathbf{I} \right) \mathbf{a}, \tag{9.49}$$

where $\mathbf{P}_{\mathbf{A}'} = \mathbf{A}'(\mathbf{A}'^H\mathbf{A}')^{-1}\mathbf{A}'^H$ is a projector to the subspace generated by columns of $\mathbf{A}'$ and $\mathbf{A}'^\dagger = (\mathbf{A}'^H\mathbf{A}')^{-1}\mathbf{A}'^H$ is a pseudoinverse of $\mathbf{A}'$. It is also worth noting that $\mathbf{P}_{\mathbf{A}'\perp} = \mathbf{I} - \mathbf{P}_{\mathbf{A}'}$ is a projector to the null-space of $\mathbf{A}'$.

The MMSE optimized equivalent noise power $P_{w_{eq}}(\hat{\alpha}, \hat{\beta})$ can be obtained by the substitution of the MMSE coefficient solution. However, it gives much better insight if we do not substitute the final result but the interim one (9.47) first

$$\begin{aligned}
P_{w_{eq}}(\hat{\alpha}, \hat{\beta}) &= P_s \left\| \hat{\alpha}\mathbf{h} - \mathbf{A}'(\mathbf{A}'^H\mathbf{A}')^{-1}\mathbf{A}'^H \left(\hat{\alpha}\mathbf{h} - \mathbf{a} \right) - \mathbf{a} \right\|^2 + |\hat{\alpha}|^2 P_w \\
&= P_s \left\| \hat{\alpha}\mathbf{h} - \mathbf{P}_{\mathbf{A}'}\left(\hat{\alpha}\mathbf{h} - \mathbf{a} \right) - \mathbf{a} \right\|^2 + |\hat{\alpha}|^2 P_w \\
&= P_s \left\| \hat{\alpha}(\mathbf{I} - \mathbf{P}_{\mathbf{A}'})\mathbf{h} - (\mathbf{I} - \mathbf{P}_{\mathbf{A}'})\mathbf{a} \right\|^2 + |\hat{\alpha}|^2 P_w \\
&= P_s \left\| \hat{\alpha}\mathbf{h}_{eq} - \mathbf{a}_{eq} \right\|^2 + |\hat{\alpha}|^2 P_w
\end{aligned} \tag{9.50}$$

where $\mathbf{h}_{eq} = (\mathbf{I} - \mathbf{P}_{\mathbf{A}'})\mathbf{h}$ and $\mathbf{a}_{eq} = (\mathbf{I} - \mathbf{P}_{\mathbf{A}'})\mathbf{a}$ are projections of $\mathbf{h}$ and $\mathbf{a}$ into the null-space of the auxiliary coefficients subspace. The optimal $\hat{\alpha}$ solution (9.48) can be also manipulated utilizing the properties of the projector, namely $(\mathbf{I} - \mathbf{P}_{\mathbf{A}'}) = (\mathbf{I} - \mathbf{P}_{\mathbf{A}'})^2$ and $(\mathbf{I} - \mathbf{P}_{\mathbf{A}'})^H = (\mathbf{I} - \mathbf{P}_{\mathbf{A}'})$

$$\begin{aligned}
\hat{\alpha} &= \frac{P_s \mathbf{h}^H(\mathbf{I} - \mathbf{P}_{\mathbf{A}'})^H(\mathbf{I} - \mathbf{P}_{\mathbf{A}'})\mathbf{a}}{P_s \mathbf{h}^H(\mathbf{I} - \mathbf{P}_{\mathbf{A}'})^H(\mathbf{I} - \mathbf{P}_{\mathbf{A}'})\mathbf{h} + P_w} \\
&= \frac{P_s \mathbf{h}_{eq}^H \mathbf{a}_{eq}}{P_s \|\mathbf{h}_{eq}\|^2 + P_w}.
\end{aligned} \tag{9.51}$$

Equivalent Equalizing Transformation

When we now compare these results for $\hat{\alpha}$ and $P_{w_{eq}}(\hat{\alpha}, \hat{\beta})$ with standard CF results (5.109) and (5.106), we reach an important conclusion. The joint H-SCFD turns the system into an *equivalent* standard CF decoding with a new *equivalent channel*

$$\mathbf{h}_{eq} = (\mathbf{I} - \mathbf{P}_{\mathbf{A}'})\mathbf{h} \tag{9.52}$$

and an *equivalent target map*

$$\mathbf{a}_{eq} = (\mathbf{I} - \mathbf{P}_{\mathbf{A}'})\mathbf{a}. \tag{9.53}$$

Both of them have neat interpretation of being *null-space projections* of the auxiliary maps space $\mathbf{A}'$. This nicely shows the *equalizing* effect of the joint H-SCFD, which removes all channel and target map components *not* lying in the subspace generated by $\mathbf{A}'$. The null-space projection equivalence allows variety of the auxiliary map optimization procedures.

The null-space projection of the equalizing transformation also gives an insight into the relationship between the target map $\mathbf{a}$ and the auxiliary maps $\mathbf{A}'$. The projection (Figure 9.1) changes the desired target map. This must be carefully respected in the end-to-end solvability of the network. A *special* case of the target map $\mathbf{a}$ orthogonal to the range-space of $\mathbf{A}'$ is of notable importance (see also a variant of the algorithm with

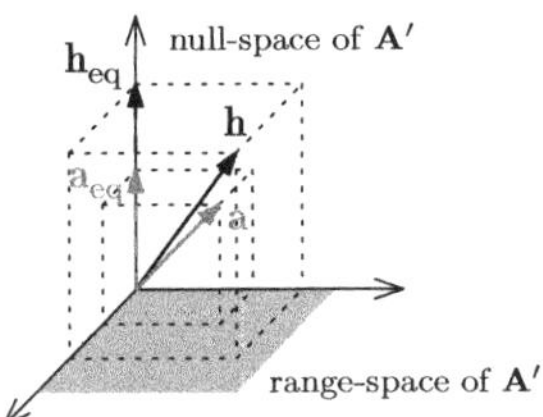

Figure 9.1 The equivalent equalizing transformation for the joint H-SCFD. The null-space projection removes all components of $\mathbf{h}$ and $\mathbf{a}$ contained in range-space of auxiliary maps $\mathbf{A}'$. It means that everything linearly dependent on auxiliary maps is removed. It applies on both the channel and the target map. The components of the target map $\mathbf{a}$ that lie in the range-space of $\mathbf{A}'$ are considered as an "interference" and (owing to the linear dependence) cannot be distinguished from the "true" interference described by the auxiliary maps.

decoupled coefficient optimization in the next section). In this case the map projection is $\mathbf{a}_{\mathrm{eq}} = \mathbf{a}$, i.e. the map is unchanged and the equalizer removes only interfering channel $\mathbf{P}_{\mathbf{A}'}\mathbf{h}$ components. Another special case is the one where the target map is a *linear combination* of auxiliary maps. Then the null-space projection is zero $\mathbf{a}_{\mathrm{eq}} = \mathbf{0}$ and subsequently also $\hat{\alpha} = 0$, and the MMSE optimization does *not* have a viable solution.

Achievable Computation Rate

Utilizing the equivalent channel and the target map, we can straightforwardly get the achievable computation rate

$$R_c(\hat{\alpha}, \hat{\boldsymbol{\beta}}) = \lg^+ \frac{P_s \|\mathbf{h}_{\mathrm{eq}}\|^2 + P_w}{P_s \left(\|\mathbf{h}_{\mathrm{eq}}\|^2 \|\mathbf{a}_{\mathrm{eq}}\|^2 - \left| \mathbf{h}_{\mathrm{eq}}^{\mathrm{H}} \mathbf{a}_{\mathrm{eq}} \right|^2 \right) + P_w \|\mathbf{a}_{\mathrm{eq}}\|^2}. \tag{9.54}$$

The optimization procedure for the choice of $\mathbf{A}'$ should clearly maximize $|\mathbf{h}_{\mathrm{eq}}^{\mathrm{H}} \mathbf{a}_{\mathrm{eq}}|^2$ by making them as collinear as possible.

9.3.4   H-SCFD with Decoupled Coefficient Optimization

We now investigate a modification of the joint H-SCFD where, instead of a joint optimization of lattice inflation coefficient α and auxiliary H-Ifc equalization coefficients $\boldsymbol{\beta}$, we use a decoupled processing.

H-Ifc Processing Step

The receiver first removes a scaled H-Ifc associated with the auxiliary maps $\mathbf{A}'$

$$\mathbf{x}' = \mathbf{x} - \sum_{\ell=1}^{L} \beta'_\ell \bar{\mathbf{s}}'_\ell. \tag{9.55}$$

In contrast to the joint H-SCFD, there is no α scaling. In order to distinguish H-Ifc processing coefficients, we denote them $\boldsymbol{\beta}' = [\beta'_1, \ldots, \beta'_L]^{\mathrm{T}}$. The coefficients are

MMSE optimized for residual interference w.r.t. the desired H-signal corresponding to the desired map $\mathbf{a}$

$$
\begin{aligned}
\hat{\boldsymbol{\beta}}' &= \arg\min_{\boldsymbol{\beta}'} \mathrm{E}\left[\left\| \mathbf{x} - \sum_{\ell=1}^{L} \beta'_\ell \bar{\mathbf{s}}'_\ell - \sum_{k=1}^{K} a_k \mathbf{s}_k \right\|^2\right] \\
&= \arg\min_{\boldsymbol{\beta}'} \mathrm{E}\left[\left\| \sum_{k=1}^{K}\left(h_k - \sum_{\ell=1}^{L} \beta'_\ell a'_{\ell,k} - a_k \right)\mathbf{s}_k + \mathbf{w} \right\|^2\right] \\
&= \arg\min_{\boldsymbol{\beta}'} N\left(P_s \|\mathbf{h} - \mathbf{A}'\boldsymbol{\beta}' - \mathbf{a}\|^2 + P_w \right).
\end{aligned}
\tag{9.56}
$$

Notice that the mean square minimization argument is equal to the equivalent noise of the joint H-SCFD where we enforced $\alpha = 1$. We can easily modify the results obtained there and we get

$$
\hat{\boldsymbol{\beta}}' = (\mathbf{A}'^{\mathrm{H}}\mathbf{A}')^{-1}\mathbf{A}'^{\mathrm{H}}(\mathbf{h} - \mathbf{a}).
\tag{9.57}
$$

After the H-Ifc processing we will have

$$
\mathbf{x}' = \sum_{k=1}^{K} h'_k \mathbf{s}_k + \mathbf{w}
\tag{9.58}
$$

where the effective equalized channel $\mathbf{h}' = [h'_1, \ldots, h'_K]^{\mathrm{T}}$ is

$$
\mathbf{h}' = \mathbf{h} - \mathbf{A}'\hat{\boldsymbol{\beta}}' = \mathbf{h} - \mathbf{P}_{\mathbf{A}'}(\mathbf{h} - \mathbf{a}).
\tag{9.59}
$$

CF Decoding on Effective Equalized Channel

After the H-Ifc equalization/cancellation we can employ a standard CF decoding. The MMSE optimized lattice inflation coefficient α' will be

$$
\hat{\alpha}' = \frac{P_s \mathbf{h}'^{\mathrm{H}}\mathbf{a}}{P_s \|\mathbf{h}'\|^2 + P_w}
\tag{9.60}
$$

and the computation rate will be

$$
R_c(\hat{\alpha}', \hat{\boldsymbol{\beta}}') = \lg^+ \frac{P_s \|\mathbf{h}'\|^2 + P_w}{P_s\left(\|\mathbf{h}'\|^2 \|\mathbf{a}\|^2 - \left|\mathbf{h}'^{\mathrm{H}}\mathbf{a}\right|^2 \right) + P_w \|\mathbf{a}\|^2}.
\tag{9.61}
$$

Target Map Orthogonal to Auxiliary Maps

It is interesting to find out the conditions under which the decoupled two-step solution produces the same result as the joint H-SCFD. To answer that, we manipulate the effective channel into

$$
\mathbf{h}' = (\mathbf{I} - \mathbf{P}_{\mathbf{A}'})\mathbf{h} + \mathbf{P}_{\mathbf{A}'}\mathbf{a} = \mathbf{h}_{\mathrm{eq}} + \mathbf{P}_{\mathbf{A}'}\mathbf{a}
\tag{9.62}
$$

where $\mathbf{h}_{\mathrm{eq}}$ is the equivalent channel of the joint H-SCFD. Clearly, if the target map is *orthogonal* to all auxiliary maps $\mathbf{a} \perp \mathbf{a}'_\ell$, for all ℓ, then the projection is zero $\mathbf{P}_{\mathbf{A}'}\mathbf{a} = \mathbf{0}$ and $\mathbf{h}' = \mathbf{h}_{\mathrm{eq}}$. Under the same condition, it also holds that $\mathbf{a}_{\mathrm{eq}} = (\mathbf{I} - \mathbf{P}_{\mathbf{A}'})\mathbf{a} = \mathbf{a}$. For the target map being orthogonal to the auxiliary map space, the two methods are equivalent and have the same computation rate.

9.4 Joint Hierarchical Interference Cancellation for Isomorphic Layered NCM

The joint H-Ifc cancellation can be applied also on the *isomorphic layered* NCM. Let us assume that the NCM is isomorphic and layered w.r.t. *two distinct HNC map* pairs $b = \chi(\tilde{b})$, $c = \chi_c(\tilde{c})$ and $\bar{b} = \bar{\chi}(\tilde{b})$, $\bar{c} = \bar{\chi}_c(\tilde{c})$. We name them the *target* and the *auxiliary* map, respectively. The corresponding H-codebooks are $c^N = \mathcal{C}(b)$, $\bar{c}^N = \bar{\mathcal{C}}(\bar{b})$, and the component product codebook is $\tilde{c}^N = \tilde{\mathcal{C}}(\tilde{b})$. The layered assumption allows us to work with discrete H-codewords that form the H-constellation. We also assume that the component channel constellation symbols are symbol-wise mapped on component codesymbols.

The core idea is to decode first the auxiliary H-codeword $\bar{c}^N$ and subsequently to use this auxiliary code structure knowledge to help the decoding of the target H-codeword c^N. If the NCM is regular w.r.t. auxiliary H-code, then any auxiliary H-code with H-rate $\bar{R} < I(\bar{C}; X)$ can be reliably decoded. Since the auxiliary H-codeword is a many-to-one function of the component codewords $\tilde{c}$, we cannot fully resolve all its components. But, we can reduce the subspace of all $\{\tilde{c}\}$ into the one that is consistent with auxiliary decoded H-codeword, i.e. $\{\tilde{c} : \bar{c}\}$. This can help in reducing the candidate decoding components of the target H-codeword $\{\tilde{c} : c\}$ because they *must also* be consistent with the auxiliary H-codeword $\{\tilde{c} : \bar{c}\}$. The reduced candidate codespace is thus $\{\tilde{c} : (c, \bar{c})\}$ and the decoding thus becomes H-decoding with the H-Ifc cancellation. The achievable rate for this decoding strategy will, however, depend on the target H-code properties when conditioned by the knowledge of the auxiliary H-codeword. The achievability proof will require an extension of the regularity principle into the conditional regularity. We now provide a detailed treatment.

9.4.1 Equivalent Hierarchical Channel with Joint H-Ifc Cancellation

Assume that the NCM is layered isomorphic and regular w.r.t. the auxiliary HNC map. If the auxiliary H-rate is $\bar{R} < I(\bar{C}; X)$ we can use a standard H-decoding (see Section 5.7.4) and the auxiliary H-codeword can be reliably decoded. The decoding ignores any other HNC map, apart from the auxiliary map. The reliably decoded $\bar{c}^N$ is then used as side-information in the target H-decoder c^N.

Following a similar way as we used for the equivalent hierarchical channel in Section 5.7.4, we can form an *equivalent hierarchical channel with H-Ifc cancellation* (Figure 9.2). The H-Ifc cancellation can be also interpreted as the H-Ifc *constraint* since we rather constrain the set of candidate component sets $\tilde{c}$. The equivalent model still contains the H-const self-dispersion, as a consequence of a many-to-one HNC target map, but on top of it, there is additional constraint by the H-Ifc side-information. The self-dispersion part of the equivalent model is constrained by the auxiliary H-codeword knowledge. This reduces the residual self-dispersion randomness and thus we expect an improvement in the target H-rate.

9.4.2 Achievable H-rate with H-Ifc Cancellation

The equivalent H-Ifc cancellation channel model is now used to find the achievable H-rate for the target H-code. The core idea is to utilize a similarity with the standard

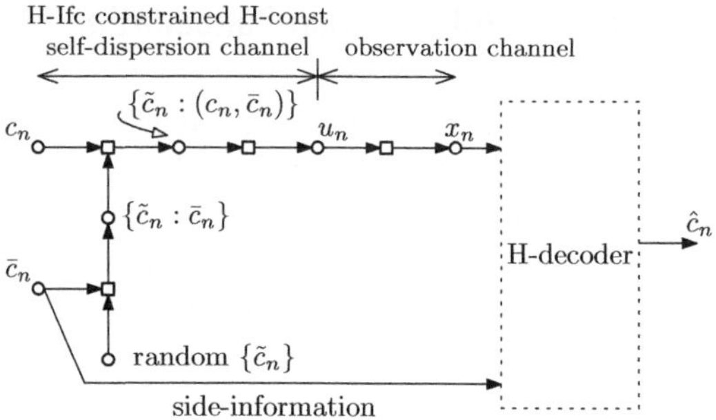

Figure 9.2 Equivalent hierarchical channel with H-Ifc cancellation.

H-decoding (Section 5.7.4) with one important difference. The difference is that instead of an independent random $\{\tilde{c}\}$ causing the self-dispersion (see Figure 5.12), the randomness source is now $\{\tilde{c} : \bar{c}\}$ and depends (it is conditioned by) on the auxiliary H-codeword that is available as side-information to the decoder.

DEFINITION 9.1 (Conditionally Regular H-Codebook)   Assume the NCM that is isomorphic and layered for two distinct codebooks, the target and the auxiliary H-codebooks $c^N = \mathcal{C}(b)$ and $\bar{c}^N = \bar{\mathcal{C}}(\bar{b})$. The target codebook $\mathcal{C}$ is called *conditionally regular* w.r.t. the auxiliary codebook $\bar{\mathcal{C}}$ if all target H-codewords are conditionally independent

$$\forall b' \neq b, \bar{b} : \left(C^N(b') \perp\!\!\!\perp C^N(b)\right) | \bar{C}^N(\bar{b}). \tag{9.63}$$

The definition means that the target H-codebook behaves as if the codewords were drawn randomly according to some distribution even when the auxiliary H-codeword is fixed and known. In other words, the auxiliary H-codeword structure must not impose the structure constraint on the target H-codebook.

THEOREM 9.2 (Achievable H-Rate with H-Ifc Cancellation)   *Assume the H-MAC with K components and the layered NCM isomorphic w.r.t. the target and the auxiliary HNC map. Component codes are $c_k^N = \mathcal{C}_k(b_k)$ with one-to-one symbol-wise channel symbol mappers $s_{k,n} = \mathcal{A}_k(c_{k,n})$ for all $k \in [1 : K]$, $n \in [1 : N]$. Component codebooks are randomly generated according to $S_k^N \sim \prod_{n=1}^N p(s_{k,n})$ for component messages $b_k \in [1 : 2^{NR_k}]$. The* target *isomorphic H-code $c^N = \mathcal{C}(b)$ is defined by the H-message map $b = \chi(\tilde{b})$, $b \in [1 : 2^{NR}]$, and the symbol-wise H-code symbol HNC map $c_n = \chi_c(\tilde{c}_n)$. The* auxiliary *isomorphic H-code is defined by $\bar{c}^N = \bar{\mathcal{C}}(\bar{b})$, $\bar{b} = \bar{\chi}(\tilde{b})$, $\bar{c}_n = \bar{\chi}_c(\tilde{c}_n)$.*

If the target isomorphic H-codebook is conditionally regular w.r.t. the auxiliary codebook and the auxiliary H-codeword $\bar{c}^N$ is available at the receiver then the target H-rate R is achievable for any $R < I(C; X|\bar{C})$. Also it holds that $I(C; X|\bar{C}) \leq I(C; U|\bar{C})$ and $I(C; X|\bar{C}) \leq I(U; X|\bar{C})$.

Proof   The proof follows similar lines to the proof of Theorem 5.10, so we will proceed in a more compact form.

The decoder is a joint typicality decoder using the available side-information – the auxiliary H-message $\bar{b}$ and the corresponding H-codeword $\bar{C}^N(\bar{b})$. The auxiliary H-message and the corresponding codeword $\bar{C}^N(\bar{b})$ are assumed to be known erroneously (they are equal to the transmitted ones). The decoder declares the decoded target H-message $\hat{b}$ if

$$\left(C^N(\hat{b}), \bar{C}^N(\bar{b}), X^N \right) \in \mathcal{T} \tag{9.64}$$

at the first occurrence of suitable $\hat{b}$, ignoring the others. If there is no such message, it declares $\hat{b} = \emptyset$. We assume a random generation of the component codebooks and thus we can fix the true transmitted message b and evaluate the probability of the error event over random codebooks.

The joint distribution of all involved variables that will be used later in the evaluation of the error event probability is

$$\left(C^N(b'), \bar{C}^N(\bar{b}), X^N, C^N(b) \right) \sim \prod_{n=1}^{N} p\left(x_n | c_n(b), \bar{c}_n(\bar{b}) \right) p\left(c_n(b), c_n(b'), \bar{c}_n(\bar{b}) \right) \tag{9.65}$$

where $b, \bar{b}$ are true transmitted H-messages and b' is a candidate message of the target H-decoder. Under the conditional regularity assumption, it holds that

$$p\left(c_n(b), c_n(b'), \bar{c}_n(\bar{b}) \right) = \begin{cases} p\left(c_n(b) | \bar{c}_n(\bar{b}) \right) \delta \left(c_n(b) - c_n(b') \right) p\left(\bar{c}_n(\bar{b}) \right); & b' = b \\ p\left(c_n(b) | \bar{c}_n(\bar{b}) \right) p\left(c_n(b') | \bar{c}_n(\bar{b}) \right) p\left(\bar{c}_n(\bar{b}) \right); & b' \neq b \end{cases} \tag{9.66}$$

for any $\bar{b}$.

The distribution of the variables involved in the joint typicality test is obtained by a marginalization

$$\left(C^N(b'), \bar{C}^N(\bar{b}), X^N \right) \sim$$

$$\sim \begin{cases} \prod_{n=1}^{N} p\left(x_n | c_n(b'), \bar{c}_n(\bar{b}) \right) p\left(c_n(b') | \bar{c}_n(\bar{b}) \right) p\left(\bar{c}_n(\bar{b}) \right); & b' = b \\ \prod_{n=1}^{N} p\left(x_n | \bar{c}_n(\bar{b}) \right) p\left(c_n(b') | \bar{c}_n(\bar{b}) \right) p\left(\bar{c}_n(\bar{b}) \right); & b' \neq b \end{cases}. \tag{9.67}$$

The test candidate codeword $C^N(b')$ for $b' \neq b$ is *conditionally independent* with X^N when the test is conditioned by the knowledge of $\bar{C}^N(\bar{b})$.

The decoding failure event Ω_{e0} means that the decoder does not recognize $(C^N(\hat{b}), \bar{C}^N(\bar{b}), X^N)$ as a typical set member even if the test message is correct $\hat{b} = b$. If $b' = b$ then random variables in (9.67) are drawn from a joint PDF and by the properties of typical set $P_{e0} = \Pr\{\Omega_{e0}\} \approx 0$ for $N \to \infty$.

The erroneous decision event is

$$\Omega_{e1}(b') = \left\{ \left(C^N(b'), \bar{C}^N(\bar{b}), X^N \right) \in \mathcal{T}, b' \neq b \right\} \tag{9.68}$$

for some given $\bar{b}$. For $b' \neq b$ it holds that $\left(C^N(b') \perp\!\!\!\perp X^N \right) | \bar{C}^N(\bar{b})$ and we can use the joint typicality lemma (A.108), and the probability of error event is

$$\Pr\{\Omega_{e1}(b')\} \approx 2^{-NI(C;X|\bar{C})}. \tag{9.69}$$

There are all together $(2^{NR} - 1)$ possible erroneous b' messages and the union bound gives

$$P_{e1} = \Pr\left\{\cup_{b' \neq b}\Omega_{e1}(b')\right\}$$
$$\lesssim 2^{NR}2^{-NI(C;X|\bar{C})}. \tag{9.70}$$

If $R < I(C;X|\bar{C})$ then $P_{e1} \approx 0$ for $N \to \infty$.

It clearly holds that $(C, \bar{C}) \mapsto U \mapsto X$. As a consequence the variables also form a conditional Markov chain $(C \mapsto U \mapsto X)|\bar{C}$, i.e.

$$p(x, u, c|\bar{c}) = p(x|u, \bar{c})p(u|c, \bar{c})p(c|\bar{c}). \tag{9.71}$$

All entropy and mutual information chain rules, when additionally conditioned by $\bar{C}$, then hold the same way as for the unconditioned case. Thus also the conditioned processing inequality gives

$$I(C;X|\bar{C}) \leq I(C;U|\bar{C}) \tag{9.72}$$

and

$$I(C;X|\bar{C}) \leq I(U;X|\bar{C}). \tag{9.73}$$

$\square$

REMARK 9.1 (Conditional Regularity)   As in Theorem 5.10, the conditional regularity is *critical* and it is not automatically fulfilled.

REMARK 9.2 (Conditionally Self-Folded NCM)   Conditional processing mutual information inequalities $I(C;X|\bar{C}) \leq I(C;U|\bar{C})$ and $I(C;X|\bar{C}) \leq I(U;X|\bar{C})$ can, as for the single-map NCM, be used to find the processing bottleneck. If the constraint by $\bar{C}$ reduces the conditional self-dispersion $\{\tilde{c} : (c, \bar{c})\}$ to zero, then U becomes *conditionally self-folded* and $I(C;U|\bar{C})$ is maximized.

9.4.3   Conditional Regularity for Linear GF HNC Maps

The following theorem shows the conditions under which a linear GF-based pair of target and auxiliary HNC maps fulfill the conditional regularity condition.

THEOREM 9.3 (Conditional Regularity for Linear GF HNC Maps)   *Assume layered NCM isomorphic w.r.t. the target and the auxiliary maps and components codebook symbols* $c_{k,n}$, $k \in [1:K]$, *from a common finite alphabet* $c_{k,n} \in \mathcal{A}_c$ *of the size* $|\mathcal{A}_c| = M_c$ *using symbol-wise constellation space mappers. All component codebooks are randomly drawn IID sequences* $C_k^N \sim \prod_{n=1}^{N} p_C(c_{k,n})$ *with uniform PMF* $p_{C_k}(c_{k,n}) = 1/M_c$. *The HNC code symbol-wise target map* $c_n = \chi_c(\tilde{c}_n)$ *and auxiliary map* $\bar{c}_n = \bar{\chi}_c(\tilde{c}_n)$ *are linear maps on GF* $\mathbb{F}_{M_c} = (\mathcal{A}_c, +, \times)$

$$c_n = \sum_{k\in[1:K]} a_k c_{k,n}, \tag{9.74}$$

$$\bar{c}_n = \sum_{k\in[1:K]} \bar{a}_k c_{k,n} \tag{9.75}$$

where $a_k, \bar{a}_k \in \mathbb{F}_{M_c}$.

If HNC map coefficient vectors $\mathbf{a} = [a_1, \ldots, a_K]^{\mathrm{T}}$ *and* $\bar{\mathbf{a}} = [\bar{a}_1, \ldots, \bar{a}_K]^{\mathrm{T}}$ *are* linearly independent *then the H-codebook is* conditionally regular.

Proof   The core idea of the proof stands on the fact that the auxiliary HNC map constraints the degrees of freedom in the many-to-one target map, but there still must remain the randomness caused by component codesymbols. Effectively it creates a new map inherently capturing the conditioning inside. This will be again linear (more strictly affine) and we can use Theorem 5.11.

Assume, without the loss of generality (we can always reindex the components), that $\bar{a}_1 \neq 0$. The auxiliary map condition implies that

$$c_{1,n} = \frac{1}{\bar{a}_1} \left(\bar{c}_n - \sum_{k \neq 1} \bar{a}_k c_{k,n} \right). \tag{9.76}$$

We substitute that into the target map

$$c_n = \frac{a_1}{\bar{a}_1} \bar{c}_n + \sum_{k \neq 1} a_k' c_{k,n} \tag{9.77}$$

where

$$a_k' = a_k - \frac{a_1}{\bar{a}_1} \bar{a}_k, \quad k \neq 1. \tag{9.78}$$

We need a_k' only for $k \neq 1$. The value $a_1' = 0$ simply means that $c_{1,n}$ was completely eliminated as a degree of freedom by the auxiliary map constraint substitution. We can generally say $\mathbf{a}' = \mathbf{a} - \alpha\bar{\mathbf{a}}$, where $\alpha = a_1/\bar{a}_1$. The vector $\mathbf{a}' = [a_1', \ldots, a_K']^{\mathrm{T}}$ together with $c_0 = \alpha\bar{c}_n$, which is a constant under the conditioning by the target map, will define a new effective linear (affine) map *inherently containing* the auxiliary conditioning inside. The effective map describes the remaining degrees of freedom in the self-dispersion after the auxiliary map conditioning.

If $\mathbf{a}$ and $\bar{\mathbf{a}}$ are linearly dependent then $\mathbf{a}' = \mathbf{0}$ and the effective map does not have any degrees of freedom. If they are independent then at least one coefficient $a_k' \neq 0$ and the map preserves a randomness from that given component codesymbol. If the component codesymbols are IID with a discrete uniform PMF, then we can use Theorem 5.11. The additional constant c_0 does not influence the result. $\square$

REMARK 9.3   Notice that both the target and auxiliary maps must be over the same GF $\mathbb{F}_{M_c}$. The proof depends on it, namely in calculating the effective coefficients $\mathbf{a}'$.

10 WPNC in Complex Stochastic Networks

10.1 Principles of Wireless Cloud Coding

The term Wireless Cloud Coding (WCC) will denote the coding concept suited for large and complex networks with a stochastic structure and connectivity. A simple low-scale WPNC network with a known deterministic structure can be designed in a "hand-crafted" style. We usually also easily identify the critical points of performance or design aspects that need to be optimized. On the other hand, a complex large stochastic network adds some specific problems.

Stochastic Network Structure

The network structure and connectivity may be stochastic and unknown. A standard NCM design needs to know the participating signals forming the H-constellation in the H-MAC stage, and also the full end-to-end hierarchical encapsulation of all involved HNC maps to guarantee the end-to-end solvability. In a large randomly changing network, this creates a significant problem for (a) *identifying* and *estimating* the system/channel state, (b) making a *decision on the processing form*, and (c) *signaling* this decision to all involved nodes. Instead of using this traditional solution, we can try to devise the NCM that is *blind* to the network structure that will not require a separate NCM/WPNC control protocol but will treat the unknown network structure as an additional degree of freedom *fully integrated with the actual payload data* (see Section 10.2).

Clustered, Nested, and Modular Cloud Framework

The number of the design degrees of freedom may be prohibiting in a large network. A nested modular encapsulation of some of its fragments/clusters can help. This has two aspects. (a) It simplifies the overall design by *clustering* the nodes and *nesting* the clusters. (b) Each cluster performs some processing as a "black-box" *self-contained virtual super-node*. It does not reveal its internal structure to the outside world and simply provides some functionality as a super-node. Its global level of clustering can even comprise the whole cloud. It can serve its outside terminals as if it was some super-relay performing some complex internal hidden processing (see Section 10.3).

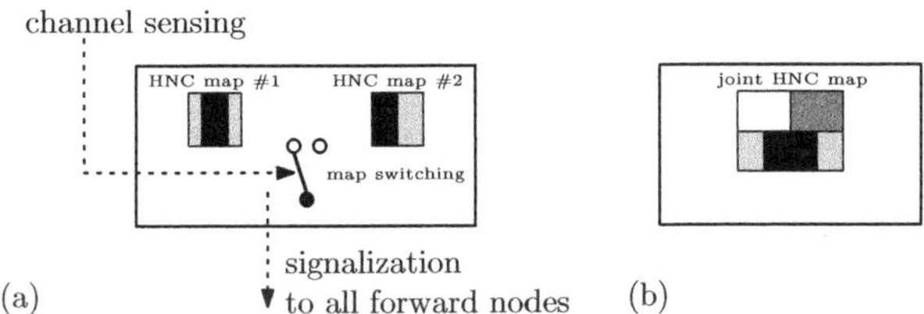

Figure 10.1 (a) HNC map switching for a classical signaling solution. (b) Joint HNC map for payload data and channel class.

10.2 Wireless Cloud-Coding-Based Design of NCM

10.2.1 Random Channel Class H-MAC and Joint HNC Map

Joint HNC map vs. Classical Signaling

A classical signaling solution (Figure 10.1a) is based on the channel sensing and then *switching* the NCM map at the node depending on the channel class (connection/disconnection). This sensing and switching is relatively easily solved for links ending at the node where the affected map needs to be switched. It can be combined with the channel estimation using a suitable preamble, which would be required anyway. Much more important and harder to avoid is the fact that this map must be made available to all succeeding nodes and stages of a large multi-hop network. This creates at least a *latency*. Because the signaling information must be forwarded reliably it must be coded by long codewords (even if it is itself very short). This cannot be done at the channel symbol level. Scenarios with multiple hops, connectivity loops, and short packet services with a rapidly changing channel state/structure are particularly vulnerable to this. In some cases, reliable signaling is not possible at all.

Our approach with the *joint* HNC map (Figure 10.1b) inherently containing the channel class (details are given in the next paragraphs) does not need explicit signaling. Of course, it is at the expense of a larger output cardinality. But as the performance bottleneck is usually the H-MAC stage, it may not restrict the overall capacity. Moreover, the channel class is contained in each payload channel symbol. It thus dissolves the small cardinality channel class in a much larger cardinality of the payload data codebook and the overall rate increase is negligible.

System Model for Random Channel Class H-MAC

Assume (as in Section 5.7.4) an H-MAC channel with K components. The kth component ($k \in [1 : K]$) transmits the constellation space coded NCM signal s_k^N, which is symbol-wise one-to-one mapped $s_{k,n} = \mathcal{A}_k(c_{k,n})$ to the codewords c_k^N. The component messages b_k are encoded by codebooks $\mathcal{C}_k$, $c_k^N = \mathcal{C}_k(b_k)$. We also define $\tilde{s}^N = \{s_1^N, \ldots, s_K^N\}$ and similarly for $\tilde{c}^N$ and $\tilde{b}$.

The channel-combined signal depends not only on the continuous channel state (inherently modeled inside the $\tilde{h}$ function) but also on a *discrete-valued random* channel class $a \in \mathcal{A}_a$, $|\mathcal{A}_a| = M_a$

$$u_n = \tilde{h}\left(\tilde{s}_n(\tilde{c}_n), a\right). \tag{10.1}$$

We have intentionally separated the continuous and discrete (the channel class) parametrization. A comparison of this model with the one used in Section 5.7.4 reveals that we have added the discrete channel class that can *switch* the form of how the component signals are combined. The channel class itself can describe a variety of channel characteristics, e.g. structure/connectivity changes, high/low SNR regime, switching between various stochastic characteristics of the fading, etc. However, the most important and simple interpretation for the channel class is a binary modeling of the connectivity (link present or absent). The connectivity can be lost as a result of the fading. The separation of this discrete style of the parametrization out of the continuous one, which is kept as usual, allows an elegant modeling of the network structure and also will play an important role in incorporating it into the global WPNC description.

Example 10.1 The simplest possible application uses the channel class for a switching on–off of the individual components, which in turn models the possible changes of the topology. For example, the two-component linear channel with a random connectivity of one component can be modeled as

$$u_n = h_A s_{A,n} + h_B s_{B,n}; \text{ for } a = 1, \tag{10.2}$$

$$u_n = h_A s_{A,n}; \text{ for } a = 0. \tag{10.3}$$

We assume that the components have a common channel constellation alphabet. In this example, the distinct channel classes imply distinct H-alphabets.

Example 10.2 A slightly more complicated example shows two configurations of a three-component H-MAC where only two of the components actively participate in the transmission. For a linear observation, the channel-combined signal is

$$u_n = h_A s_{A,n} + h_B s_{B,n}; \text{ for } a = 1, \tag{10.4}$$

$$u_n = h_B s_{B,n} + h_C s_{C,n}; \text{ for } a = 2. \tag{10.5}$$

Assuming again a common component channel alphabet, this example shows that the observed H-alphabet does not need to change with the channel class.

The overall observation model is a memoryless two-stage Markov chain

$$(\tilde{C}_n, A) \mapsto U_n \mapsto X_n \tag{10.6}$$

where the first stage is a random channel class enabled H-constellation dispersion channel and the second stage is the observation channel $x_n = x_n(u_n)$. We can amend the system model and the equivalent model used in Section 5.7.4 and Figure 5.12 to obtain the *joint equivalent model* for a channel class enabled H-MAC channel (Figure 10.2). Clearly, the only modification is in allowing the channel class a to modify the H-alphabet and H-constellation.

The channel class is generally an unknown random variable with some PMF $p(a)$. Although it generally does not need to be, we will assume a special case here – a *block-constant* channel class. The channel class is a constant for the whole duration of the $\tilde{c}^N$ codeword and it randomly changes in between the codewords.

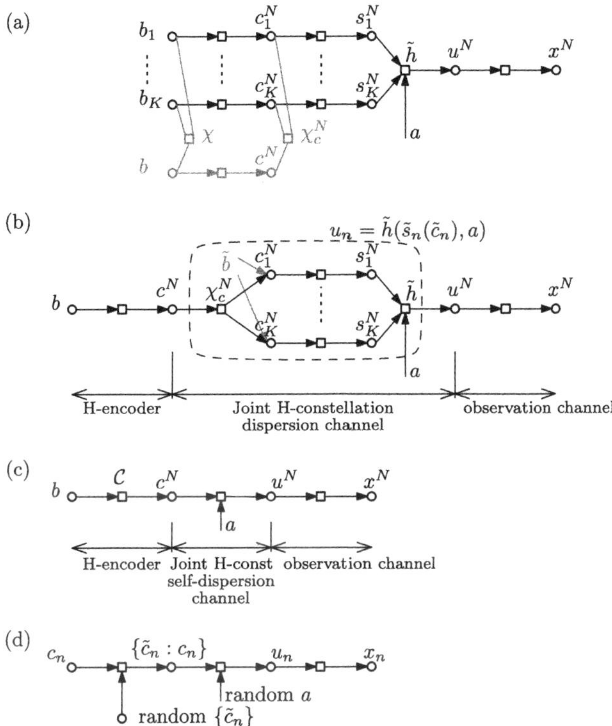

Figure 10.2 Joint equivalent model for channel class enabled H-MAC channel.

Joint HNC Map and WCC NCM

The H-alphabet and the H-constellation are additionally affected by the discrete random variable a – the channel class. As such, it creates additional discrete-valued degrees of freedom in the received signal that are similar to the way in which the discrete code symbols $\tilde{c}_n$ affect the received signal. This is a fundamental observation that motivates the *joint data and channel class* NHC map design. It allows us to view the situation as if the H-codebook size was enlarged to accommodate the channel class together with the payload data.

DEFINITION 10.1 (Joint HNC Map and Joint H-Constellation) The symbol-wise joint HNC map for the channel class a and channel code symbols $\tilde{c}_n$ is defined as

$$\bar{c}_n = \bar{\chi}_c(\tilde{c}_n, a). \tag{10.7}$$

The joint H-constellation is defined as

$$\mathcal{U}(\bar{c}_n) = \left\{ u_n : u_n = \tilde{h}\left(\tilde{s}_n(\tilde{c}_n), a\right) | \bar{c}_n = \bar{\chi}_c(\tilde{c}_n, a) \right\} \tag{10.8}$$

and the joint H-alphabet is the set $\bar{\mathcal{A}}_u$, $u_n \in \bar{\mathcal{A}}_u$.

The definition can be easily extended for any other entities, such as data symbols or data/code vectors. Notice also that the channel class plays an equal role as coded symbols, as opposed to the continuous valued channel state that is on par with the constellation space description.

The coding construction strategy using this joint HNC map will be called *Wireless Cloud Coding (WCC) NCM*. It jointly solves the problem of the payload data coding over the network with the random structure/properties described by the channel class. The channel class affects the observation similarly as the payload data and we will incorporate both and treat them as discrete joint information carrying data. The information is jointly the payload data and the "network structure" data described by the channel class. At the expense of slight cardinality increase of the map, it allows us to avoid the explicit and latency increasing signaling of the maps used in all involved nodes.

WCC NCM in the Context of Multi-Map and H-Ifc NCM

As we saw, the WCC NCM treats the HNC map in a new form including some additional input. It is worth putting it into the context with other forms of NCM with more complex HNC map handling. The structured NCM for multi-map H-MAC (Chapter 8, Figure 10.3a) defines multiple HNC maps over the *independent* data payloads (e.g. obtained by splitting the node data stream) and processes these maps in parallel. The main purpose was to allow some asymmetry in the rates of participating node codebooks dictated, e.g. by asymmetric channels.

The NCM with joint H-Ifc processing (Chapter 9, Figure 10.3b) used multiple HNC maps on *common* payload component data. These maps can support each other in reducing the hierarchical interference and thus can provide higher rates or can better match the NCM for a given channel parametrization.

The WCC NCM treats the third form of more complex HNC map handling (Figure 10.3c) where the amended map is used to solve the unknown network structure (named "channel class"). The joint map extension is thus done not by adding component payload data but by adding the channel class of the given H-MAC.

10.2.2 Coding Theorems for WCC NCM

WCC NCM Codebook Construction Principle

The channel class discrete random variable plays a similar role to that played by the payload data. Effectively, the channel class adds additional information about the H-MAC structure to the payload data. Both the channel class and the payload data are random

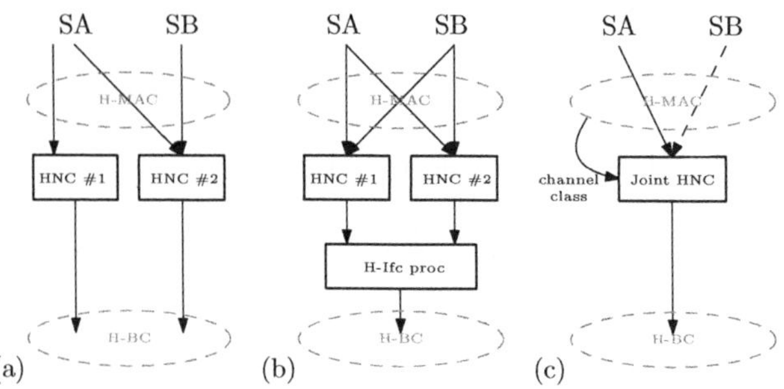

Figure 10.3 WCC NCM in the context of multi-map and H-Ifc NCM.

and unknown on the receiver side and they both influence the observation signal. From the point of view of the received signal observation, this appears as a joint codebook composed from the payload-only part and the channel class part. The channel class is assumed to be constant per payload codeword frame. As a consequence, the cardinality of the channel class set is much lower than the cardinality of the payload codebook size. The WCC concept does not impose any special requirements on the payload data codebook structure. It is applicable to any form of single-user, classical multi-user, or NCM hierarchical data system. In the context of this text, we, however, concentrate on the *hierarchical* case.

A payload-only isomorphic layered NCM design (Section 4.2.4) requires that the hierarchical data $b = \chi(\tilde{b})$ map is a one-to-one map to the hierarchical codeword $c^N = C(b)$, which, in turn, is observed through the equivalent hierarchical channel (Figure 5.12) with the self-dispersion part $u_n = u(c_n)$. In order to prove the achievability of the hierarchical rate, we showed (Section 5.7.4) that the H-codebook must be *regular*, i.e. it must behave as a standard random codebook having independent codewords $C^N(b') \perp\!\!\!\perp C^N(b)$, $b' \neq b$. This principle will be now extended to the WCC NCM with the joint HNC map.

DEFINITION 10.2 (Jointly Isomorphic Layered NCM)   The layered NCM consisting of component codes C_k, $c_k^N = C_k(b_k)$, and the joint HNC data $\bar{b} = \bar{\chi}(\tilde{b}, a)$ and code symbol-wise maps $\bar{c}_n = \bar{\chi}_c(\tilde{c}_n, a)$, where a is the channel class, is called a *jointly isomorphic* layered NCM if there exists a valid one-to-one equivalent *jointly isomorphic* hierarchical codebook $\bar{C}$, such that $\bar{c}^N = \bar{C}(\bar{b})$, i.e.

$$\forall \tilde{b}, a : \bar{c}^N = \bar{\chi}_c^N\left(\tilde{C}(\tilde{b}), a\right) = \bar{C}\left(\bar{\chi}(\tilde{b}, a)\right). \tag{10.9}$$

The jointly isomorphic NCM describes the situation where the channel class a is *decodable* as a part of *joint* HNC map $\bar{b}$ *without* being itself encoded by any component codebook. The channel class only influences the joint H-codeword by affecting the way in which it is observed in the self-dispersion equivalent model. If the alternations of how the payload-only H-codewords appear produce *distinct* joint H-codewords for distinct joint H-data (payload and channel class), the channel class will be *decodable* through $\bar{b}$. Clearly, the fact that the channel class is not itself encoded, but needs to appear inside the channel-combined symbols u_n as if it was encoded, narrows significantly the possible solutions.

Joint and Conditional Regularity

The jointly isomorphic H-codebook must form a decodable codebook. In the achievability theorems relying on a random codebook construction with a joint typicality decoding, it means that all codewords must be symbol-wise IID and mutually independent. This is now formalized by defining joint and conditional regularity.

DEFINITION 10.3 (Joint and Conditional H-codebook Regularity)   The joint isomorphic hierarchical codebook $\bar{c}^N = \bar{C}(\bar{b})$ is called *jointly regular* if

$$\forall \bar{b}' \neq \bar{b} : \bar{C}^N(\bar{b}') \perp\!\!\!\perp \bar{C}^N(\bar{b}) \tag{10.10}$$

where $\bar{b}, \bar{b}'$ are particular distinct joint H-messages $\bar{b} = \bar{\chi}(\tilde{b}, a)$.

The joint isomorphic hierarchical codebook $\bar{c}^N = \bar{\mathcal{C}}(\bar{b})$ is called *conditionally regular* if

$$\forall \bar{b}' \neq \bar{b} : \ \left(\bar{C}^N(\bar{b}') \perp\!\!\!\perp \bar{C}^N(\bar{b})\right)\,|a \tag{10.11}$$

for some fixed a.

REMARK 10.1 (Factorized Joint HNC Map)   The simplest and the most directly interpretable form is the factorized form of joint maps $\bar{b} = \bar{\chi}(\tilde{b}, a)$, $\bar{c}_n = \bar{\chi}_c(\tilde{c}_n, a)$. It separates the standard hierarchical payload-only map and the channel class

$$\bar{b} = (b, a), \ \bar{c} = (c, a), \ \ b = \chi(\tilde{b}), \ c = \chi_c(\tilde{c}). \tag{10.12}$$

The joint regularity condition is then

$$\forall b' \neq b \vee a' \neq a : \ \bar{C}^N(b', a') \perp\!\!\!\perp \bar{C}^N(b, a). \tag{10.13}$$

Notice the "or" in the condition. The codewords must be independent even for $b' = b$ and $a' \neq a$.

REMARK 10.2 (Conditional Regularity)   The conditionally regularity condition means that the H-code is regular w.r.t. payload, in a standard sense of Section 5.7.4, for *any particular* channel class a. This allows us to treat the conditional joint H-decoding as usual for a given known channel class. The channel class, however, becomes a part of the map and Theorem 5.10 holds with a minor formal change of including the known channel class a in the HNC map.

REMARK 10.3 (No Natural Random Coding for the Channel Class)   Generally, the problem of having independent H-codewords for joint HNC map $\bar{b} = \bar{\chi}(\tilde{b}, a)$ and $\bar{b}' = \bar{\chi}(\tilde{b}, a')$ that differ *only* by the channel class $a' \neq a$ is the missing underlying random codebook construction, which is present in the component codes. The channel class influences directly the observation. It will have to be imposed by some "scrambling" randomization of other components in the H-MAC.

Example 10.3   Assume the model from Example 10.1 where $c_{A,n}, c_{B,n} \in \mathbb{F}_2$ with BPSK constellations $s_{A,n}, s_{B,n} \in \{\pm 1\}$, $s_{k,n}(0) = -1$, $s_{k,n}(1) = +1$, and unity channels $h_A = h_B = 1$. The joint HNC will be the factorized one, $\bar{c}_n = (c_n, a)$, where $c_n = c_{A,n} + c_{B,n}$ on $\mathbb{F}_2$. The data and the channel class are assumed to have uniform PMFs. The map table and the observable channel combined symbols are as follows.

Clearly, it holds for PMF $p(\bar{c}_n) \neq p(\bar{c}_n|a)$ and thus the codewords are not independent w.r.t. the channel class a. In the factorized map, where a directly participates in $\bar{\chi}_c$, the randomizing effect of the random component codebook construction is missing. However, for a fixed channel class they are independent (see Theorem 5.11). The joint HNC map is thus not jointly regular; however, it is conditionally regular. We also see

$c_{A,n}$	$c_{B,n}$	a	c_n	$\bar{c}_n$	u_n
0	0	0	0	(0,0)	-1
0	1	0	1	(1,0)	-1
1	0	0	1	(1,0)	$+1$
1	1	0	0	(0,0)	$+1$
0	0	1	0	(0,1)	-2
0	1	1	1	(1,1)	0
1	0	1	1	(1,1)	0
1	1	1	0	(0,1)	$+2$

that the H-alphabet changes with the channel class. Notice that the linear factorized HNC map exhibits unresolved singular states, particularly for $u_n \in \{-1, +1\}$.

Example 10.4　A modification of the previous example that avoids unresolved singular fading must use a nonlinear $\bar{c}$ map. Clearly, the first component of $\bar{c}_{n,1}$ is $\bar{c}_{n,1} = c_{A,n} + ac_{B,n}$ and the map is thus quadratic in the standard $\mathbb{F}_2$ arithmetic with symbols "0" and "1" having a standard meaning.

$c_{A,n}$	$c_{B,n}$	a	c_n	$\bar{c}_n$	u_n
0	0	0	0	(0,0)	-1
0	1	0	1	(0,0)	-1
1	0	0	1	(1,0)	$+1$
1	1	0	0	(1,0)	$+1$
0	0	1	0	(0,1)	-2
0	1	1	1	(1,1)	0
1	0	1	1	(1,1)	0
1	1	1	0	(0,1)	$+2$

Achievable Joint H-Rate

The achievable joint H-rate can be evaluated by using the random channel coding, the joint typicality decoding, and the regularity conditions. The joint H-rate defines the size of product H-codebook involving both the hierarchical payload data and the channel class.

THEOREM 10.4 (Achievable Joint H-Rate)　*Assume H-MAC with random channel class $a \in \mathcal{A}_a$, $|\mathcal{A}_a| = M_a$, and K components NCM defined by component codes $c_k^N = \mathcal{C}_k(b_k)$ with one-to-one symbol-wise channel symbol mappers $s_{k,n} = \mathcal{A}_k(c_{k,n})$ for all $k \in [1 : K]$, $n \in [1 : N]$. Component codebooks are randomly generated according to $S_k^N \sim \prod_{n=1}^{N} p(s_{k,n})$ for component messages $b_k \in [1 : 2^{NR_k}]$. Assume that there exists a jointly isomorphic hierarchical codebook $\bar{C}$, such that $\bar{c}^N = \bar{C}(\bar{b})$, where $\bar{b} = \bar{\chi}(\tilde{b}, a)$, $\bar{b} \in [1 : 2^{N\bar{R}}]$, and code symbol-wise maps are $\bar{c}_n = \bar{\chi}_c(\tilde{c}_n, a)$.*

If the jointly isomorphic H-codebook is jointly regular *with* IID symbols *then the joint hierarchical rate $\bar{R}$ is achievable for any $\bar{R} < I(\bar{C}; X)$.*

Proof   The observation in joint equivalent model with memoryless observation part is X^N; see Figure 10.2. The decoder is the joint typicality decoder. It declares joint H-message $\hat{\bar{b}}$ as decoded if

$$\left(\bar{C}^N(\hat{\bar{b}}), X^N \right) \in \mathcal{T} \tag{10.14}$$

at the first occurrence of $\hat{\bar{b}}$ ignoring the other ones. If no such message exists, it declares $\hat{\bar{b}} = \emptyset$.

The decoding error probability analysis follows similar lines as in the proof of Theorem 5.10, so will proceed in a compact form. However, there is one important difference. In Theorem 5.10, the IID property of the H-code symbols was automatically implied by symbol-wise IID component codebooks and symbol-wise mappers. However, here, because of the presence of the channel class constant over the N symbol long codeword frame, we must *explicitly* require the IID property for joint H-codebook. It will be fulfilled if the component codes perform an IID randomizing scrambling on the channel class and make it appear as if it was random IID symbol encoded.

Under the *symbol IID* assumption, we have

$$p(\bar{c}^N(\tilde{c}^N, a)) = \prod_{n=1}^{N} p(\bar{c}_n(\tilde{c}_n, a)). \tag{10.15}$$

We denote by $\bar{C}(\bar{b})$ the true transmitted joint H-codeword and by $\bar{C}(\bar{b}')$ the test codeword used by the decoder in the joint typicality decoding. The test codeword and the received observation with true transmitted codeword are drawn according to the joint density

$$\left(\bar{C}^N(\bar{b}'), X^N, \bar{C}^N(b) \right) \sim \prod_{n=1}^{N} p(x_n | \bar{c}_n(\bar{b})) p(\bar{c}_n(\bar{b}), \bar{c}_n(\bar{b}')). \tag{10.16}$$

Under the *regularity* of the joint H-codebook, it holds that

$$p(\bar{c}_n(\bar{b}), \bar{c}_n(\bar{b}')) = \begin{cases} p(\bar{c}_n(\bar{b}))\delta(\bar{c}_n(\bar{b}) - \bar{c}_n(\bar{b}')); & \bar{b}' = \bar{b} \\ p(\bar{c}_n(\bar{b}))p(\bar{c}_n(\bar{b}')); & \bar{b}' \neq \bar{b} \end{cases} \tag{10.17}$$

and, as a consequence, it is $\bar{C}^N(\bar{b}') \perp\!\!\!\perp X^N$ for $\bar{b}' \neq \bar{b}$.

The decoding failure event means that the decoder does not recognize the pair $(\bar{C}^N(\hat{\bar{b}}), X^N)$ as a typical set member even if the test message is the true transmitted one $\hat{\bar{b}} = \bar{b}$. Using similar arguments as in Theorem 5.10, it has the probability $P_{e0} \approx 0$ as $N \to \infty$.

The erroneous decision event is

$$\Omega_{e1}(\bar{b}') = \left\{ \left(\bar{C}^N(\bar{b}'), X^N \right) \in \mathcal{T}, \, \bar{b}' \neq \bar{b} \right\}. \tag{10.18}$$

Since $\bar{C}^N(\bar{b}')$ and $X^N(\bar{b})$ are *independent*, then we can use joint typicality lemma (A.112) and

$$\Pr\{\Omega_{e1}(\bar{b}')\} \approx 2^{-NI(\bar{C};X)}. \tag{10.19}$$

There are all together $(2^{N\bar{R}} - 1)$ messages $\bar{b}'$ and the probability of the union of events is upper bounded by

$$P_{e1} = \Pr\left\{\cup_{\bar{b}' \neq \bar{b}}\Omega_{e1}(\bar{b}')\right\}$$
$$\lesssim 2^{N\bar{R}}2^{-NI(\bar{C};X)}. \tag{10.20}$$

If $\bar{R} < I(\bar{C};X)$ then $P_{e1} \approx 0$ for $N \to \infty$.

Notice that, although the proof is formally almost identical with the proof of Theorem 5.10, the core enabler is the *joint* regularity and *IID* symbols of the joint H-codebook, which is much harder to be fulfilled. In particular, the missing natural randomization of the channel class itself is a major problem. $\square$

Conditional H-Rates

The case where the joint H-codebook fulfills *only* the *conditional* regularity condition can no longer be interpreted as a joint *decoding*. It rather becomes a *detection* problem[1] with a composite hypothesis. The detection of the channel class exhibits degrees of freedom (sometimes also called nuisance parameters) represented by random payload H-messages.

Because we are interested in both the channel class and payload data, the problem becomes *joint hypothesis testing* and *decoding*. The situation becomes similar to the joint synchronization and decoding with the difference that instead of a continuous-valued synchronization task we detect discrete channel class hypotheses. The problem can be approached by many strategies, as with the joint synchronization problem.

- *Separate detection and then decoding*   We can first perform the channel class detection (as a composite hypothesis testing with payload data degrees of freedom) and then perform the decoding conditioned by the detected class estimate. Once the channel class is detected with some fidelity measure (e.g. detection vs. false alarm probability), we can employ a standard reliable H-decoding conditioned by the channel class. A clear advantage for the detection process is the long observation of a relatively low-cardinality channel class variable. The channel class was assumed to be a constant over the codeword frame with the length N. This long observation easily allows a good detection probability performance essentially with no additional resource requirements (e.g. dedicated channel class sensing pilots, etc.) and with no need of their separate signaling to the rest of the network (it is an inherent part of the joint map).

---

[1] We use the term *decoding* to denote the situation where the receiver exploits some internal structure of the codebook. In this case the coding theorems identify the conditions for reliable (i.e. asymptotically having no errors) decoding. On the other hand, the term *detection* is usually related to identifying some hypothesis from the observation of some signals that do *not* need to have any internal code-like structure. A typical performance measure is the probability of the correct detection under the given false alarm probability.

- *Joint detection and decoding*   Both operations can be performed together as a joint detection and decoding problem.
- *Various forms of cooperative detection and decoding algorithms*   This includes all possible forms of mutually cooperating detection and decoding, exchanging the aiding information in any form (soft-information, decisions, etc.), in many scheduling variants (e.g. single-shot exchange, iterative operation, etc.).

The conditional decoding performance is measured by the conditional H-rates and the mutual information. The simplest situation is the one with *factorized* joint HNC maps.

THEOREM 10.5 (Conditional H-Rates for Factorized Joint HNC Map)   *Assume jointly isomorphic layered NCM with symbol-wise IID random component codebooks that is conditionally regular and has factorized joint HNC maps* $\bar{b} = (b, a)$, $\bar{c}_n = (c_n, a)$, *where the payload-only H-codebook is* $c^N = \mathcal{C}(b)$, $b \in [1 : 2^{NR}]$. *The channel class is constant over the code frame N and has a known value* $A = a$. *Then, the conditional payload-only H-rate R for the given* $A = a$ *is achievable if* $R < I(C; X | A = a)$.

Proof   The observation model for some given known a is

$$
\prod_{n=1}^{N} p(x_n | c_n, a) p(c_n). \tag{10.21}
$$

The joint typicality decoder finds the message $\hat{b}$ such that $\left(\bar{C}^N(\hat{b}, a), X^N \right)$ are jointly typical for a given known a. The observation model and the decoder are essentially identical to the one used in Theorem 5.10 with the *exception* that here we have the observation conditioned by a *particular* value of $A = a$. It is just enough to reuse all statements in Theorem 5.10 where we additionally parametrize everything by the conditioning $A = a$.

The fact that we must use the condition with a *particular* value $A = a$ is important. The joint typicality lemma, particularly (A.108), requires that all involved sequences are symbol-wise IID, including the one used for conditioning in the conditional independence. This allows us to use an *"average"* conditioning in the conditional mutual information $I(C; X | A)$ as opposed to conditioning by a particular value $I(C; X | A = a)$, in the evaluation of the decoding error probability. The symbol-wise IID property implies that all variables fully develop their randomness in the frame of the length N and thus the average conditioning justifies. However, this does not apply to the channel class in our setup. It is constant over the payload codeword frame. $\qquad \square$

10.3   Clustered, Nested, and Modular Cloud Framework

The clustering and modular nesting in complex WPNC cloud networks is an important concept that simplifies the design and allows some form of an abstraction from the detailed cloud structure. The concept is quite wide and does not have any strict rules or boundaries. However, we can somewhat isolate two main mutually related aspects.

10.3.1 Clustered Cloud

A large cloud network can contain many nodes and many radio links. Treating the network as a whole can be an extremely difficult design task with many degrees of freedom. Even the performance analysis itself (for some given defined WPNC coding/processing) can be out of practical feasibility. We can *cluster* the network into a set of smaller subnetworks where we *fully* take into consideration only those signals and processing that *originate* or *terminate* at nodes *inside* the cluster. All signals and processing from/to the *outside* of the cluster are simply *completely ignored* or treated only in some *simplistic manner* (e.g. by ignoring the codebook structure or contents and considering them only as some simply modeled interference). The only outside interaction from/to other clusters is through the node *common* to both of them. The clustering can be viewed as a side-by-side split of the cloud into smaller subnetworks ignoring the outside world. There are many ways to cluster the network. Some of the particular clusterings may, of course, affect both the performance and/or the desired overall end-to-end functionality/connectivity of the whole cloud network. The need of the proper way of clustering is the price paid for simplifying the design.

An example of a clustered cloud is shown in Figure 10.4. Figure 10.4a shows the full network with all radio links. Figure 10.4b shows cluster 1, consisting of nodes S_A, S_B, R_1, D_A, D_B. Radio visibility links that are on both the Tx and Rx sides outside the cluster (e.g. S_C–R_2) do not influence the cluster nodes directly and they are ignored. These links and corresponding nodes are in a gray color. Some links from the outside influence the receivers in the cluster (e.g. S_C–R_1, dashed lines). These are either ignored or treated as some simple interference without attempting to utilize the information content. Also some radio links lead outside and influence the rest of the network (dashed lines). These are also ignored, in the sense that the cluster does not rely on their presence. For example the chain of links S_A–R_2–D_A has some potential to help cluster 1, but it is ignored, or treated as a simple interference. In the example, we define two clusters Figure 10.4b,c. This particular clustering is also an example of how the overall end-to-end functionality/connectivity of the cloud is affected by the cluster choices. Clearly, the original cloud (Figure 10.4a) has the end-to-end connectivity (or capability of solving the global HNC map) for source S_C at destination D_A. However, the clustering in Figures 10.4b,c has lost this capability.

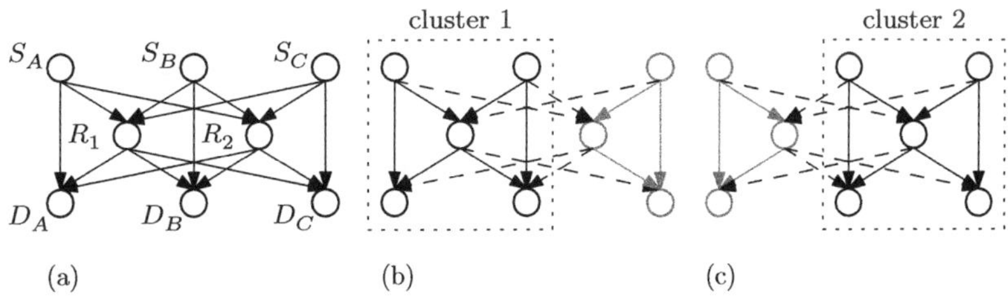

Figure 10.4 Clustered cloud network.

10.3.2   Nested Cloud

The nested cloud is a specific form of creating a *hierarchical nesting* of clusters. The inner cluster is modeled, from the outside perspective, as one "super-node" in the outer cluster. The properties and the performance of the super-node are given by the properties of the inner cluster with all the assumptions about ignoring the outside links as explained above for the "pure" clustering. In comparison with "pure" clustering, which is a "side-by-side" construction, this relies on the encapsulation. Clearly, both approaches can be combined. The advantage of modular nesting is the increased level of the abstraction (hence increased modularity) as we move up in the level of the encapsulation. Each encapsulation step hides the internal structure of the super-node and makes its behavior self-contained. The highest level of the encapsulation is the whole cloud, which can be then viewed as a super-super-node that provides services to the end terminals as if there was just one relay.

An example of the nested cloud is shown in Figure 10.5. The numbers at the links denote the stages (e.g. dictated by the half-duplex constraint) where the links belong. The butterfly subnetwork cluster $(R_1, R_2, R_3, R_4, R_5)$ is nested and modeled as super-node R_0 in Figure 10.5b. From the perspective of the new network $(S_A, S_B, R_0, D_A, D_B)$, it simply behaves as if R_0 was one node receiving both S_A, S_B at stage 1 and transmitting at stage 4. Notice that it formally looks like H-MAC and H-BC so we can apply all global network analysis (e.g. end-to-end solvability, the HNC maps, etc.) as usual in WPNC. However, the actual implementation of the virtual super-node achieves that by separate processing on individual cluster-internal nodes. This does not need to be revealed to the cluster-outside world. Of course it affects the *performance* (e.g. rates) and this must be reflected by the outside network. But the internal *structure* itself does not need to be revealed. Notice that, in this example, one nesting of the butterfly subnetwork leads again to the butterfly topology at a higher nesting level. Clearly this could have been represented again as another super-node in a more complex network. There also many ways in which to do the nesting/clustering. The example in Figure 10.6 shows a two-level nesting utilizing a two-way relay subnetwork at an interim step.

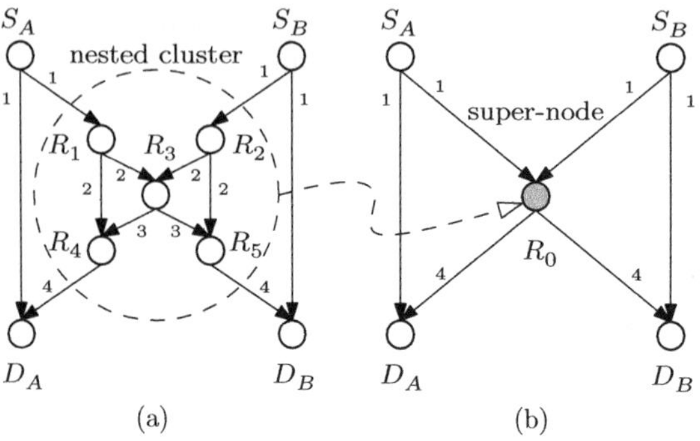

Figure 10.5 Nested cloud network.

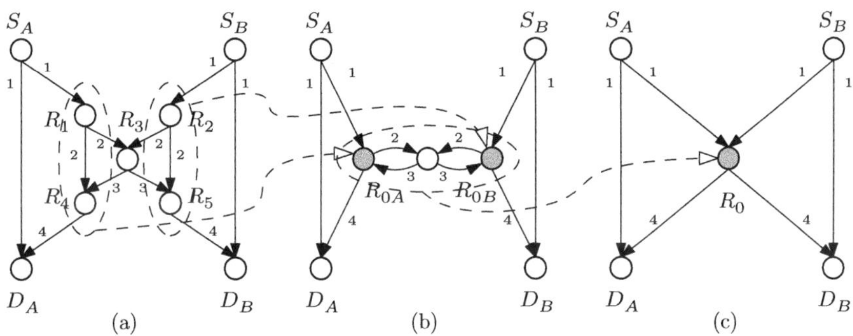

Figure 10.6 Nested cloud network with alternative clustering.

10.3.3   Modular Cloud Framework

Cloud clustering and nesting can be used as a practically feasible modular design framework. A large network is decomposed into smaller building blocks, each with a manageable design and analysis complexity. This procedure has obviously many options and does not have to be globally optimal because the local clusters simplify or ignore the wider context in the network. However, this is a traditional trade-off between the optimality and the complexity of the design.

The procedure itself should first identify the particular fragments of the network that can be modeled as clusters in both a "pure" side-by-side or nested hierarchical encapsulated manner. We can readily use the hierarchical network transfer function (Chapter 5) to capture the topology and the structure of processing in the cloud. Then, by searching for a given pattern, we can identify clusters to be replaced by the super-nodes. We assume that the set of available building blocks is given with all their performance properties.

Appendix A   Background Theory and Selected Fundamentals

A.1   Basic Mathematical Definitions

Gaussian Distribution

A complex-valued scalar zero-mean rotationally invariant Gaussian PDF is defined as

$$p_w(w) = \frac{1}{\pi \sigma_w^2} \exp\left(-\frac{1}{\sigma_w^2}|w|^2\right). \tag{A.1}$$

A complex-valued n-dimensional vector zero-mean rotationally invariant Gaussian PDF with variance σ_w^2 per dimension is defined as

$$p_{\mathbf{w}}(\mathbf{w}) = \frac{1}{\pi^n \sigma_w^{2n}} \exp\left(-\frac{1}{\sigma_w^2}\|\mathbf{w}\|^2\right). \tag{A.2}$$

Fourier Transform

A Discrete Fourier Transform (DFT) $\mathbf{S} = \text{DFT}[\mathbf{s}]$, where $\mathbf{s} = [s_0, \ldots, s_{N-1}]^{\mathrm{T}}$ and $\mathbf{S} = [S_0, \ldots, S_{N-1}]^{\mathrm{T}}$, is defined as

$$S_k = \sum_{n=0}^{N-1} s_n e^{-\mathrm{j}\frac{2\pi}{N}kn}, \tag{A.3}$$

$$s_n = \frac{1}{N} \sum_{k=0}^{N-1} S_k e^{\mathrm{j}\frac{2\pi}{N}kn}. \tag{A.4}$$

A.2   Linear Algebra

A.2.1   Algebraic Structures

Number theory, and algebraic structures in particular, is a well-founded mathematical area covered by many excellent books. Algebraic structures formalize the handling of basic mathematical operations and their properties, and this lays the ground for the coding theory (see e.g. [8]).

Group

A group is an algebraic structure $(\mathcal{A}, \circ)$ defining one operation $\circ$ and the set of elements $\mathcal{A}$ with the following properties

$$\text{(closure)} \quad \forall a, b \in \mathcal{A} : \ (a \circ b) \in \mathcal{A}, \tag{A.5}$$

$$\text{(associativity)} \quad \forall a, b, c \in \mathcal{A} : \ (a \circ b) \circ c = a \circ (b \circ c), \tag{A.6}$$

$$\text{(neutral element)} \quad \exists e \in \mathcal{A}, \forall a \in \mathcal{A} : \ a \circ e = e \circ a = a, \tag{A.7}$$

$$\text{(inverse)} \quad \forall a \in \mathcal{A}, \exists \bar{a} \in \mathcal{A} : \ a \circ \bar{a} = \bar{a} \circ a = e. \tag{A.8}$$

If, on top of these properties, the following holds

$$\text{(commutativity)} \quad \forall a, b \in \mathcal{A} : \ a \circ b = b \circ a \tag{A.9}$$

the group is called a commutative (or Abelian) group.

Ring

A ring is an algebraic structure $\mathbb{S} = (\mathcal{A}, \square, \circ)$, defining two operations $\square, \circ$, and the set of elements $\mathcal{A}$ with the following properties. The ring is a commutative group w.r.t. $\square$

$$\text{(closure)} \quad \forall a, b \in \mathcal{A} : \ (a \square b) \in \mathcal{A}, \tag{A.10}$$

$$\text{(associativity)} \quad \forall a, b, c \in \mathcal{A} : \ (a \square b) \square c = a \square (b \square c), \tag{A.11}$$

$$\text{(neutral element)} \quad \exists e \in \mathcal{A}, \forall a \in \mathcal{A} : \ a \square e = e \square a = a, \tag{A.12}$$

$$\text{(inverse)} \quad \forall a \in \mathcal{A}, \exists \bar{a} \in \mathcal{A} : \ a \square \bar{a} = \bar{a} \square a = e, \tag{A.13}$$

$$\text{(commutativity)} \quad \forall a, b \in \mathcal{A} : \ a \square b = b \square a. \tag{A.14}$$

The neutral element w.r.t. $\square$ is usually called "zero" and also denoted as such: $e = 0$. The *commutative* ring is a commutative group w.r.t. $\circ$ but not all non-zero elements have an inverse

$$\text{(closure)} \quad \forall a, b \in \mathcal{A} : \ (a \circ b) \in \mathcal{A}, \tag{A.15}$$

$$\text{(associativity)} \quad \forall a, b, c \in \mathcal{A} : \ (a \circ b) \circ c = a \circ (b \circ c), \tag{A.16}$$

$$\text{(neutral element)} \quad \exists i \in \mathcal{A}, \forall a \in \mathcal{A} : \ a \circ i = i \circ a = a, \tag{A.17}$$

$$\text{(commutativity)} \quad \forall a, b \in \mathcal{A} : \ a \circ b = b \circ a. \tag{A.18}$$

The neutral element w.r.t. $\circ$ is usually called "unity" and also denoted as such: $i = 1$. Finally the $\circ$ operation must distribute over the $\square$ operation

$$\text{(distributive law)} \quad \forall a, b, c \in \mathcal{A} : \ a \circ (b \square c) = (a \circ b) \square (a \circ c). \tag{A.19}$$

A ring may contain at least one non-zero *zero divisor* element a such that for some non-zero element b, $a \circ b = 0$.

Field

A field is an algebraic structure $\mathbb{F} = (\mathcal{A}, \square, \circ)$ defining two operations $\square, \circ$, and the set of elements $\mathcal{A}$ with the properties given below. It differs from the ring in that all non-zero elements have an inverse w.r.t. the $\circ$ operation. Thus the field, like the ring, is a commutative group w.r.t. $\square$

$$\text{(closure)} \quad \forall a, b \in \mathcal{A} : (a \square b) \in \mathcal{A}, \tag{A.20}$$

$$\text{(associativity)} \quad \forall a, b, c \in \mathcal{A} : (a \square b) \square c = a \square (b \square c), \tag{A.21}$$

$$\text{(neutral element)} \quad \exists e \in \mathcal{A}, \forall a \in \mathcal{A} : a \square e = e \square a = a, \tag{A.22}$$

$$\text{(inverse)} \quad \forall a \in \mathcal{A}, \exists \bar{a} \in \mathcal{A} : a \square \bar{a} = \bar{a} \square a = e, \tag{A.23}$$

$$\text{(commutativity)} \quad \forall a, b \in \mathcal{A} : a \square b = b \square a. \tag{A.24}$$

The neutral element e w.r.t. $\square$ is usually called "zero" and also denoted as such: $e = 0$. The field is a commutative group w.r.t. $\circ$ with an inversion for all non-zero elements

$$\text{(closure)} \quad \forall a, b \in \mathcal{A} : (a \circ b) \in \mathcal{A}, \tag{A.25}$$

$$\text{(associativity)} \quad \forall a, b, c \in \mathcal{A} : (a \circ b) \circ c = a \circ (b \circ c), \tag{A.26}$$

$$\text{(neutral element)} \quad \exists i \in \mathcal{A}, \forall a \in \mathcal{A} : a \circ i = i \circ a = a, \tag{A.27}$$

$$\text{(inverse)} \quad \forall a \in \mathcal{A} \setminus \{0\}, \exists \bar{a} \in \mathcal{A} : a \circ \bar{a} = \bar{a} \circ a = i, \tag{A.28}$$

$$\text{(commutativity)} \quad \forall a, b \in \mathcal{A} : a \circ b = b \circ a. \tag{A.29}$$

The neutral element i ($i \neq e$) w.r.t. $\circ$ is usually called "unity" and also denoted as such: $i = 1$. Finally the $\circ$ operation must distribute over the $\square$ operation

$$\text{(distributive law)} \quad \forall a, b, c \in \mathcal{A} : a \circ (b \square c) = (a \circ b) \square (a \circ c). \tag{A.30}$$

If the cardinality of the elements set is finite, we call it a *finite field* or *Galois field* (GF). The *finite* field exists only if its size is p^m where p is a prime number and $m \in \mathbb{N}$. It is denoted by $\mathbb{F}_{p^m}$. If $m > 1$, it is called an extended GF. All finite fields with the same size are mutually isomorphic. If $m = 1$, we can conveniently define $\square$ and $\circ$ as modulo p addition and multiplication. However, for an extended field with $m > 1$, we must use other approaches, e.g. a polynomial representation and modulo-polynomial operations (see e.g. [8]).

A.2.2 Matrix Analysis

Matrix analysis, vector spaces, vector space with inner product, eigensystems, and many other topics from applied linear algebra form the basic tools for communication theory. A detailed treatment is covered by many textbooks, e.g. [40].

Orthogonal Projection

Assume $\mathbf{A} \in \mathbb{C}^{N \times L}$, where $\text{rank}(\mathbf{A}) = L$, is a matrix with columns being a basis of an L-dimensional subspace of $\mathbb{C}^N$. A linear operator represented by the matrix $\mathbf{P_A} \in \mathbb{C}^{N \times N}$ is called an orthogonal projector on the subspace generated by $\mathbf{A}$ if $\mathbf{P_A}^2 = \mathbf{P_A}$, and it can be constructed as

$$\mathbf{P_A} = \mathbf{A}(\mathbf{A}^{\mathrm{H}}\mathbf{A})^{-1}\mathbf{A}^{\mathrm{H}}. \tag{A.31}$$

For the projector, it holds that $\mathbf{P_A}\mathbf{A} = \mathbf{A}$ and $\mathbf{P_A}^{\mathrm{H}} = \mathbf{P_A}$.

The orthogonal projector on null-space of $\mathbf{A}$, i.e. on the subspace perpendicular to the subspace generated by $\mathbf{A}$, is

$$\mathbf{P}_{\mathbf{A}^\perp} = \mathbf{I} - \mathbf{P}_{\mathbf{A}}. \tag{A.32}$$

Pseudoinverse

Assume $\mathbf{A} \in \mathbb{C}^{N \times L}$ with independent columns, i.e. $\mathrm{rank}(\mathbf{A}) = L$; then the left pseudoinverse is

$$\mathbf{A}^\dagger = (\mathbf{A}^H \mathbf{A})^{-1} \mathbf{A}^H. \tag{A.33}$$

Kronecker Matrix Product

The Kronecker matrix product is defined as

$$\mathbf{A} \boxtimes \mathbf{B} = \begin{bmatrix} a_{11}\mathbf{B} & \cdots & a_{1n}\mathbf{B} \\ \vdots & \ddots & \vdots \\ a_{m1}\mathbf{B} & \cdots & a_{mn}\mathbf{B} \end{bmatrix}. \tag{A.34}$$

A.2.3 Miscellaneous

Modulo Operation

Assume arbitrary $x, y \in \mathbb{Z}$ and $p \in \mathbb{N}$. The modulo operation is defined as a reminder after division

$$x \bmod p = x_p \in [0 : p-1] : \ x = x_0 p + x_p, \ x_0 \in \mathbb{Z}. \tag{A.35}$$

The modulo operation distributes over the addition and the multiplication

$$(x + y) \bmod p = ((x \bmod p) + (y \bmod p)) \bmod p, \tag{A.36}$$

$$(xy) \bmod p = ((x \bmod p)(y \bmod p)) \bmod p \tag{A.37}$$

and it holds also in a semi-distribution form

$$(x + y) \bmod p = (x + (y \bmod p)) \bmod p, \tag{A.38}$$

$$(xy) \bmod p = (x(y \bmod p)) \bmod p. \tag{A.39}$$

A.3 Detection, Decoding, and Estimation Theory

Detection, decoding, and estimation theory (and the closely related statistical signal processing) is a wide and long-established field. There are many excellent books and tutorials covering the topic from various angles. Books [22], [25], [26], [48], [51], [52] can be used as a general reference and starting point for any more specific problem solution.

A.3.1 Bayesian Estimators

A Bayesian estimation is a generic approach that forms a whole range of estimators that differ by the choice of particular error measure function (the loss function). Because the Bayesian estimation optimizes a *mean* error measure function by averaging over *all* random influences in the system (including the target parameter itself), it is applicable only on stochastic parameters with *known* a priori PDF/PMF.

We define a so-called Bayesian *loss function*, which quantitatively determines the consequences of the estimation error. Assume the vector parameter θ observed in observation $\mathbf{x}$ with the observation model $p(\mathbf{x}|\theta)$. The loss function is $L(\theta,\check{\theta}(\mathbf{x}))$ where $\check{\theta}(\mathbf{x})$ is a tentative (trial) estimator, i.e. a function of observation $\mathbf{x}$ providing value $\check{\theta}(\mathbf{x})$. Typically it is assumed that $L(\theta,\check{\theta}(\mathbf{x})) = 0$ for $\theta = \check{\theta}(\mathbf{x})$, i.e. the estimator has a zero error. The loss function is random due to the randomness of the observation and the parameter itself. The mean value of the loss, called the mean Bayesian risk, is

$$R(\check{\theta}(\mathbf{x})) = \mathrm{E}_{\theta,\mathbf{x}}\left[L(\theta,\check{\theta}(\mathbf{x}))\right]. \tag{A.40}$$

The Bayesian estimator is the one that minimizes the mean risk over all possible trial tentative estimators

$$\hat{\theta} = \arg\min_{\check{\theta}(\mathbf{x})} R(\check{\theta}(\mathbf{x})). \tag{A.41}$$

An important aspect of the Bayesian estimation is the fact that the above-stated form of the estimator is directly *unrealizable*. The problem lies in the averaging over the random observation. In order to find the risk-minimizing tentative estimator, we would have to repeat the observation to obtain the stochastic mean. However, a practical implementation has only *one* realization of the observation available. Given this single realization, we must find such estimator that would minimize the mean risk *if* we repeated the observation many times. Fortunately, there is a systematic way to guarantee this. It is based on the *conditional a posteriori* risk.

For a notation simplicity, we now assume a *discrete-valued* case (the continuous or mixed-valued cases follow a similar route). The mean risk is

$$R(\check{\theta}(\mathbf{x})) = \sum_{\mathbf{x}}\sum_{\theta} L(\theta,\check{\theta}(\mathbf{x}))p(\theta|\mathbf{x})p(\mathbf{x})$$
$$= \sum_{\mathbf{x}} R(\check{\theta}|\mathbf{x})p(\mathbf{x}) \tag{A.42}$$

where

$$R(\check{\theta}|\mathbf{x}) = \sum_{\theta} L(\theta,\check{\theta}(\mathbf{x}))p(\theta|\mathbf{x}) \tag{A.43}$$

is a conditional risk evaluated for one *given realization* of $\mathbf{x}$. As $p(\mathbf{x})$ is non-negative, the minimization of conditional risk for any given particular realization of $\mathbf{x}$ also *guarantees* that the mean risk is minimized, as is required by the Bayesian strategy. As an additional bonus, owing to the fixed $\mathbf{x}$, we do not need to minimize over all estimator *functions* $\check{\theta}(\mathbf{x})$, but only over the *values* $\check{\theta}$. The Bayesian estimator is thus equivalent to

$$\hat{\theta} = \arg\min_{\check{\theta}} R(\check{\theta}|\mathbf{x}). \tag{A.44}$$

MAP

Notable widespread examples of the Bayesian strategy are maximum a posteriori probability (MAP) and minimum mean square error (MMSE) estimators. The MAP estimator uses a uniform loss function

$$L(\boldsymbol{\theta}, \check{\boldsymbol{\theta}}(\mathbf{x})) = 1 - \delta(\boldsymbol{\theta} - \check{\boldsymbol{\theta}}(\mathbf{x})). \tag{A.45}$$

The uniform loss should be interpreted as an error measure that does not pay any attention to the size of the error. The conditional risk then becomes

$$R(\check{\boldsymbol{\theta}}|\mathbf{x}) = 1 - p(\check{\boldsymbol{\theta}}|\mathbf{x}). \tag{A.46}$$

Its minimization is clearly equivalent to the maximization of a posteriori probability

$$\hat{\boldsymbol{\theta}} = \arg\max_{\check{\boldsymbol{\theta}}} p(\check{\boldsymbol{\theta}}|\mathbf{x}). \tag{A.47}$$

MMSE

The MMSE estimator uses a quadratic loss function

$$L(\boldsymbol{\theta}, \check{\boldsymbol{\theta}}(\mathbf{x})) = \|\boldsymbol{\theta} - \check{\boldsymbol{\theta}}(\mathbf{x})\|^2 \tag{A.48}$$

and the corresponding mean risk is thus a mean square error

$$R(\check{\boldsymbol{\theta}}(\mathbf{x})) = \mathrm{E}_{\boldsymbol{\theta},\mathbf{x}}\left[\|\boldsymbol{\theta} - \check{\boldsymbol{\theta}}(\mathbf{x})\|^2\right]. \tag{A.49}$$

The minimization of the conditional risk can be solved utilizing the generalized derivative

$$\frac{\tilde{\partial}}{\tilde{\partial}\check{\boldsymbol{\theta}}}R(\check{\boldsymbol{\theta}}|\mathbf{x}) = \frac{\tilde{\partial}}{\tilde{\partial}\check{\boldsymbol{\theta}}}\sum_{\boldsymbol{\theta}}\|\boldsymbol{\theta} - \check{\boldsymbol{\theta}}\|^2 p(\boldsymbol{\theta}|\mathbf{x}) \triangleq \mathbf{0} \tag{A.50}$$

where

$$\begin{aligned}
\frac{\tilde{\partial}}{\tilde{\partial}\check{\boldsymbol{\theta}}}\|\boldsymbol{\theta} - \check{\boldsymbol{\theta}}(\mathbf{x})\|^2 &= \frac{\tilde{\partial}}{\tilde{\partial}\check{\boldsymbol{\theta}}}\left(\boldsymbol{\theta} - \check{\boldsymbol{\theta}}\right)^{\mathrm{H}}\left(\boldsymbol{\theta} - \check{\boldsymbol{\theta}}\right) \\
&= \frac{\tilde{\partial}}{\tilde{\partial}\check{\boldsymbol{\theta}}}\left(\boldsymbol{\theta}^{\mathrm{H}}\boldsymbol{\theta} + \check{\boldsymbol{\theta}}^{\mathrm{H}}\check{\boldsymbol{\theta}} - \boldsymbol{\theta}^{\mathrm{H}}\check{\boldsymbol{\theta}} - \check{\boldsymbol{\theta}}^{\mathrm{H}}\boldsymbol{\theta}\right) \\
&= \check{\boldsymbol{\theta}}^{*} - \boldsymbol{\theta}^{*} \tag{A.51}
\end{aligned}$$

and thus the MMSE estimator is a conditional mean

$$\hat{\boldsymbol{\theta}} = \sum_{\boldsymbol{\theta}}\boldsymbol{\theta}p(\boldsymbol{\theta}|\mathbf{x}) = \mathrm{E}\left[\boldsymbol{\theta}|\mathbf{x}\right]. \tag{A.52}$$

A.3.2   Maximum Likelihood Estimator

The maximum likelihood (ML) estimator is a popular choice for estimating parameters with unknown statistics. It finds the value that best matches the actual observed measurement $\mathbf{x}$ for a given stochastic observation input–output model $p(\mathbf{x}|\boldsymbol{\theta})$

$$\hat{\boldsymbol{\theta}} = \arg\max_{\check{\boldsymbol{\theta}}} p(\mathbf{x}|\check{\boldsymbol{\theta}}). \tag{A.53}$$

The ML estimate is asymptotically *unbiased* and *efficient*, i.e. it attains CRLB. Moreover, the estimation error has asymptotically Gaussian distribution

$$\hat{\boldsymbol{\theta}} \sim \mathcal{N}(\boldsymbol{\theta}, \mathbf{J}^{-1}(\boldsymbol{\theta})) \tag{A.54}$$

where $\boldsymbol{\theta}$ is the true parameter value and $\mathbf{J}(\boldsymbol{\theta})$ is a Fisher information matrix.

A.3.3   MAP Sequence and Symbol Decoding

It is easy to show that the mean MAP risk is equal to the mean probability of the detection/decoding error for discrete data when all error types are treated uniformly. The MAP criterion is thus a standard data detection/decoding strategy. However, we must distinguish between applying the MAP criterion to the whole data/codeword *vector (sequence)* or applying it to the individual data/code *symbols*.

In the case of applying the MAP criterion to the whole data/codeword vector/sequence, we talk about *MAP sequence* decoding

$$\hat{\mathbf{c}} = \arg\max_{\check{\mathbf{c}}} p(\check{\mathbf{c}}|\mathbf{x}). \tag{A.55}$$

When applying the criterion symbol-wise, it becomes

$$\hat{c}_n = \arg\max_{\check{c}_n} p(\check{c}_n|\mathbf{x}) \tag{A.56}$$

and we call it *MAP symbol decoding*. The symbol-wise PMF/PDF can be obtained by a *marginalization* of the joint PMF/PDF. It marginalizes all the other symbols apart from the one of our interest. But this marginalization must respect the internal structure (if there is any, e.g. dictated by the codebook) of the codeword. The generic way of capturing this is to use a "consistent with" notation $\mathbf{c} : c_n$. The marginalized PMF/PDF is then

$$p(c_n|\mathbf{x}) = \sum_{\mathbf{c}:c_n} p(\mathbf{c}|\mathbf{x}). \tag{A.57}$$

Regardless of whether we use a sequence or a symbol MAP decoding, it properly utilizes the a priori statistics of the data. In a *special* case of *uniform* a priori data distribution $p(\mathbf{c}) = \text{const}$, it becomes (we show the sequence MAP case, the symbol MAP case is similar)

$$\begin{aligned}
\hat{\mathbf{c}} &= \arg\max_{\check{\mathbf{c}}} p(\check{\mathbf{c}}|\mathbf{x}) \\
&= \arg\max_{\check{\mathbf{c}}} \frac{p(\mathbf{x}|\check{\mathbf{c}})p(\check{\mathbf{c}})}{p(\mathbf{x})} \\
&= \arg\max_{\check{\mathbf{c}}} p(\mathbf{x}|\check{\mathbf{c}}).
\end{aligned} \tag{A.58}$$

The decoder becomes identical with the ML decoding criterion.

A.3.4   Pairwise Error Union Upper Bound

The pairwise error probability is an important and frequently used tool for analyzing error performance and for forming the codebook design targets. Although it is only an upper bound, it has one fundamental feature that has very useful practical consequences for the codebook design. It closely brings together the error performance with the decoder's metric. The decoder metric (e.g. MAP or ML), in turn, directly depends on the transmitted signals and the observation model. It allows us to connect the error performance with the form of transmitted codebook signals and the observation model, and we can use this in the optimal codebook design. A classical example is the "maximize the free Euclidean distance" design rule for the AWGN observation model.

We now briefly review the concept of the pairwise error union upper bound technique for the sequence decoder (the symbol-wise case is similar) with some *generic* metric $\rho(\mathbf{x}, \mathbf{c})$ and the decoding rule $\hat{\mathbf{c}} = \arg\min_{\mathbf{c}} \rho(\mathbf{x}, \mathbf{c})$. For example, it can be the minimization of Euclidian distance for ML decoding in AWGN channel. The pairwise error event describes that the incorrect test sequence $\mathbf{c}^{(i)}$ has a better metric than the correct one $\mathbf{c}^{(k)}$, i.e. the one that was actually transmitted

$$\mathbb{E}_2(i|k) = \left\{ \rho(\mathbf{x}, \mathbf{c}^{(i)}) < \rho(\mathbf{x}, \mathbf{c}^{(k)}) \,\middle|\, \mathbf{c}^{(k)} \right\}. \tag{A.59}$$

It means that the decoder *does not* select the correct message. However, it does not mean that the decoder would select the message $\mathbf{c}^{(i)}$. The events $\mathbb{E}_2(i|k)$ are *not* disjunct. The pairwise error probability is

$$P_2(i|k) = \Pr\left\{\mathbb{E}_2(i|k)\right\} = \Pr\left\{ \rho(\mathbf{x}, \mathbf{c}^{(i)}) < \rho(\mathbf{x}, \mathbf{c}^{(k)}) \,\middle|\, \mathbf{c}^{(k)} \right\}. \tag{A.60}$$

The union bound on the probability of the decoding error is

$$\begin{aligned}
P_e &= \sum_k \Pr\left\{ \hat{\mathbf{c}} \neq \mathbf{c}^{(k)} \,\middle|\, \mathbf{c} = \mathbf{c}^{(k)} \right\} \Pr\left\{ \mathbf{c}^{(k)} \right\} \\
&= \sum_k \Pr\left\{ \bigcup_{i \neq k} \left(\rho\left(\mathbf{x}, \mathbf{c}^{(i)}\right) < \rho\left(\mathbf{x}, \mathbf{c}^{(k)}\right) \right) \,\middle|\, \mathbf{c}^{(k)} \right\} \Pr\left\{ \mathbf{c}^{(k)} \right\} \\
&\stackrel{\text{(ub)}}{\leq} \Pr\{\mathbf{c}\} \sum_k \sum_{i \neq k} \Pr\left\{ \rho\left(\mathbf{x}, \mathbf{c}^{(i)}\right) < \rho\left(\mathbf{x}, \mathbf{c}^{(k)}\right) \,\middle|\, \mathbf{c}^{(k)} \right\} \\
&= \Pr\{\mathbf{c}\} \sum_k \sum_{i \neq k} P_2(i|k) \tag{A.61}
\end{aligned}$$

where (ub) comes from the union of *non-disjunct* events $\Pr\left\{\bigcup_i \mathbb{A}_i\right\} \leq \sum_i \Pr\left\{\mathbb{A}_i\right\}$ and where we have assumed uniformly distributed messages $\Pr\{\mathbf{c}^{(k)}\} = \Pr\{\mathbf{c}\} = \text{const}$.

The pairwise error probability upper bound can be further simplified by considering only the most probable one and ignoring the rest of them. Let the most probable pairwise error have the probability

$$P_{2\max} = \max_{i,k} P_2(i|k). \tag{A.62}$$

The pairwise events with this probability usually appear in multiple instances and we denote this multiplicity K_m. The resulting *most probable pairwise event* upper bound approximation is

$$P_e \lesssim K_m P_{2\max}. \tag{A.63}$$

This approximation is frequently used for specifying the codebook design targets. For example, when applied on the AWGN observation model, it leads to the well-known Euclidean free distance design rule.

A.3.5   Complex-Valued Optimization

In the estimation theory, we frequently need to find the minimum or the maximum (the stationary point) of a *real-valued* function $f(z) \in \mathbb{R}$ of a *complex-valued* parameter $z \in \mathbb{C}$, with real x and imaginary y parts, $z = x + jy$. Unfortunately, a real-valued function of a complex-valued parameter is *not* differentiable in the sense of Cauchy–Riemann conditions (apart from a trivial case of a constant). But because we are not generally interested in complete differentiability but only in finding the stationary point, we can split the evaluation into real and imaginary parts. However, there is a more elegant solution for finding the stationary points using a *generalized (Wirtinger) derivative* [25], [52], [59].

The generalized derivative is defined as

$$\frac{\tilde{\partial} f}{\tilde{\partial} z} = \frac{1}{2}\frac{\partial f}{\partial x} - j\frac{1}{2}\frac{\partial f}{\partial y} \tag{A.64}$$

and

$$\frac{\tilde{\partial} f}{\tilde{\partial} z^*} = \frac{1}{2}\frac{\partial f}{\partial x} + j\frac{1}{2}\frac{\partial f}{\partial y}. \tag{A.65}$$

Major properties of the generalized derivative follow.

(1) If f is Cauchy–Riemann differentiable (analytic function) then

$$\frac{\tilde{\partial} f}{\tilde{\partial} z} = \frac{\partial f}{\partial z}. \tag{A.66}$$

(2) If f is pure real-valued $f \in \mathbb{R}$ then

$$\frac{\tilde{\partial} f}{\tilde{\partial} z} = 0 \quad \Leftrightarrow \quad \frac{\partial f}{\partial x} = 0 \wedge \frac{\partial f}{\partial y} = 0, \tag{A.67}$$

i.e. we can find the stationary point using the generalized derivative.

(3) The generalized derivative shares most of the usual properties (e.g. a linearity, see [25], [52], [59] for details) with the ordinary derivative with one important exception related to the *complex conjugation*. In fact, we could treat the generalized derivative as if z and z^* were *two different variables*

$$\frac{\tilde{\partial} z}{\tilde{\partial} z} = 1, \quad \frac{\tilde{\partial}(z^*)}{\tilde{\partial} z} = 0, \quad \frac{\tilde{\partial}(|z|^2)}{\tilde{\partial} z} = z^*. \tag{A.68}$$

(4) The generalized derivative can be also extended to vectors and gradients (see [25], [52], [59]). The generalized derivative over a vector is

$$\frac{\tilde{\partial} f}{\tilde{\partial} \mathbf{z}} = \left[\frac{\tilde{\partial} f}{\tilde{\partial} z_1}, \ldots, \frac{\tilde{\partial} f}{\tilde{\partial} z_N}\right]^{\mathrm{T}} \tag{A.69}$$

where $\mathbf{z} = [z_1, \ldots, z_N]^{\mathrm{T}} \in \mathbb{C}^{N \times 1}$. If $\mathbf{A} \in \mathbb{C}^{N \times N}$ and $\mathbf{a} \in \mathbb{C}^{N \times 1}$ then it holds that

$$\frac{\tilde{\partial}}{\tilde{\partial} \mathbf{z}}\left(\mathbf{a}^{\mathrm{H}}\mathbf{z}\right) = \mathbf{a}^*, \tag{A.70}$$

$$\frac{\tilde{\partial}}{\tilde{\partial}\mathbf{z}}\left(\mathbf{z}^{\mathrm{H}}\mathbf{a}\right) = \mathbf{0}, \tag{A.71}$$

$$\frac{\tilde{\partial}}{\tilde{\partial}\mathbf{z}}\left(\mathbf{z}^{\mathrm{H}}\mathbf{A}\mathbf{z}\right) = \mathbf{A}^{\mathrm{T}}\mathbf{z}^{*}, \tag{A.72}$$

and

$$\frac{\tilde{\partial}}{\tilde{\partial}\mathbf{z}^{*}}\left(\mathbf{a}^{\mathrm{H}}\mathbf{z}\right) = \mathbf{0}, \tag{A.73}$$

$$\frac{\tilde{\partial}}{\tilde{\partial}\mathbf{z}^{*}}\left(\mathbf{z}^{\mathrm{H}}\mathbf{a}\right) = \mathbf{a}, \tag{A.74}$$

$$\frac{\tilde{\partial}}{\tilde{\partial}\mathbf{z}^{*}}\left(\mathbf{z}^{\mathrm{H}}\mathbf{A}\mathbf{z}\right) = \mathbf{A}\mathbf{z}. \tag{A.75}$$

A.3.6   Cramer–Rao Lower Bound

The Cramer-Rao lower bound (CRLB) sets the fundamental limits on MSE performance of any unbiased parameter estimator. Assume the observation $\mathbf{x}$ of the vector deterministic parameter $\boldsymbol{\theta} = [\theta_1, \ldots, \theta_N]^{\mathrm{T}}$. The observation model is described by the likelihood $p(\mathbf{x}|\boldsymbol{\theta})$. The estimate $\hat{\boldsymbol{\theta}}$ is said to be unbiased if $\mathrm{E}[\hat{\boldsymbol{\theta}}] = \boldsymbol{\theta}$.

A classical form of CRLB theorem for a *real-valued* parameter vector $\boldsymbol{\theta}$ follows [25]. Assume $\boldsymbol{\theta} \in \mathbb{R}^{N \times 1}$. If the regularity condition holds

$$\mathrm{E}\left[\frac{\partial \ln p(\mathbf{x}|\boldsymbol{\theta})}{\partial \boldsymbol{\theta}}\right] = 0 \tag{A.76}$$

then the error covariance matrix $\mathbf{C} = \mathrm{E}\left[(\hat{\boldsymbol{\theta}} - \boldsymbol{\theta})(\hat{\boldsymbol{\theta}} - \boldsymbol{\theta})^{\mathrm{T}}\right]$ of any *unbiased* estimator obeys

$$\mathbf{C} \geq \mathbf{J}^{-1}(\boldsymbol{\theta}) \tag{A.77}$$

where $\mathbf{J}$ is a Fisher information matrix with elements

$$[\mathbf{J}]_{k,\ell} = -\mathrm{E}\left[\frac{\partial^2 \ln p(\mathbf{x}|\boldsymbol{\theta})}{\partial \theta_k \partial \theta_\ell}\right]. \tag{A.78}$$

An extension of CRLB for *complex-valued* parameters is possible and requires to use the generalized derivative [41], [52], [61]. Assume that $\boldsymbol{\theta} \in \mathbb{C}^{N \times 1}$. If the regularity condition holds

$$\mathrm{E}\left[\frac{\tilde{\partial} \ln p(\mathbf{x}|\boldsymbol{\theta})}{\tilde{\partial}\boldsymbol{\theta}^{*}}\right] = \mathbf{0} \tag{A.79}$$

then the error covariance matrix $\mathbf{C} = \mathrm{E}\left[(\hat{\boldsymbol{\theta}} - \boldsymbol{\theta})(\hat{\boldsymbol{\theta}} - \boldsymbol{\theta})^{\mathrm{H}}\right]$ of any *unbiased* estimator obeys

$$\mathbf{C} \geq \mathbf{J}^{-1}(\boldsymbol{\theta}) \tag{A.80}$$

where $\mathbf{J}$ is a *complex* Fisher information matrix with elements

$$[\mathbf{J}(\boldsymbol{\theta})]_{k,\ell} = -\,\mathrm{E}\left[\frac{\tilde{\partial}^2 \ln p(\mathbf{x}|\boldsymbol{\theta})}{\tilde{\partial}\theta_k^* \tilde{\partial}\theta_\ell}\right] \tag{A.81}$$

or equivalently

$$\mathbf{J}(\boldsymbol{\theta}) = \mathrm{E}\left[\left(\frac{\tilde{\partial} \ln p(\mathbf{x}|\boldsymbol{\theta})}{\tilde{\partial}\boldsymbol{\theta}^*}\right)\left(\frac{\tilde{\partial} \ln p(\mathbf{x}|\boldsymbol{\theta})}{\tilde{\partial}\boldsymbol{\theta}}\right)^{\mathrm{T}}\right]. \tag{A.82}$$

The most important application is related to the diagonal elements of the error covariance matrix which describe the MSE for individual parameters $[\mathbf{C}]_{k,k} = \mathrm{E}[|\hat{\theta}_k - \theta_k|^2]$. Since any positive semi-definite matrix has real-valued non-negative diagonal elements, it implies that

$$\mathrm{E}[|\hat{\theta}_k - \theta_k|^2] \geq [\mathbf{J}^{-1}(\boldsymbol{\theta})]_{k,k}. \tag{A.83}$$

A.3.7   Sufficient Statistic

Assume the observation model $p(\mathbf{x}|\boldsymbol{\theta})$ for the parameter $\boldsymbol{\theta}$. A sufficient statistic is a function $T(\mathbf{x})$ of the observation that contains all available information contained in the original observation $\mathbf{x}$ about $\boldsymbol{\theta}$ necessary for the estimator $\hat{\boldsymbol{\theta}}$. The estimate obtained from the original observation must equal the one obtained from the sufficient statistic $\hat{\boldsymbol{\theta}}(\mathbf{x}) = \hat{\boldsymbol{\theta}}(T(\mathbf{x}))$. The sufficient statistic usually serves to reduce the dimensionality of other simplifications of the estimator input.

The Neyman–Fisher factorization theorem [25] allows us to verify whether the function $T(\mathbf{x})$ is a sufficient statistic. $T(\mathbf{x})$ is a sufficient statistic if and only if we can factorize the observation likelihood as

$$p(\mathbf{x}|\boldsymbol{\theta}) = g(T(\mathbf{x}), \boldsymbol{\theta})h(\mathbf{x}). \tag{A.84}$$

A.4   Information Theory

Information theory and network information theory form a theoretical framework for many fundamental concepts treated in the book. The reader might find an abundant amount, depth, and rigor of the knowledge in several excellent landmark tutorial books [13], [18], [62]. Here we state only the most important principles in a streamlined form. This section will use strict notation as is used in information theory. A random variable will be denoted by a capital X, its realization by a lower-case letter x, and its PDF/PMF by $p_X(x)$ or, if no ambiguity is possible, by $p(x)$. A sequence of variables will be denoted as $X^K = \{X_1, X_2, \ldots, X_K\}$. A notation $x(\mathcal{S})$ means a tuple containing all x_k such that $k \in \mathcal{S}$.

A.4.1   Basic Concepts

Conditional Independence

Random variables A and B are said to be conditionally independent given C, in notation $A \perp\!\!\!\perp B | C$ or $(A \perp\!\!\!\perp B)|C$, if any of the following equivalent forms hold

$$p(a,b|c) = p(a|c)p(b|c), \tag{A.85}$$

$$p(a|b,c) = p(a|c), \tag{A.86}$$

$$p(b|a,c) = p(b|c), \tag{A.87}$$

$$p(a,b,c) = p(a|c)p(b,c), \tag{A.88}$$

$$p(a,b,c) = p(a,c)p(b|c). \tag{A.89}$$

Markov Chain

Random variables $X \mapsto Y \mapsto Z$ are said to form a Markov chain if $p(x,y,z) = p(z|y)p(y|x)p(x)$. If $X \mapsto Y \mapsto Z$ then also $Z \mapsto Y \mapsto X$.

Entropy and Mutual Information

The entropy is defined as

$$\mathcal{H}[X] = -\mathrm{E}[\lg p_X(x)] \tag{A.90}$$

and the mutual information is defined as

$$I(X;Y) = \mathcal{H}[X] - \mathcal{H}[X|Y] = \mathcal{H}[Y] - \mathcal{H}[Y|X]. \tag{A.91}$$

The conditional entropy is defined as

$$\mathcal{H}[X|Z] = -\mathrm{E}_{p_{X,Z}(x,z)}[\lg p_{X|Z}(x|z)]. \tag{A.92}$$

Notice that the expectation is over a *joint* PDF whereas the argument contains a *conditional* PDF. The conditional mutual information is defined as

$$I(X;Y|Z) = \mathcal{H}[X|Z] - \mathcal{H}[X|Y,Z] = \mathcal{H}[Y|Z] - \mathcal{H}[Y|X,Z]. \tag{A.93}$$

Binary Entropy Function

Binary entropy function describes the entropy of a binary random variable with instance probabilities $\{p, 1-p\}$

$$H(p) = -p \lg p - (1-p)\lg(1-p), \quad p \in [0,1]. \tag{A.94}$$

Chain Rules

Entropy

$$\mathcal{H}[X_1,\ldots,X_N] = \sum_{n=1}^{N} \mathcal{H}[X_n|X_{n-1},\ldots,X_1]. \tag{A.95}$$

Mutual information

$$I(X_1,\ldots,X_N;Y) = \sum_{n=1}^{N} I(X_n;Y|X_{n-1},\ldots,X_1]. \tag{A.96}$$

Data Processing Inequality

If $X \mapsto Y \mapsto Z$ form the Markov chain then

$$I(X;Y) \geq I(X;Z). \tag{A.97}$$

Fano's Inequality

Discrete random variables $X, \hat{X} \in \mathcal{A}$ are drawn according to $(X, \hat{X}) \sim p(x, \hat{x})$. We define probability of the error $P_e = \Pr\{X \neq \hat{X}\}$. Then

$$\mathcal{H}[X|\hat{X}] \leq \mathrm{H}(P_e) + P_e \lg |\mathcal{A}| \leq 1 + P_e \lg |\mathcal{A}|. \tag{A.98}$$

Typical Set

A typical set plays a fundamental role in many information theoretic solutions. It has a particularly important role in coding theorem proofs. A typical set formalizes a very natural concept that a sequence of experimentally drawn random IID variables X^N approach their true stochastic properties as the sequence size N increases to infinity.

Assume the IID random sequence drawn according to $X^N \sim \prod_{n=1}^{N} p_X(x_i)$. The true stochastic entropy of the random variable is $\mathcal{H}[X] = -\mathrm{E}[\lg p_X(x)]$. Now imagine that, instead of using the stochastic expectation operator, we calculate an *empirical* averaging on a sequence of drawn random variables X^N

$$\mathrm{h}_N[X] = -\frac{1}{N} \sum_{n=1}^{N} \lg p_X(X_n). \tag{A.99}$$

We sometimes call this an empirical entropy. As one can naturally anticipate, empirical entropy approaches the true one. This is called the *asymptotic equipartition property* theorem

$$\lim_{N \to \infty} \mathrm{h}_N[X] \overset{(\mathrm{prob})}{=} \mathcal{H}[X]. \tag{A.100}$$

It converges in probability, i.e. $\forall \epsilon > 0$, $\lim_{N \to \infty} \Pr\{|\mathrm{h}_N[X] - \mathcal{H}[X]| > \epsilon\} = 0$, and it is formally a consequence of the weak law of large numbers [13].

A *typical set* of IID sequences X^N is defined as

$$\mathcal{T}_\epsilon^{(N)}(X) = \left\{ X^N : \mathcal{H}[X] - \epsilon \leq \mathrm{h}_N[X] \leq \mathcal{H}[X] + \epsilon \right\}. \tag{A.101}$$

$\mathcal{T}_\epsilon^{(N)}(X)$ thus defines a set of sequences X^N that have the empirical entropy within ϵ toleration around the true stochastic entropy $\mathcal{H}[X]$. A typicality property classifies all possible sequences X^N into two classes: the typical set members and the non-typical set members.

Typical sets for a *discrete-valued* X have a number of useful properties

$$\forall X^N \in \mathcal{T}_\epsilon^{(N)}(X) : 2^{-N(\mathcal{H}[X]+\epsilon)} \leq p_{X^N}(X_1, \ldots, X_N) \leq 2^{-N(\mathcal{H}[X]-\epsilon)}, \tag{A.102}$$

$$\forall \epsilon > 0, \exists N : \Pr\left\{ \mathcal{T}_\epsilon^{(N)}(X) \right\} > 1 - \epsilon, \tag{A.103}$$

$$\forall \epsilon > 0, \exists N : (1 - \epsilon)2^{N(\mathcal{H}[X]-\epsilon)} \leq \left| \mathcal{T}_\epsilon^{(N)}(X) \right| \leq 2^{N(\mathcal{H}[X]+\epsilon)}. \tag{A.104}$$

These properties mean that, for a sufficiently large N, (1) all sequences in the typical set are almost equally probable with probability $2^{-N\mathcal{H}[X]}$, (2) the probability of X^N being a typical sequence is almost one, and (3) the cardinality of the typical set is asymptotically $2^{N\mathcal{H}[X]}$.

In the case of *continuous-valued* X most of the properties hold similarly, with a notable exception that, instead of the cardinality of the typical set, we have a volume

$$V\left(\mathcal{T}_\epsilon^{(N)}(X)\right) = \int_{\mathcal{T}_\epsilon^{(N)}(X)} \mathrm{d}x^N \approx 2^{N\mathcal{H}[X]}. \tag{A.105}$$

The typicality property is also straightforwardly applicable to multiple random variables by using the joint entropy.

In the case when no ambiguity is possible and all is clearly defined by the context, we may use a simplified notation omitting explicit identification of the length of the sequence and the width of the toleration range, and the typical set is simply denoted as $\mathcal{T}(X)$. Also, we may use the notation $h_N[X] \approx \mathcal{H}[X]$ to denote equality in the $\pm\epsilon$ tolerance sense. We also define asymptotic inequalities in the ϵ tolerances sense as $\gtrsim$ and $\lesssim$.

Joint Typicality Lemma

The joint typicality lemma in various variants (see e.g. [18] Section 2.5.1) plays an important role in achievability proofs of coding theorems. Here, we state the lemma in several forms and in a slightly relaxed notation. In particular, we drop the explicit ϵ and $\delta(\epsilon)$ asymptotic $N \to \infty$ notation. We will use $\approx$, $\lesssim$, and $\gtrsim$ to denote asymptotically equal, less than, and greater than. All should be understood in the sense that for any suitably chosen ϵ tolerance range around the condition there exists a large enough N that the condition is fulfilled. For example, $a(N) \approx b(N)$ denotes that for any $\epsilon > 0$ there exists N such that $|a(N) - b(N)| < \epsilon$, or $a(N) \lesssim b(N)$ denotes $a(N) \leq b(N)(1 + \epsilon)$. We should, however, warn against too liberal extrapolations of this notation. Many results in information theory depend on the proper order of the convergence in the asymptotic statements and we need frequently to make sure that one entity approaches faster than the other (e.g. by having two tolerances $\epsilon < \epsilon'$). One should always carefully understand the internal meaning of the notation and turn back to the explicit notation when needed. More details can be found in [13], [18], [39], [62].

LEMMA A.1 (Joint Typicality Lemma) *Assume that random variables X, Y, Z are jointly described by PDF $p(x, y, z)$.*

(1) If (x^N, y^N) is a pair of arbitrary given realizations of sequences, and the random sequence Z^N is drawn according to $Z^N \sim \prod_n p_{Z|X}(z_n|x_n)$ then

$$\Pr\left\{(x^N, y^N, Z^N) \in \mathcal{T}(X, Y, Z)\right\} \lesssim 2^{-NI(Y;Z|X)}. \tag{A.106}$$

(2) If the pair of sequence realizations is jointly typical $(x^N, y^N) \in \mathcal{T}(X, Y)$, and $Z^N \sim \prod_n p_{Z|X}(z_n|x_n)$ then

$$\Pr\left\{(x^N, y^N, Z^N) \in \mathcal{T}(X, Y, Z)\right\} \gtrsim 2^{-NI(Y;Z|X)}. \tag{A.107}$$

(3) If the pair of sequences is randomly drawn according to $(X^N, Y^N) \sim \prod_n p_{X,Y}(x_n, y_n)$, and random sequence Z^N is drawn conditionally given X^N independent of Y^N, i.e. $Z^N|\{X^N = x^N, Y^N = y^N\} \sim \prod_n p_{Z|X}(z_n|x_n)$, then

$$\Pr\left\{(X^N, Y^N, Z^N) \in \mathcal{T}(X, Y, Z)\right\} \approx 2^{-NI(Y;Z|X)}. \tag{A.108}$$

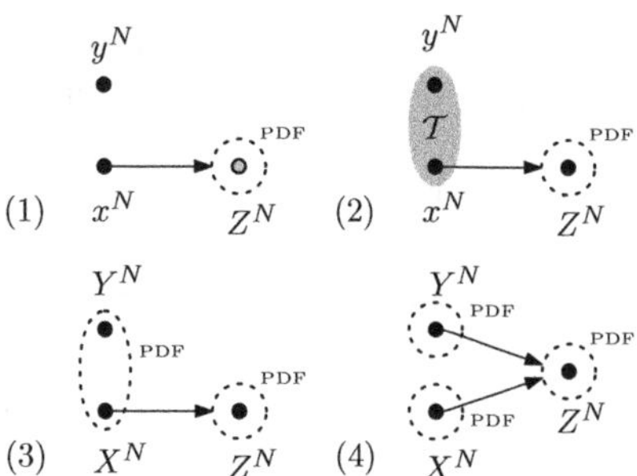

Figure A.1 Joint typicality lemma. The relationships among the involved sequences in four variants of the lemma. Particular realizations are denoted by black dots, random sequences drawn according to their PDF/PMF by dotted circle/ellipsis, and the typical set is denoted by the gray region. The conditional relationship is denoted by arrows.

(4) If sequences are randomly drawn according to
$$(X^N, Y^N, Z^N) \sim \prod_n p_X(x_n) p_Y(y_n) p_{Z|X,Y}(z_n|x_n, y_n) \; then$$

$$\Pr\left\{(X^N, Y^N, Z^N) \in \mathcal{T}(X, Y, Z)\right\} \approx 2^{-NI(X;Y)}. \tag{A.109}$$

Proof   See [18] Section 2.5.1. The mutual relationships among the variables are shown in Figure A.1 for all four variants.   □

While skipping the exact proof, we instead provide the insight for the above stated four variants of the joint typicality lemma. First of all, and this is common to all of them, there is an assumption of *IID sequences*, i.e. individual symbols in the sequence are IID. This is expressed formally by a PDF having a form of $\prod_n p_X(x_n)$. This assumption is closely related to the underlying properties of the typical sets that require the sequence being IID. If the sequence was not the IID sequence, we could not infer anything about the probabilities of being inside the typical set. From the information-theoretic point of view it means that the rate performance is given by the capabilities of the *single-letter* channel symbol alphabet, i.e. the single-letter channel characterization. In a practical application for proving the coding theorems it means that the codewords must be IID sequences.

Variant (1) of the lemma describes the situation when random sequence Z^N is drawn according to the conditional distribution for some given particular x^N. For example, it is the noisy observation Z^N of the transmitted particular sequence x^N. Z^N can be viewed as typically "matched" to x^N. The third sequence y^N is an arbitrary fixed sequence (see Figure A.1). The probability that this triple will be "jointly matched" is very low. It is essentially given by the stochastic connection between random $Z^N|\{x^N\}$ and y^N. The stochastic connection is given by PDF/PMF and it is quantified by the mutual information. If there is a strong connection, i.e. the behavior of $Z^N|\{x^N\}$ is strongly connected to Y^N, then if we choose some arbitrary y^N there is quite a small chance that we will hit exactly the one that would correspond the typically "matched" one to the $Z^N|\{x^N\}$. On the other side, if the connection is zero (the mutual information is zero) the arbitrary choice is as good as the "matched" one, and the probability is high. The sequence y^N is an arbitrary one. It is *not* drawn according to its distribution. Therefore the properties derived from its stochastic behavior (i.e. respecting the distribution), in this case the mutual information, present "the best" we can expect. The probability of being in the typical set will be, at the best, the one given by the mutual information. This explains the upper bound on the probability $\Pr\{(x^N, y^N, Z^N) \in \mathcal{T}(X, Y, Z)\}$.

Variant (2) of the lemma differs from variant (1) by assuming that $(x^N, y^N) \in \mathcal{T}$ belong to the joint typical set. The behavior described by the stochastic description, i.e. the mutual information, thus describes the worst case. Thus (x^N, y^N) cannot be the low probability pair that would violate the behavior dictated by their stochastic distribution. They are typical representatives of the given PDF/PMF. This explains the lower bound on the probability $\Pr\{(x^N, y^N, Z^N) \in \mathcal{T}(X, Y, Z)\}$.

Variant (3) assumes that the pair (X^N, Y^N) is drawn according to the joint PDF/PMF and thus the behavior described by the mutual information "exactly" (in the ϵ sense) describes the probability $\Pr\{(X^N, Y^N, Z^N) \in \mathcal{T}(X, Y, Z)\}$.

The last variant (4) assumes that Z^N is drawn from the PDF/PMF jointly conditioned by (x^N, y^N) while X^N and Y^N are drawn *independently*. Therefore Z^N always "matches" to (x^N, y^N) but X^N and Y^N themselves may *not* be matched. The probability $\Pr\{(X^N, Y^N, Z^N) \in \mathcal{T}(X, Y, Z)\}$ is thus dictated only by the stochastic connection between X^N and Y^N described by $I(X; Y)$ and the result does not depend on Z^N. Also, since all sequences are drawn according to the PDF/PMF, the result is "exact" in the ϵ sense.

Special forms of the lemma can be obtained by removing the random sequence X^N, or equivalently setting it as a degenerative $X = \emptyset$. It implies that $I(Y; Z|X) = I(Y; Z)$. Particularly, variant (3) in this special form is useful. Notice that this, in fact, corresponds to the properties of the jointly typical pairs (see [13] Theorem 7.6.1). Similarly, we can derive special forms by dropping Y (or setting it to $Y = \emptyset$), which implies $I(Y; Z|X) = 0$ and again leads to some classical results.

LEMMA A.2 (Joint Typicality Lemma (Special Forms))   *Assume that random variables Y, Z are jointly described by PDF $p(y, z)$.*

(1) If (y^N) is an arbitrary *given realization of the sequence, and the random sequence Z^N is drawn according to $Z^N \sim \prod_n p_Z(z_n)$ then*

$$\Pr\{(y^N, Z^N) \in \mathcal{T}(Y, Z)\} \lesssim 2^{-NI(Y;Z)}. \tag{A.110}$$

(2) If the sequence realizations is typical *$y^N \in \mathcal{T}(Y)$, and $Z^N \sim \prod_n p_Z(z_n)$ then*

$$\Pr\{(y^N, Z^N) \in \mathcal{T}(Y, Z)\} \gtrsim 2^{-NI(Y;Z)}. \tag{A.111}$$

(3) If the sequence is randomly drawn according to $(Y^N) \sim \prod_n p_Y(y_n)$, and random sequence Z^N is drawn independent of Y^N, i.e. $Z^N|\{Y^N = y^N\} \sim \prod_n p_Z(z_n)$ then

$$\Pr\{(Y^N, Z^N) \in \mathcal{T}(Y, Z)\} \approx 2^{-NI(Y;Z)}. \tag{A.112}$$

(4) If the sequence realization is typical *$(x^N) \in \mathcal{T}(X)$, and $Z^N \sim \prod_n p_{Z|X}(z_n|x_n)$ then*

$$\Pr\{(x^N, Z^N) \in \mathcal{T}(X, Z)\} \approx 1. \tag{A.113}$$

Random Codebook

A random coding using random codebooks is a widely used principle in many information-theoretic assessments, namely the error probability analysis in achievability proofs of coding theorems. The idea stands on the following. The codebook $\mathcal{C}$ defining mapping between the message $b \in [1 : 2^{NR}]$ and the codeword c^N is generated randomly for each b symbol-wise IID according to $C^N(b) \sim \prod_{n=1}^{N} p(c_n)$. The probability of defining a particular codebook $\mathcal{C}^{(i)}$ is $\Pr\{\mathcal{C}^{(i)}\}$ and it is clearly related to the codeword distribution.

The error probability analysis in the achievability proofs evaluates the average probability of the error over all codebooks that could be generated according to the given distribution

$$\bar{P}_e = \sum_i \Pr\{\mathcal{C}^{(i)}\} P_e(\mathcal{C}^{(i)}). \tag{A.114}$$

The probability of the error for a given particular codebook $\mathcal{C}^{(i)}$ is in turn an average over all codewords in that codebook

$$P_e(\mathcal{C}^{(i)}) = \sum_{b=1}^{2^{NR}} \Pr\{B = b\} P_e(\mathcal{C}^{(i)}|b) \tag{A.115}$$

where $P_e(\mathcal{C}^{(i)}|b)$ is the probability of making an error for a particular transmitted message b using a particular codebook $\mathcal{C}^{(i)}$

$$P_e(\mathcal{C}^{(i)}|b) = \Pr\{\hat{B} \neq b | c^N = \mathcal{C}^{(i)}(b)\}. \tag{A.116}$$

The beauty of the random codebook principle is three-fold. First, because of a random generation symmetry, we can fix the transmitted message to any suitable value, typically $b_0 = 1$, and the result remains unaffected, i.e. $P_e(\mathcal{C}^{(i)}|b) = P_e(\mathcal{C}^{(i)}|b_0)$ for any b. Then

$$P_e(\mathcal{C}^{(i)}) = \sum_{b=1}^{2^{NR}} \Pr\{B = b\} P_e(\mathcal{C}^{(i)}|b_0)$$

$$= P_e(\mathcal{C}^{(i)}|b_0). \tag{A.117}$$

This simplifies the evaluation.

Second, the averaging over all codebooks guarantees that the error probability $\bar{P}_e$ will have evaluated properties, typically $\bar{P}_e \to 0$ for $N \to \infty$, in the *average*. Of course, there can be some bad codebooks that would not obey this behavior. But, on the other hand, there *must exist at least one* that would obey. So the evaluation of the average guarantees the existence of at least one good codebook. We can interpret the random codebook principle as the indirect tool for proving the existence of a good code.

Third, the evaluation of the average $\bar{P}_e$ is much *simpler* than the evaluation for one particular codebook $P_e(\mathcal{C}^{(i)})$. The average error is

$$\bar{P}_e = \sum_i \Pr\{\mathcal{C}^{(i)}\} P_e(\mathcal{C}^{(i)}|b_0)$$

$$= \Pr\left\{\hat{B} \neq b_0 | C^N(b_0) \sim \prod_{n=1}^{N} p(c_n)\right\}. \tag{A.118}$$

The probability of error is thus evaluated as if the transmitted codewords were *drawn randomly* at the moment of the evaluation.[1] This enables us to use the codebook randomness and the stochastic behavior, namely the *IID sequence* and the *codeword independence*, in the evaluation. In contrast, one particular codebook $C^{(i)}$ defines a one-to-one mapping of some very particular values c^N for each b and thus cannot be treated as random. We should stress that the random codebook and the equivalence of averaging over all codebooks are primarily used as a theorem-proving tool and one should not misinterpret this with the properties of some particular codebook defining a non-random mapping.

A.4.2   Capacity Region and Bounds

Capacity Region and its Inner and Outer Bounds

The *capacity region* is a convex hull of *all achievable* information rate K-tuples $[R_1, \ldots, R_K]$ over *all possible transmission strategies*

$$\mathcal{R} = \bigcup [R_1, \ldots, R_K] \tag{A.119}$$

where R_k are all rates relevant to the analyzed system. If the rates are associated with independent codebooks with corresponding channel symbols $S_1, \ldots, S_K$ we may use the notation $\mathcal{R}(S_1, \ldots, S_K)$.

It is important to stress two important aspects of the capacity region that are usually proved in the coding theorem for the given system. The rates must be *achievable*. It means that there exists some coding strategy (i.e. a codebook construction with given stochastic properties of codesymbols) and some decoding strategy (i.e. a procedure defining what to do with the received signal in order to find the desired data), for which we can prove that the decoding error probability approaches zero. Second, as a consequence of the union over *all* possible strategies, there is no other achievable rate point outside the region. In other words, if some coding/decoding strategy has a decoding error approaching zero, it must have the rate coordinates inside the capacity region. This is usually called the *converse* part of the coding theorem.

The time (or proportional) *sharing theorem* states that, given any two achievable rate region points $\mathbf{R}', \mathbf{R}''$ with strategies C', C'', the point

$$\mathbf{R} = t\mathbf{R}' + (1 - t)\mathbf{R}'', \quad t \in [0, 1] \tag{A.120}$$

is also achievable by using the strategy C' in proportion t and the strategy C'' in proportion $1 - t$. A simple consequence of this theorem is the *convexity* of any achievable rate region. Also, a zero rate on any coordinate of the rate vector is clearly "achievable." It implies that the achievable rate region must contain all points on the line connecting any point on the boundary by a perpendicular line to any of the axes.

Finding the capacity region is generally a complicated task, particularly the converse part showing that there is no other achievable rate outside. It is much easier to find inner and outer bounds. Notice that we call them inner and outer, not "lower" and "upper,"

---

[1] Essentially we use the fact that for any function $g(a)$ it holds that $\sum_a p(a)g(a) = \mathrm{E}_{A \sim p(a)}[g(A)]$.

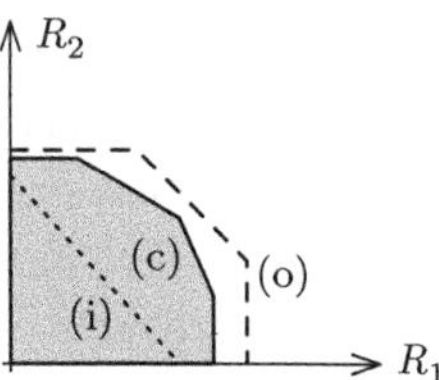

Figure A.2 Capacity region (c) and its inner (i) and outer (o) bounds illustration for $K = 2$.

since they generally confine multidimensional space region either from the inside or from the outside (Figure A.2).

The *inner bound* is achievable, but there might exist some other better coding strategy with the achievable rate point outside the bound. The inner bounds are generally easy to find. We simply set some coding and decoding strategy, analyze its decoding error probability and find the rates for which it approaches the zero.

The *outer bound* might not be achievable (we do not know it) but we are sure that no achievable point is outside. The outer bound is usually constructed by assuming some idealizing assumptions that cannot be beaten by any real coding strategy. The inner and outer bounds thus create at least the tolerance region in which the capacity region boundary must lie. The most comfortable way of finding the capacity region is the situation when we find some inner and outer bound and these two *coincide*. Then they also coincide with the capacity region.

Cut-Set Bound

The cut-set bound is a particular example of an easily and quite generally applicable outer bound. Here we state it in the basic unicast form where each node has a dedicated message for a given one-destination node.

THEOREM A.3 (Cut-Set Bound – Unicast) *Assume the network with the set of nodes S_k, $k \in [1 : K]$ where each node transmits the signal x_k and receives the signal y_k. Assume that source node i has the message b_{ij} to be transferred to the destination node j with the rate R_{ij}, i.e. $b_{ij} \in [1 : 2^{NR_{ij}}]$, where N is the common length of the codewords. All messages are independent and uniformly distributed. Any node can be the source, or the destination, or both, or none of them. If there is no message between some given pair of nodes, we simply set $R_{ij} = 0$. We denote the node set two-part partition, called a cut-set, as $(\mathcal{S}, \bar{\mathcal{S}})$, where $\mathcal{S} \subset \{S_k\}_{k=1}^K$, $\bar{\mathcal{S}} = \{S_k\}_{k=1}^K \setminus \mathcal{S}$.*

Then there exists a joint distribution $p(x_1, \ldots, x_K)$ such that any achievable rate R_{ij} is bounded by

$$\sum_{i \in \mathcal{S}, j \in \bar{\mathcal{S}}} R_{ij} \leq I\left(x(\mathcal{S}); y(\bar{\mathcal{S}})|x(\bar{\mathcal{S}})\right) \tag{A.121}$$

for all cut-sets $(\mathcal{S}, \bar{\mathcal{S}})$ where $x(\mathcal{S}) = \{x_i : i \in \mathcal{S}\}$, $y(\bar{\mathcal{S}}) = \{y_j : j \in \bar{\mathcal{S}}\}$, $x(\bar{\mathcal{S}}) = \{x_j : j \in \bar{\mathcal{S}}\}$.

For a rigorous proof see e.g. [13]. Here, we rather focus on the insight, proper understanding, and an intuitive interpretation of the theorem. First, the theorem assumes generally full-duplex nodes, i.e. nodes transmitting and receiving the signal at the same time. Networks with explicit half-duplex imposed stages must use an advanced form of the theorem that will be shown later.

A major idea that provides the insight for the theorem is the principle of assuming some idealizing assumptions that lead to the unbeatable performance bound. No real system can surpass this. The idealized system uses the idea of the *super-transmitter* and the *super-receiver* with perfect *interference cancellation*. Imagine the idealized hypothetical situation that all nodes in an S form a super-Tx allowing then to arbitrarily and fully coordinate their transmitted super-codeword, as if they were just one physical node with all resources (e.g. power) of S available. Similarly on the $\bar{S}$ side, the super-Rx assumes that all receiving nodes behave as if they were just one node with multiple observations and a common decode processing. The joint virtual message communicated by this super-Tx to the super-Rx is then a combination of all possible values of all messages originating in S and terminating in $\bar{S}$. The overall cardinality is then $\prod_{i \in S, j \in \bar{S}} 2^{NR_{ij}}$ and the corresponding rate is thus $\sum_{i \in S, j \in \bar{S}} R_{ij}$. Clearly the luxury of collecting all transmit resources on one side and behaving as one super-Tx node with joint reception processing of multiple collected observations is unbeatable by any set of *separately* operating nodes in S and $\bar{S}$. Moreover, this idea has turned the system into a virtual single super-user system, for which we can easily set the bound on achievable rates $\sum_{i \in S, j \in \bar{S}} R_{ij} \leq I(x(S); y(\bar{S}))$.

However, there is still one additional problem. There are also interfering transmitters in the $\bar{S}$ affecting the receivers in $\bar{S}$. The most idealizing measure would be to switch them off. From the information-theoretic point of view, this is equivalent to knowing the interfering signals at receivers. This, in turn, corresponds to conditioning the mutual information in the above-stated single super-user rate bound, i.e. $\sum_{i \in S, j \in \bar{S}} R_{ij} \leq I\left(x(S); y(\bar{S}) | x(\bar{S})\right)$, which is the final form of the cut-set bound.

Another aspect of the theorem that is worth mentioning is the formulation "there exists a joint distribution $p(x_1, \ldots, x_K)$." First, the *joint* distribution reference is a mathematical form of allowing the joint resource utilization and coordination in a super-Tx. Second, the fact that *there exists* a distribution means practically the maximization of the RHS of (A.121) that upper bounds the LHS sum rate.

The rates R_{ij} that appear in (A.121) are end-to-end rates, i.e. from the source to destination node, over the given cut-set $(S, \bar{S})$ boundary where the message source is in S and its destination is in $\bar{S}$. There does not need to be an explicit radio link between a given node pair crossed by the cut-set boundary. Both the source and the destination may be buried deeply inside the cut-sets. One given node may be a source for multiple destinations, or a destination for multiple sources. If the node is the source, or the destination, for some message that does not terminate on, or originates from, the "other side" of the cut, it does not appear in the sum of rates on the LHS of the equation.

If there are some radio links across the cut boundary missing between some pair of points, we may drop them from the mutual information expression, since they do not affect its value anyway. This fact is implied by the Markov property. For example, if $x_i \mapsto x_j \mapsto y_k$ form a Markov chain, then $I(x_i, x_j; y_k) = I(x_j; y_k)$. Similarly, if $x_i \mapsto y_j \mapsto y_k$ form a Markov chain, it implies that also $y_k \mapsto y_j \mapsto x_i$ is a Markov chain, and then $I(x_i; y_j, y_k) = I(x_i; y_j)$. Also, in the conditioning part $x(\bar{S})$ (A.121), we can consider only those components that actually affect $y(\bar{S})$ by existing link pairs. Figure A.3 shows the example.

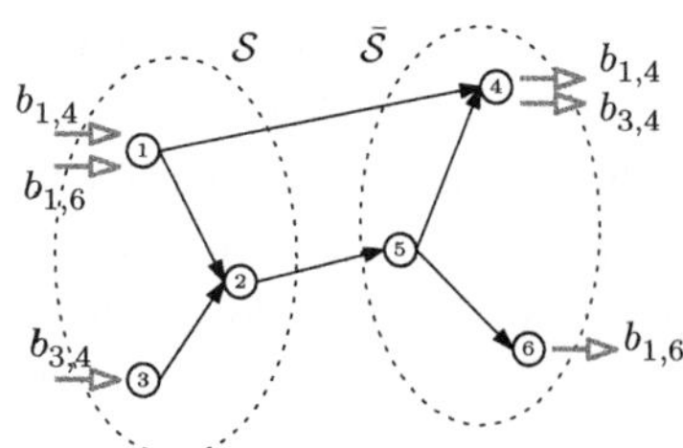

Figure A.3 Cut-set bound example. This particular cut implies $R_{14} + R_{16} + R_{34} \leq I(x_{123}; y_{456}$ $|x_{456}) = I(x_{12}; y_{45}|x_5)$. The simplification of mutual information utilizes that $x_3 \mapsto (x_1, x_2) \mapsto (y_4, y_5) \mapsto y_6$.

Cut-Set Bound for Multi-Stage Network

Many practical networks operate in multiple stages (forced e.g. by the half-duplex constraint) where the topology, or the connectivity, or the channel state properties, differ in each state. We set aside the problems of the causality and the end-to-end solvability of sequentially applied stages of the network. These need to be addressed separately. Apart from this, we can view the multi-stage network as a set of parallel networks corresponding to each stage and where each stage ℓ is used in some proportion t_ℓ, $\sum_{\ell=1}^{L} t_\ell = 1$. The resulting multi-stage cut-set bound is then given by the following

$$\sum_{i \in \mathcal{S}, j \in \bar{\mathcal{S}}} R_{ij} \leq \sup_{\{t_\ell\}_{\ell=1}^{L}: \sum_{\ell=1}^{L} t_\ell = 1} \sum_{\ell=1}^{L} t_\ell I\left(x_\ell(\mathcal{S}); y_\ell(\bar{\mathcal{S}})|x_\ell(\bar{\mathcal{S}})\right) \tag{A.122}$$

where L is the number of stages, x_ℓ, y_ℓ are transmitted and received signals in the ℓth stage, and the rest of the notation is identical to the one used in the standard cut-set bound.

A.5 Lattice Coding

This section summarizes selected lattice coding fundamentals. More details can be found in e.g. [63], which can serve as a good starting point and which lists numerous additional references on specific aspects of lattice coding. In this section, we constrain all treatment to *real-valued* lattices and lattice codes. The generalization to *complex* spaces is achieved in the simplest case by considering two separate lattice subspaces for real and imaginary parts, having identical properties. However, there are also more involved designs based on Gaussian and Einstein integers.

A.5.1 Lattices

Lattice

A lattice in N-dimensional *real-valued* space is defined as

$$\Lambda = \left\{ \lambda = \sum_{n=1}^{N} i_n \mathbf{g}_n; \; i_n \in \mathbb{Z} \right\} = \left\{ \lambda = \mathbf{Gi}; \; \mathbf{i} \in \mathbb{Z}^N \right\} \tag{A.123}$$

where $\mathbf{G} \in \mathbb{R}^{N \times N}$ is a generator $N \times N$ real-valued matrix composed from column basis vectors $\mathbf{g}_n$. The lattice Λ is a subgroup of $\mathbb{R}^N$ closed under the reflection, $\forall \lambda \in \Lambda : (-\lambda) \in \Lambda$, and real addition, $\forall \lambda_1, \lambda_2 \in \Lambda : (\lambda_1 + \lambda_2) \in \Lambda$.

Lattice Determinant

A lattice determinant is defined as

$$\det(\Lambda) = |\det(\mathbf{G})|. \tag{A.124}$$

Nearest Neighbor Quantizer

A Euclidean nearest neighbor quantizer is defined as

$$Q_\Lambda(\mathbf{x}) = \arg\min_{\lambda \in \Lambda} \|\mathbf{x} - \lambda\|. \tag{A.125}$$

The problem with border points is usually solved by systematically including the points in one axis direction (e.g. the negative direction, "South-West").

Voronoi Cell

For some lattice point λ, the region

$$\mathcal{V}_\lambda(\Lambda) = \left\{ \mathbf{x} \in \mathbb{R}^N : Q_\Lambda(\mathbf{x}) = \lambda \right\} \tag{A.126}$$

is called a Voronoi cell of the lattice Λ and it is associated with λ lattice point. A *fundamental* Voronoi cell $\mathcal{V}_0(\Lambda)$ is associated with the origin $\lambda = \mathbf{0}$. If the associated lattice point does not play any role in some given context, we may drop it from the notation $\mathcal{V}(\Lambda)$. The Voronoi cell is usually (unless explicitly stated otherwise) defined by the Euclidean distance quantizer. There are also more general approaches to defining cells using a general metric (see [63]). For the purpose of this text, we will, however, use a Euclidean-based approach.

Cell Volume

A volume of the cell is defined as

$$V(\Lambda) = \int_{\mathcal{V}(\Lambda)} d\mathbf{x} = \det(\Lambda) = |\det(\mathbf{G})|. \tag{A.127}$$

Modulo Lattice Operation

A modulo lattice operation is defined as follows. For any point $\mathbf{a} \in \mathbb{R}^N$

$$\mathbf{a} \bmod \Lambda = \mathbf{a} - Q_\Lambda(\mathbf{a}) \in \mathcal{V}_0(\Lambda). \tag{A.128}$$

The properties of a modulo lattice operation are

$$(\mathbf{a} + \mathbf{b}) \bmod \Lambda = ((\mathbf{a} \bmod \Lambda) + (\mathbf{b} \bmod \Lambda)) \bmod \Lambda, \tag{A.129}$$

$$(m\mathbf{a}) \bmod \Lambda = (m(\mathbf{a} \bmod \Lambda) \bmod \Lambda, \tag{A.130}$$

$$\left(Q_{\Lambda_c}(\mathbf{a})\right) \bmod \Lambda = \left(Q_{\Lambda_c}(\mathbf{a} \bmod \Lambda)\right) \bmod \Lambda, \tag{A.131}$$

where $\mathbf{a}, \mathbf{b} \in \mathbb{R}^N$, $m \in \mathbb{Z}$, $\Lambda \subset \Lambda_c$. The outer mod Λ operation on the RHS of (A.131) practically affects only a proper association of the points on the border of $\mathcal{V}_0(\Lambda)$. It holds that $\mathbf{a} \bmod \Lambda \in \mathcal{V}_0(\Lambda)$ but the inner Q_{Λ_c} quantizer might shift the points close to the cell boundary to the boundary point while violating a proper boundary association of $\mathcal{V}_0(\Lambda)$.

Normalized Second-Order Moment

The second-order moment per dimension (the power) for $\mathbf{u} \sim \mathcal{U}(\mathcal{V}_0(\Lambda))$ is

$$P(\Lambda) = \sigma^2(\Lambda) = \frac{1}{N} \mathrm{E}\left[\|\mathbf{u}\|^2\right] = \frac{1}{N} \frac{1}{V(\Lambda)} \int_{\mathcal{V}_0(\Lambda)} \|\mathbf{u}\|^2 \, \mathrm{d}\mathbf{u}, \tag{A.132}$$

and the normalized second-order moment is

$$G(\Lambda) = \frac{\sigma^2(\Lambda)}{V^{2/N}(\Lambda)}. \tag{A.133}$$

The normalized moment for a lattice product is

$$G(\Lambda \times \Lambda) = G(\Lambda). \tag{A.134}$$

For integer lattice it is $G(\mathbb{Z}^N) = 1/12$, and for *good quantizer* lattice it is $G(\Lambda) \to 1/(2\pi\, e)$ for $N \to \infty$.

A.5.2   Lattice Coding

Nested Lattice Code

A lattice code in N-dimensional *real* space is defined as a bounded subset of a so-called *fine*, or *code*, lattice $\Lambda_c \subset \mathbb{R}^N$. If the bounding region of the lattice codebook is related to some subset of the fine lattice, we talk about a *nested lattice code*. The sublattice $\Lambda_s \subset \Lambda_c$ is then called a *coarse*, or *shaping*, lattice. The nested lattice code is defined as a codebook

$$\mathcal{C} = \{\mathbf{c} \in (\Lambda_c \cap \mathcal{V}_0(\Lambda_s))\}. \tag{A.135}$$

The code rate of the nested lattice code is

$$R = \frac{\lg\left(V(\Lambda_s)/V(\Lambda_c)\right)}{N} = \frac{\lg |\Lambda_c/\Lambda_s|}{N} \tag{A.136}$$

where $M = |\Lambda_c/\Lambda_s|$ is the size of the quotient group.

Nested Lattice Code Based on Construction A

Under some limitations, we can construct a nested lattice code based on a linear GF block code using the so-called "Construction A" [10], [15], [37]. Construction A cannot be used for coarse lattices with irrational ratios of base vector lengths. The construction is done in the following steps.

Linear GF Block Code

Assume that we have a linear (N, N_b) block code over a prime-sized GF $\mathbb{F}_M$ with integer alphabet[2] $\{0, 1, \ldots, M - 1\}$

$$\mathcal{C}_0 = \left\{ \mathbf{c}_0 \in \mathbb{F}_M^N : \mathbf{c}_0 = \mathbf{G}_0 \mathbf{b}, \ \mathbf{b} \in \mathbb{F}_M^{N_b} \right\} \tag{A.137}$$

where $\mathbf{G}_0 \in \mathbb{F}_M^{N \times N_b}$ is a full-rank code generating matrix.

Scale the Codebook for Embedding into the Unit Cube $[0, 1)^N$

A scaling by $1/M$ will embed N-dimensional codebook points $\frac{1}{M}\mathcal{C}_0$ inside the $[0, 1)^N$ unit cube.

Expand by Repetition into $\mathbb{R}^N$

The basic cube of points is expanded into $\mathbb{R}^N$ by repeating the pattern with a $\mathbb{Z}^N$ shift

$$\Lambda = \frac{1}{M}\mathcal{C}_0 + \mathbb{Z}^N. \tag{A.138}$$

Shaping Lattice Transform

Assume that $\mathbf{G}_s \in \mathbb{R}^{N \times N}$ is a generating matrix of the coarse shaping lattice Λ_s, i.e. its columns are basis vectors $\mathbf{g}_{s,n}$ of Λ_s. Clearly,

$$\mathbf{G}_s(\frac{1}{M}\mathcal{C}_0) = \left\{ \lambda : \lambda = \mathbf{G}_s\left(\frac{1}{M}\mathbf{c}_0\right), \ \mathbf{c}_0 \in \mathcal{C}_0 \right\} \tag{A.139}$$

will transform the unit cube $[0, 1)^N$ bounded points $\frac{1}{M}\mathcal{C}_0$ into the points inside the volume of the set $\mathcal{P}_0 = \{\mathbf{x} : \mathbf{x} = \sum_{n=1}^{N} \mathbf{g}_{s,n} a_n, \ a_n \in [0, 1)\}$.

Fine Lattice

A fine lattice of the code is then

$$\Lambda_c = \mathbf{G}_s \Lambda = \left\{ \lambda : \lambda = \mathbf{G}_s\left(\frac{1}{M}\mathbf{G}_0 \mathbf{b} + \mathbf{z}\right), \ \mathbf{b} \in \mathbb{F}_M^{N_b}, \ \mathbf{z} \in \mathbb{Z}^N \right\}. \tag{A.140}$$

Nested Lattice Code

The final nested lattice code is

$$\mathcal{C} = \Lambda_c \cap \mathcal{V}_0(\Lambda_s). \tag{A.141}$$

Notice that both the set $\mathcal{P}_0$ and the fundamental Voronoi cell $\mathcal{V}_0$ have the same volume, and contain exactly the same number of fine lattice points. Points inside the fundamental Voronoi cell are modulo Λ_s shifted version of the points in $\mathcal{P}_0$, then

$$\mathcal{C} = \Lambda_c \bmod \Lambda_s. \tag{A.142}$$

The nested lattice codebook is thus

$$\mathcal{C} = \left\{ \mathbf{c} : \mathbf{c} = \left(\mathbf{G}_s\left(\frac{1}{M}\mathbf{G}_0 \mathbf{b} + \mathbf{z}\right) \right) \bmod \Lambda_s, \ \mathbf{b} \in \mathbb{F}_M^{N_b}, \ \mathbf{z} \in \mathbb{Z}^N \right\}. \tag{A.143}$$

---

[2] Generally, the GF alphabet can have many formal forms. The fact that we have chosen the set of integers allows us to avoid the need of a "conversion" function when embedding GF into $\mathbb{R}^N$.

Nested Lattice Code with Euclidean Lattice Decoding

The nested lattice code can be used in the AWGN channel for the transmission of the information. However, the *transmission strategy* does not send the lattice points directly but it requires a modification by a continuous-valued random dithering and a modulo lattice operation. Both operations are needed to have an equivalent channel with an *independent* noise distribution. Since the dithering is done over the fundamental Voronoi cell, the scheme is also sometimes named the Voronoi lattice encoder.

The transmitted signal is

$$\mathbf{s} = (\mathbf{c} + \mathbf{u}) \bmod \Lambda_s \tag{A.144}$$

where $\mathbf{c} \in \mathcal{C}$ is the nested lattice code and $\mathbf{u}$ is the random dither with a uniform distribution $\mathbf{u} \sim \mathcal{U}(\mathcal{V}_0(\Lambda_s))$ over the fundamental Voronoi cell. First, the dither, together with the modulo lattice operation, randomizes the distribution of the transmitted signal to be also $\mathbf{s} \sim \mathcal{U}(\mathcal{V}_0(\Lambda_s))$. As a consequence, if the transmitted signal is used many times in a super-block of length LN, the distribution of $\mathbf{s}$ truly becomes uniform and thus the capacity is achieved (as will be explained later). Second, the modulo lattice and the dither preserve the transmitted mean power, which is the second moment of $\mathbf{s}$, i.e.

$$P_s = P(\Lambda_s) = \frac{1}{N} \mathrm{E}\left[\|\mathbf{s}\|^2\right]. \tag{A.145}$$

Third, the dither and the modulo lattice operation will play an important role in the modulo lattice equivalent channel properties, as will be explained later.

The received signal in the AWGN channel is

$$\mathbf{x} = \mathbf{s} + \mathbf{w} \tag{A.146}$$

where $\mathbf{w}$ is IID N-dimensional real-valued Gaussian zero mean noise with variance $P_w = \frac{1}{N} \mathrm{E}\left[\|\mathbf{w}\|^2\right]$ per dimension.

The *decoding strategy* is called Euclidean lattice decoding. It consists of (1) a pre-processing (scaling) of the received signal, (2) a removal of the dither, (3) a fine lattice quantization based on the Euclidean nearest neighbor, and (4) a modulo coarse (shaping) lattice operation, which provides a decision of the codeword. The preprocessing operation is a linear scaling by $\alpha \in \mathbb{R}$, which has a strong effect on the decoding and will be discussed later in detail. The decoding strategy first calculates the decision metric

$$\mathbf{y} = \alpha \mathbf{x} - \mathbf{u} \tag{A.147}$$

and then it makes the decision

$$\hat{\mathbf{c}} = \left(Q_{\Lambda_c}(\mathbf{y})\right) \bmod \Lambda_s. \tag{A.148}$$

Equivalent Modulo Lattice Channel

If we ignored the scaling, the dither, and the modulo lattice operation in the communication strategy above, everything would look like an ordinary minimum Euclidean distance receiver in the AWGN channel. However, it is not. In order to evaluate the performance of lattice coding, we need to define the *equivalent modulo lattice channel*, which would capture all the phenomena mentioned but, at the same time, would have a structure that could be handled more easily.

We start by realizing that

$$\hat{\mathbf{c}} = \left(Q_{\Lambda_c}(\mathbf{y})\right) \bmod \Lambda_s = \left(Q_{\Lambda_c}(\mathbf{y} \bmod \Lambda_s)\right) \bmod \Lambda_s, \tag{A.149}$$

i.e. an additional modulo lattice operation before the quantization does not change the result. We define

$$\mathbf{y}' = (\alpha\mathbf{x} - \mathbf{u}) \bmod \Lambda_s. \tag{A.150}$$

A substitution of the received signal and subsequent manipulations using properties of $\bmod\,\Lambda_s$ give

$$\begin{aligned}
\mathbf{y}' &= (\alpha\,((\mathbf{c} + \mathbf{u}) \bmod \Lambda_s + \mathbf{w}) - \mathbf{u}) \bmod \Lambda_s \\
&= (\mathbf{c} - \mathbf{c} + \alpha\,((\mathbf{c} + \mathbf{u}) \bmod \Lambda_s) + \alpha\mathbf{w} - \mathbf{u}) \bmod \Lambda_s \\
&= (\mathbf{c} + \alpha\,((\mathbf{c} + \mathbf{u}) \bmod \Lambda_s) + \alpha\mathbf{w} - (\mathbf{c} + \mathbf{u}) \bmod \Lambda_s) \bmod \Lambda_s \\
&= (\mathbf{c} + (\alpha - 1)\,((\mathbf{c} + \mathbf{u}) \bmod \Lambda_s) + \alpha\mathbf{w}) \bmod \Lambda_s.
\end{aligned} \tag{A.151}$$

Since $\mathbf{u} \sim \mathcal{U}(\mathcal{V}_0(\Lambda_s))$ and thus also $(\mathbf{c} + \mathbf{u}) \bmod \Lambda_s \sim \mathcal{U}(\mathcal{V}_0(\Lambda_s))$ for an arbitrary $\mathbf{c}$ (i.e. $(\mathbf{c} + \mathbf{u}) \bmod \Lambda_s \perp \mathbf{c}$), we can substitute the expression $(\mathbf{c} + \mathbf{u}) \bmod \Lambda_s$ by the equivalent one that has the same stochastic properties $\mathbf{u}_{\mathrm{eq}} \sim \mathcal{U}(\mathcal{V}_0(\Lambda_s))$ and which is *independent* of $\mathbf{c}$. The equivalent dither is a zero mean random vector and has the same power as the transmitted signal, $\frac{1}{N}\,\mathrm{E}\left[\|\mathbf{u}_{\mathrm{eq}}\|^2\right] = P(\Lambda_s) = P_s$.

The *equivalent channel* is

$$\mathbf{y}_{\mathrm{eq}} = \left(\mathbf{c} + \mathbf{w}_{\mathrm{eq}}\right) \bmod \Lambda_s \tag{A.152}$$

where the *equivalent noise* is

$$\mathbf{w}_{\mathrm{eq}} = (\alpha - 1)\mathbf{u}_{\mathrm{eq}} + \alpha\mathbf{w}. \tag{A.153}$$

The resulting equivalent modulo lattice channel will have the same stochastic input–output description ($p_{\mathbf{y}_{\mathrm{eq}}|\mathbf{c}}(\mathbf{y}|\mathbf{c}) = p_{\mathbf{y}'|\mathbf{c}}(\mathbf{y}|\mathbf{c})$) as the original one and, as a consequence,

$$\hat{\mathbf{c}} = \left(Q_{\Lambda_c}(\mathbf{y})\right) \bmod \Lambda_s = \left(Q_{\Lambda_c}(\mathbf{y}')\right) \bmod \Lambda_s = \left(Q_{\Lambda_c}(\mathbf{y}_{\mathrm{eq}})\right) \bmod \Lambda_s. \tag{A.154}$$

The equivalent channel is the additive noise channel with the modulo lattice operation. Notice the importance of the *uniform random dither* for making the equivalent noise *independent* on the transmitted signal for *arbitrary* scaling α. Of course, the special case of no scaling, $\alpha = 1$, trivially removes any impact of the dither and only a pure intact AWGN noise remains. We will use the equivalent channel when investigating the information-theoretic properties like the capacity and the rate achievability of the lattice coding.

The additive equivalent noise is, however, no longer Gaussian. It is a zero mean mixture of the uniform and Gaussian distributions. Its variance is

$$P_{w_{\mathrm{eq}}} = \frac{1}{N}\,\mathrm{E}\left[\|\mathbf{w}_{\mathrm{eq}}\|^2\right] = (\alpha - 1)^2 P_s + \alpha^2 P_w. \tag{A.155}$$

The distribution of equivalent noise is close to Gaussian but with "lighter" tails (see [63] Section 9 for details). This light-tail noise is more favorable than Gaussian noise from the probability of the decoding error point of view. The equivalent noise PDF can be upper bounded, up to the fixed scaling, by the Gaussian PDF with the same variance (see details in [45]). All these statements refer to the noise $\mathbf{w}_{eq}$ itself. There is still the outer modulo lattice operation in the channel model that additionally influences the result.

Optimization of the Preprocessor Scaling

The equivalent noise power $P_{w_{eq}}$ depends, apart from the given signal and true AWGN noise powers P_s, P_w, also on the choice of the scaling coefficient α. The analysis of the α scaling optimization provides a nice insight and we show that in more detail.

The minimization of equivalent noise power by finding the stationary point

$$\frac{\partial P_{w_{eq}}}{\partial \alpha} = 2(\alpha - 1)P_s + 2\alpha P_w = 0 \tag{A.156}$$

gives

$$\hat{\alpha} = \frac{P_s}{P_s + P_w}. \tag{A.157}$$

The resulting minimized equivalent noise power is

$$P_{w_{eq}}(\hat{\alpha}) = \frac{P_s P_w}{P_s + P_w}. \tag{A.158}$$

Clearly, this solution is in fact an MMSE preprocessing filter minimizing the mean square error between the desired value $\mathbf{c}$ and the observed output of the equivalent channel ($\mathbf{c} + \mathbf{w}_{eq}$). In fact, it is a trivial single-tap scalar Wiener filter. It is worth mentioning that the optimization is done for the processing before mod Λ_s operation.

The MMSE scaling coefficient is $\hat{\alpha} \in (0, 1)$; the higher the SNR P_s/P_w we have, the closer we get to unity. For SNR values where the noise power is comparable to signal power, the optimal value can be significantly smaller than one. From the perspective of the true received signal $\mathbf{x}$, it creates an *intentional* misalignment of the received lattice points w.r.t. lattice quantizer Q_{Λ_c} and mod Λ_s operation in the lattice decoder. Although this intentional distortion might look counter-intuitive and harmful, it proves to be exactly the opposite. The core of understanding why it is so is in realizing (as will be shown later) that the *mean second-order error* properties are more important from the achievable rate point of view than the actual bare lattice alignment.

Capacity of Equivalent Modulo Lattice Channel

Now we investigate the capacity of the equivalent channel (A.152) and (A.153). The capacity is, by definition,

$$C_{\Lambda_s} = \frac{1}{N} \max_{p(\mathbf{c})} I(\mathbf{c}; \mathbf{y}_{eq}) = \frac{1}{N} \max_{p(\mathbf{c})} \left(\mathcal{H}[\mathbf{y}_{eq}] - \mathcal{H}[\mathbf{y}_{eq}|\mathbf{c}] \right). \tag{A.159}$$

We first evaluate $\mathcal{H}[\mathbf{y}_{eq}]$ entropy. The variable $\mathbf{y}_{eq}$ has a *finite* support $\mathcal{V}_0(\Lambda_s)$ and the entropy is thus maximized by the *uniform* distribution. As a consequence of the mod Λ_s

operation, the uniform channel input $\mathbf{c} \sim \mathcal{U}(\mathcal{V}_0(\Lambda_s))$ guarantees the uniform $\mathbf{y}_{eq}$ and the entropy is

$$\max_{p(\mathbf{c})} \mathcal{H}[\mathbf{y}_{eq}] = \lg V(\Lambda_s)$$

$$= \lg \left(\frac{\sigma^2(\Lambda_s)}{G(\Lambda_s)} \right)^{\frac{N}{2}}$$

$$= \frac{N}{2} \lg P_s - \frac{N}{2} \lg G(\Lambda_s) \qquad (A.160)$$

where we used a second-order normalized moment of the shaping lattice $G(\Lambda_s)$.

The conditional entropy is

$$\mathcal{H}[\mathbf{y}_{eq}|\mathbf{c}] = \mathcal{H}[\mathbf{w}_{eq} \bmod \Lambda_s] \qquad (A.161)$$

and it does not depend on the input distribution. The noise $\mathbf{w}_{eq}$ is not Gaussian; neither is $\mathbf{w}_{eq} \bmod \Lambda_s$ Gaussian. However, the entropy $\mathcal{H}[\mathbf{w}_{eq} \bmod \Lambda_s]$ can be upper bounded by the entropy of the Gaussian noise with variance equal to the variance of noise $\mathbf{w}_{eq} \bmod \Lambda_s$. Also, because the $\bmod \Lambda_s$ operation *does not* increase the second moment ([63] Section 9.5.1, Problem 9.6), it can in turn be upper bounded by the entropy of Gaussian noise with variance $\mathrm{E}[\|\mathbf{w}_{eq}\|^2]$

$$\mathcal{H}[\mathbf{w}_{eq} \bmod \Lambda_s] \leq \frac{N}{2} \lg \left(2\pi \, \mathrm{e} \, \frac{1}{N} \mathrm{E}\left[\|\mathbf{w}_{eq} \bmod \Lambda_s\|^2 \right] \right)$$

$$\leq \frac{N}{2} \lg \left(2\pi \, \mathrm{e} \, \frac{1}{N} \mathrm{E}\left[\|\mathbf{w}_{eq}\|^2 \right] \right)$$

$$= \frac{N}{2} \lg \left(2\pi \, \mathrm{e} \, P_{w_{eq}} \right). \qquad (A.162)$$

The resulting *equivalent modulo lattice channel capacity* is lower bounded

$$C_{\Lambda_s} \geq \frac{1}{2} \lg \frac{P_s}{P_{w_{eq}}} - \frac{1}{2} \lg \left(2\pi \, \mathrm{e} \, G(\Lambda_s) \right). \qquad (A.163)$$

The capacity is achieved by the uniform input $\mathbf{c}$ distribution. If the shaping lattice is a good quantizer, $G(\Lambda_s) \to 1/(2\pi \, \mathrm{e})$ for $N \to \infty$, then for $N \to \infty$

$$C_{\Lambda_s} \geq \frac{1}{2} \lg^+ \frac{P_s}{P_{w_{eq}}}. \qquad (A.164)$$

The *tightness* of the lower bound depends on how close to Gaussian is the noise $\mathbf{w}_{eq} \bmod \Lambda_s$. Generally, there is no guarantee that $P_s > P_{w_{eq}}$ and thus we use $\lg^+$ function.

The capacity effectively describes the price paid for the nonlinear modulo lattice operation in comparison with a pure AWGN linear channel. The effect of the dither and the preprocessor scaling is hidden in non-Gaussian equivalent noise properties. Clearly, as a result of the data processing inequality, the capacity of equivalent modulo lattice channel is upper bounded by AWGN channel capacity and thus

$$\frac{1}{2} \lg \left(1 + \frac{P_s}{P_w} \right) \geq C_{\Lambda_s} \geq \frac{1}{2} \lg^+ \frac{P_s}{P_{w_{eq}}}. \qquad (A.165)$$

The equivalent modulo lattice channel capacity lower bound will be called a *modulo lattice channel rate*

$$R_{\Lambda_s} = \frac{1}{2}\lg^+ \frac{P_s}{P_{W_{eq}}}. \tag{A.166}$$

Achievability of Modulo Lattice Channel Rate

The modulo lattice channel rate is *achievable* by the *nested lattice code* as the transmission strategy, and the *Euclidean lattice decoding* as the decoding strategy. This statement can be declared in two forms. The first one takes the α scaling as it is, without any attempt to optimize it ([63] Lemma 9.6.1).

THEOREM A.4   *The modulo lattice channel rate*

$$R_{\Lambda_s} = \frac{1}{2}\lg^+ \frac{P_s}{P_{W_{eq}}} \tag{A.167}$$

where the equivalent noise power is $P_{W_{eq}} = (1-\alpha)^2 P_s + \alpha^2 P_w$ is achievable by the nested lattice code *and the* Euclidean lattice decoding *for arbitrary α and* good nested lattices, *i.e. Λ_s is good for a quantization and Λ_c is a good channel code for the equivalent noise.*

The second form optimizes the α scaling and shows that the achievable modulo lattice channel rate approaches the $\frac{1}{2}\lg(1 + P_s/P_w)$ AWGN channel achievable rate ([63] Theorem 9.6.1). In principle, it is obtained by simply plugging in the MMSE scaler solution (A.158) into the modulo lattice channel rate (A.166).

THEOREM A.5   *The AWGN channel capacity*

$$C = \frac{1}{2}\lg\left(1 + \frac{P_s}{P_w}\right) \tag{A.168}$$

where the AWGN noise power is P_w is achievable by the nested lattice code *and the* Euclidean lattice decoding *with an* optimized Wiener (MMSE) *scaling*

$$\hat{\alpha} = \frac{P_s}{P_s + P_w} \tag{A.169}$$

and good nested lattices, *i.e. Λ_s is good for a quantization and Λ_c is a good channel code for the equivalent noise.*

The coding and decoding strategy ignores (1) the granularity of actual lattice transmitted points $\mathbf{c} \in \Lambda_c$, whereas the entropy maximization requires the uniform distribution, (2) single realization of encoding and particularly the dither, whereas the achievability requires random behavior, and (3) the mismatch in the decoding where the decoder uses the Euclidean metric, whereas the actual equivalent noise is non-Gaussian. Nevertheless, if the lattice dimension is large $N \to \infty$, these limitations do not affect the achievability (see [63] Section 9.6).

References

[1] "RFC 1122: Requirements for internet hosts – communication layers."

[2] "ISO/IEC standard 7498-1:1994," 1994.

[3] "FP7 project DIWINE (Dense Cooperative Wireless Cloud Network)," 2013. [Online]. Available: http://diwine-project.eu/public/

[4] R. Ahlswede, "Multi-way communication channels," in *2nd Int. Symp. on Information Theory*, 1971.

[5] R. Ahlswede, N. Cai, S.-Y. R. Li, and R. W. Yeung, "Network information flow," *IEEE Trans. Inf. Theory*, vol. 46, no. 4, pp. 1204–1216, 2000.

[6] A. Behboodi and P. Piantanida, "Mixed noisy network coding and cooperative unicasting in wireless networks," *IEEE Trans. Inf. Theory*, vol. 61, no. 1, pp. 189–222, 2015.

[7] S. S. Bidokhti and V. M. Prabhakaran, "Is non-unique decoding necessary?" *IEEE Trans. Inf. Theory*, vol. 60, no. 5, pp. 2594–2610, 2014.

[8] R. E. Blahut, *Algebraic Codes for Data Transmission*. Cambridge University Press, 2003.

[9] M. Conti and S. Giordano, "Multihop ad hoc networking: The theory," *IEEE Commun. Mag.*, vol. 45, no. 4, pp. 78–86, 2007.

[10] J. H. Conway and N. J. A. Sloane, *Sphere Packings, Lattices and Groups*, 3rd edn. Springer, 1999.

[11] T. M. Cover, "Broadcast channels," *IEEE Trans. Inf. Theory*, vol. 18, no. 1, pp. 2–14, 1972.

[12] T. M. Cover and A. E. Gamal, "Capacity theorems for the relay channel," *IEEE Trans. Inf. Theory*, vol. 25, no. 5, pp. 572–584, 1979.

[13] T. M. Cover and J. A. Thomas, *Elements of Information Theory*, 2nd edn. John Wiley & Sons, 2006.

[14] P. Elias, A. Feinstein, and C. E. Shannon, "A note on the maximum flow through a network," *IEEE Trans. Inf. Theory*, vol. 2, no. 4, pp. 117–119, 1956.

[15] U. Erez, S. Litsyn, and R. Zamir, "Lattices which are good for (almost) everything," *IEEE Trans. Inf. Theory*, vol. 51, no. 10, pp. 3401–3416, 2005.

[16] U. Erez and R. Zamir, "Achieving $1/2 \log(1 + \text{SNR})$ on the AWGN channel with lattice encoding and decoding," *IEEE Trans. Inf. Theory*, vol. 50, no. 10, pp. 2293–2314, 2004.

[17] C. Feng, D. Silva, and F. R. Kschischang, "An algebraic approach to physical-layer network coding," *IEEE Trans. Inf. Theory*, vol. 59, no. 11, pp. 7576–7596, 2013.

[18] A. E. Gamal and Y.-H. Kim, *Network Information Theory*. Cambridge University Press, 2011.

[19] A. Goldsmith, *Wireless Communications*. Cambridge University Press, 2005.

[20] P. Gupta and P. R. Kumar, "The capacity of wireless networks," *IEEE Trans. Inf. Theory*, vol. 46, no. 2, pp. 388–404, Mar. 2000.

[21] J. Harshan and B. S. Rajan, "On two-user Gaussian multiple access channels with finite input constellations," *IEEE Trans. Inf. Theory*, vol. 57, no. 3, pp. 1299–1327, 2011.

[22] M. H. Hayes, *Statistical Digital Signal Processing and Modeling*. John Wiley & Sons, 1996.

[23] J. Hou and G. Kramer, "Short message noisy network coding with a decode-forward option," *arXiv:1304.1692 [cs.IT]*, 2015.

[24] ITU, "The world in 2014," 2014. [Online]. Available: www.itu.int/en/ITU-D/Statistics/ Documents/facts/ ICTFactsFigures2014-e.pdf

[25] S. M. Kay, *Fundamentals of Statistical Signal Processing: Estimation Theory*. Prentice-Hall, 1993.

[26] ——, *Fundamentals of Statistical Signal Processing: Detection Theory*. Prentice-Hall, 1998.

[27] R. Koetter and M. Medard, "An algebraic approach to network coding," *IEEE Tran. on Networking*, vol. 11, no. 5, pp. 782–795, 2003.

[28] T. Koike-Akino, P. Popovski, and V. Tarokh, "Denoising maps and constellations for wireless network coding in two-way relaying systems," in *Proc. IEEE Global Telecommun. Conf. (GlobeCom)*, 2008.

[29] ——, "Adaptive modulation and network coding with optimized precoding in two-way relaying," in *Proc. IEEE Global Telecommun. Conf. (GlobeCom)*, 2009.

[30] ——, "Denoising strategy for convolutionally-coded bidirectional relaying," in *Proc. IEEE Internat. Conf. on Commun. (ICC)*, 2009.

[31] ——, "Optimized constellations for two-way wireless relaying with physical network coding," *IEEE J. Sel. Areas Commun.*, vol. 27, no. 5, pp. 773–787, 2009.

[32] J. N. Laneman, D. N. C. Tse, and G. W. Wornell, "Cooperative diversity in wireless networks: Efficient protocols and outage behavior," *IEEE Trans. Inf. Theory*, vol. 50, no. 12, pp. 3062–3080, December 2004.

[33] P. Larsson, N. Johansson, and K.-E. Sunell, "Coded bi-directional relaying," in *5th Scandinavian WS on Wireless Ad-hoc Networks (AdHoc'05)*, 2005.

[34] S.-Y. R. Li, R. W. Yeung, and N. Cai, "Linear network coding," *IEEE Trans. Inf. Theory*, vol. 49, no. 2, pp. 371–381, 2003.

[35] H. H. J. Liao, "Multiple access channels," Ph.D. dissertation.

[36] S. H. Lim, Y.-H. Kim, A. E. Gamal, and S.-Y. Chung, "Noisy network coding," *IEEE Trans. Inf. Theory*, vol. 57, no. 5, pp. 3132–3152, 2011.

[37] H.-A. Loeliger, "Averaging bounds for lattices and linear codes," *IEEE Trans. Inf. Theory*, vol. 43, no. 6, pp. 1767–1773, 1997.

[38] L. Lu, T. Wang, S. C. Liew, and S. Zhang, "Implementation of physical-layer network coding," in *Proc. IEEE Internat. Conf. on Commun. (ICC)*, 2012.

[39] R. J. McEliece, *The Theory of Information and Coding*, 2nd edn. Cambridge University Press, 2002.

[40] C. D. Meyer, *Matrix Analysis and Applied Linear Algebra*. SIAM, 2000.

[41] V. Nagesha and S. Kay, "Cramer–Rao lower bounds for complex parameters," 1991. [Online]. Available: www.ele.uri.edu/faculty/kay.html

[42] B. Nazer, "Successive compute-and-forward," in *International Zurich Seminar on Communication (IZS 2012)*, 2012.

[43] B. Nazer and M. Gastpar, "Reliable computation over multiple-access channels," in *Proc. Annual Allerton Conf. on Commun., Cont., and Comp.*, 2005.

[44] ——, "Computation over multiple-access channels," *IEEE Trans. Inf. Theory*, vol. 53, no. 10, pp. 3498–3516, 2007.

[45] ——, "Compute-and-forward: Harnessing interference through structured codes," *IEEE Trans. Inf. Theory*, vol. 57, no. 10, pp. 6463–6486, 2011.

[46] ——, "Reliable physical layer network coding," *Proc. IEEE*, vol. 99, no. 3, pp. 438–460, 2011.

[47] O. Ordentlich, U. Erez, and B. Nazer, "The approximate sum capacity of the symmetric Gaussian K-user interference channel," *IEEE Trans. Inf. Theory*, vol. 60, no. 6, pp. 3450–3482, 2014.

[48] A. Papoulis, *Probability, Random Variables, and Stochastic Processes*, 2nd edn. McGraw-Hill, 1984.

[49] P. Popovski and H. Yomo, "The anti-packets can increase the achievable throughput of a wireless multi-hop network," in *Proc. IEEE Internat. Conf. on Commun. (ICC)*, 2006.

[50] B. Rankov and A. Wittneben, "Achievable rate regions for the two-way relay channel," in *Proc. IEEE Int. Symp. on Information Theory (ISIT)*, 2006.

[51] L. L. Scharf, *Statistical Signal Processing: Detection, Estimation, and Time Series Analysis*. Addison-Wesley, 1991.

[52] P. J. Schreier and L. L. Scharf, *Statistical Signal Processing of Complex-Valued Data – The Theory of Improper and Noncircular Signals*. Cambridge University Press, 2010.

[53] N. Sommer, M. Feder, and O. Shalvi, "Low-density lattice codes," *IEEE Trans. Inf. Theory*, vol. 54, no. 4, pp. 1561–1585, 2008.

[54] J. Sykora, "Tapped delay line model of linear randomly time-variant WSSUS channel," *Electronics Letters*, vol. 36, no. 19, pp. 1656–1657, 2000.

[55] ——, "Hierarchical network transfer function and doubly-greedy half-duplex stage scheduling for WPNC networks," *IEEE Commun. Lett.*, vol. 19, no. 6, pp. 1029–1032, 2015.

[56] ——, "Hierarchical pairwise error probability for hierarchical decode and forward strategy in PLNC," *IEEE Commun. Lett.*, vol. 20, no. 1, pp. 49–52, 2016.

[57] J. Sykora and A. Burr, "Layered design of hierarchical exclusive codebook and its capacity regions for HDF strategy in parametric wireless 2-WRC," *IEEE Trans. Veh. Technol.*, vol. 60, no. 7, pp. 3241–3252, 2011.

[58] J. Sykora and J. Hejtmanek, "Joint and recursive hierarchical interference cancellation with successive compute & forward decoding," in *COST IC1004 MCM*, Aalborg, Denmark, 2014.

[59] C. W. Therrien, *Discrete Random Signals and Statistical Signal Processing*. Englewood Cliffs, New Jersey: Prentice-Hall, 1992.

[60] P. Valerio, "Internet of things: 50 billion is only the beginning," 2014. [Online]. Available: www.eetimes.com/document.asp?doc_id=1321229

[61] A. van den Bos, "A Cramer–Rao lower bound for complex parameters," *IEEE Trans. Signal Process.*, vol. 42, no. 10, p. 2859, 1994.

[62] R. W. Yeung, *Information Theory and Network Coding*. Springer, 2008.

[63] R. Zamir, *Lattice Coding for Signals and Networks*. Cambridge University Press, 2014.

[64] S. Zhang, S. C. Liew, and P. P. Lam, "Hot topic: Physical-layer network coding," in *MobiCom'06*, 2006, pp. 358–365.

Index